ASTRONAUTICAL AND AERONAUTICAL EVENTS OF 1962

REPORT

OF THE

NATIONAL AERONAUTICS AND SPACE ADMINISTRATION

TO THE

COMMITTEE ON SCIENCE AND ASTRONAUTICS
U.S. HOUSE OF REPRESENTATIVES
EIGHTY-EIGHTH CONGRESS
FIRST SESSION

COMMITTEE ON SCIENCE AND ASTRONAUTICS

GEORGE P. MILLER, California, *Chairman*

OLIN E. TEAGUE, Texas
JOSEPH E. KARTH, Minnesota
KEN HECHLER, West Virginia
EMILIO Q. DADDARIO, Connecticut
J. EDWARD ROUSH, Indiana
THOMAS G. MORRIS, New Mexico
BOB CASEY, Texas
WILLIAM J. RANDALL, Missouri
JOHN W. DAVIS, Georgia
WILLIAM F. RYAN, New York
THOMAS N. DOWNING, Virginia
JOE D. WAGGONNER, JR., Louisiana
EDWARD J. PATTEN, New Jersey
RICHARD H. FULTON, Tennessee
DON FUQUA, Florida
NEIL STAEBLER, Michigan
CARL ALBERT, Oklahoma

JOSEPH W. MARTIN, JR., Massachusetts
JAMES G. FULTON, Pennsylvania
J. EDGAR CHENOWETH, Colorado
WILLIAM K. VAN PELT, Wisconsin
R. WALTER RIEHLMAN, New York
CHARLES A. MOSHER, Ohio
RICHARD L. ROUDEBUSH, Indiana
ALPHONZO BELL, California
THOMAS M. PELLY, Washington
DONALD RUMSFELD, Illinois
JAMES D. WEAVER, Pennsylvania
EDWARD J. GURNEY, Florida
JOHN W. WYDLER, New York

CHARLES F. DUCANDER, *Executive Director and Chief Counsel*
JOHN A. CARSTARPHEN, JR., *Chief Clerk*
PHILIP B. YEAGER, *Counsel*
FRANK R. HAMMILL, JR., *Counsel*
W. H. BOONE, *Technical Consultant*
WILLIAM E. DITCH, *Technical Consultant*
HAROLD A. GOULD, *Technical Consultant*
RICHARD P. HINES, *Staff Consultant*
JOSEPH M. FELTON, *Assistant Staff Consultant*
DENIS C. QUIGLEY, *Publications Clerk*

II

FOREWORD

Days, weeks, months, and years of the space age clock on. Noteworthy events in space science and technology flash by at an ever-accelerating rate. Many events appearing of current importance become less so with the passage of time, while others emerge slowly as the significant milestones align themselves into the dynamic patterns of progress. Our comprehension of the meaning of astronautics to the future of mankind is likewise a growing thing. The great breadth of the social, political, economic, and strategic impact of man's nascent steps to explore his universe must not be ignored. Yet all of us tend to become preoccupied with our problems and responsibilities near at hand.

One of the useful tools to help gain perspective and greater appreciation is a chronology of documented events. As current meaning stems from the decisions and progress of the past, so also is the future conditioned by the comprehension and the actions of the present. Prepared from open sources, this chronology has reference value of contemporary utility and will also serve the cause of future historians and analysts.

The year 1962 was only the fifth year since the Soviet SPUTNIK opened many American eyes to the early practical significance of space science and technology. Yet it was another spectacular year in space affairs. The highlights of 1962 were many: the orbital flights of Mercury Astronauts Glenn, Carpenter, and Schirra; the successful launching of more than 61 American vehicles; the spectacular data-recording flight past the planet Venus by MARINER II, not to slight RANGER V hitting the moon and the first international satellites, ARIEL I and ALOUETTE. There was the continued contribution of Tiros weather satellites as well as the dramatic first live global telecommunications achieved by TELSTAR. The rocket-powered X–15 research airplane continued its record-making contribution to the science and technology of manned space flight. Decisions by management and progress throughout NASA's program also are reflected in this chronology, not to ignore the indispensable contributions of other governmental agencies, the aerospace industry, and the academic community to the massive research and development effort now well underway.

In his address at Rice University on September 12, 1962, President John F. Kennedy said that this Nation "means to be a leading spacefaring nation." These words sound the challenge for which a response must be fashioned by all Americans. While 1962 was an eventful year, the future of astronautics appears even more stimulating.

GEORGE L. SIMPSON, Jr.,
Assistant Administrator for Technology Utilization and Policy Planning, National Aeronautics and Space Administration.

ASTRONAUTICAL AND AERONAUTICAL EVENTS OF 1962

CONTENTS

PREFACE

A chronology is not a full-fledged history. But this chronology is a necessary beginning of the historical process of documentation, analysis, and verification concerning the activities, problems, and accomplishments of the National Aeronautics and Space Administration and its academic, industrial, governmental, and international partners in the exploration and use of space for the benefit of all mankind.

This historical report was prepared from open, public sources. Science and technology in today's world are essentially indivisible. Space-related efforts by the Department of Defense and the military services, as well as international items of a non-NASA character, have therefore been included to help retain a valid historical context. Any effort by a free society also manifests considerable public discussion and policy comment, the inclusion of which was considered pertinent to meaningful presentation of known events in the science and technology of space exploration. The NASA Historical Staff appreciates the generous support by various NASA offices and centers as well as by members of the historical community in compilation of this report.

Astronautical and Aeronautical Events of 1962 is supplemental to *Aeronautics and Astronautics, 1915–60*, published by NASA (GPO, Superintendent of Documents), as well as *Aeronautical and Astronautical Events of 1961*, a NASA Historical Report published in 1962 by the House Committee on Science and Astronautics. Appendix A: "Satellites, Space Probes, and Manned Space Flights—1962," compiled by Dr. Frank W. Anderson, Jr., Assistant NASA Historian, is a continuation of Appendix A in both of the above chronologies. Appendix B: "Chronology of Major NASA Launchings, 1958–62," originally drafted by Mr. Robert Rosholt, provides a useful catalog not previously available.

Compilation of this chronology involved the entire NASA Historical Staff with Mrs. Helen T. Wells carrying the major drafting and editorial responsibilities. Appendix A was launched by Mr. Alfred Rosenthal, Historian of the Goddard Space Flight Center, while NASA Center historians at Launch Operations Center, Manned Spacecraft Center, and Marshall Space Flight Center also made their mark in this preliminary report.

Incompleteness, perhaps errors, either by commission or omission, may require further guidance by history-makers and scholars. Comments and criticism are welcomed by the NASA Historical Staff at any time.

EUGENE M. EMME,
NASA Historian (AFEH).

JANUARY 1962

January 1: National Bureau of Standards and U.S. Naval Observatory increased the standard frequency transmissions by 2 parts in 1 billion to allow for higher precision in scientific measurements, radio communications and navigation, and satellite tracking.

• The United Nations should tax commercial ventures in outer space, the ocean depths, and polar regions to obtain financial support, said Dr. Eugene Staley of the Stanford Research Institute in a memorandum to the United Nations. He also proposed that the U.N. should be given exclusive authority to license and regulate space traffic and satellites relaying telephone and television signals.

January 2: Army announced installation of new 102-foot antenna near Fort Dix built in conjunction with DOD's Project Advent, one link in development of a microwave radio-relay system for global communications using active-repeater satellites in a 22,300-mile-high orbit.

• Walter Hahn of General Electric's Defense System Department was named NASA's Director of Management Analysis Division.

January 3: NASA announced that Mercury Mark II spacecraft would be named "Gemini," after the third constellation of the zodiac featuring the twin stars Castor and Pollux. Gemini would be a two-man spacecraft used in development of the rendezvous technique, would be 50% larger than the Mercury capsule, and launched into orbit by a Titan II booster.

• Mercury capsule installed on top of Atlas booster preparatory for MA–6 manned flight; it was also reported from Cape Canaveral that first American orbital manned flight was now unofficially scheduled for January 23.

• Vice President Johnson sent a congratulatory telegram to members of the OSCAR amateur radio satellite team: "For me this project is symbolic of the type of freedom for which this country stands—freedom of enterprise and freedom of participation on the part of individuals throughout the world." OSCAR I was launched with DISCOVERER XXXVI on December 12, 1961.

• Dartmouth College announced new graduate program leading to a doctorate in the field of molecular biology.

January 3–10: Soviet cosmonaut, Major Gherman S. Titov, visited Indonesia at the personal invitation of President Sukarno, was then scheduled to go to Burma.

January 4: Announced at the Manned Spacecraft Center that a large "innertube" or "flotation collar" may be used to keep Mercury capsule afloat after a water landing. Collar would be installed by Navy frogmen. Astronaut Alan B. Shepard took part in proving tests conducted on Chesapeake Bay.

1

January 4: NASA announced contract with the University of Texas to design and build a radio antenna at the Balcones Research Center to be used in making radiation measurements of the moon and planets. It would be 16 feet in diameter and operate effectively at 30,000 to 150,000 megacycles.

- Dr. William W. Kellogg of the NAS Space Science Board reported on the study of the planet Venus at the American Geophysical Union. Bolometer studies of the atmosphere of Venus indicated a temperature of $-40°$ C (presumably the cloud tops), while temperatures deduced from measurements by large radiotelescopes indicated temperatures of about $300°$ C ($572°$ F) (believed to be surface temperature). A planetary probe could probably answer questions raised on the thickness and nature of Venus' aerosphere.

January 5: President Kennedy released part of a report submitted earlier by Vice President Johnson, Chairman of the Space Council. The report stated that the U.S. had generated a greater rate of progress in space in 1961 than in any other year but that "it is too early to make definitive comparisons as between our newly developing competence and the capabilities of the U.S.S.R."

- NASA first made public drawings of three-man Apollo spacecraft to be used in lunar landing development program.
- Dr. J. P. Kuettner, formerly chief of the Mercury-Redstone program, was named MFSC Manager of the Saturn-Apollo System Integration Program.
- USAF Minuteman successfully fired from silo at Cape Canaveral, its third straight success in underground firing.

January 6: FAA released memorandum dated December 29, 1961, stating that Stanford Research Institute's radiotelescope was a hazard to air navigation. The first of three such telescopes to be built in the U.S. by SRI extends 90 feet too high for Moffett Field air traffic and 128 feet too high for Palo Alto Airport traffic.

January 7: National Science Foundation reported that Congress had appropriated $10.8 billion for research and development in FY 62, which included $3 billion for research, $6.7 billion for development, $1 billion for facilities, and $100 million for information. NASA spent $1.4 billion, the DOD $6.2 billion.

- Executive Director of the Space Science Board of the National Academy of Sciences, Dr. Hugh Odishaw, reviewed the prospects in space in first of a series of lectures prepared for the Voice of America. Despite the great promise of practical application of space technology, Odishaw said: "I would contend that the challenges of research dwarf those of adventure and application." Other members of the Space Science Board were slated for later lectures in this Voice of America series.
- Reported by J. Alsop that experts estimate that the U.S.S.R. could possibly test an antisatellite missile during 1962, and would probably do so in 1963.

January 8: Special hand tools for use in zero-gravity conditions were tested by personnel of the Manned Spacecraft and the Marshall Space Flight Centers. Experiments were conducted in simulated space environment to try out non-torque hand tools drawn from a number of industrial sources.

January 8–13: Five-day symposium on aerospace medicine held at School of Aerospace Medicine at Brooks AFB, Texas, which included some 500 medical school professors, industrial scientists, military, and other Government specialists.

January 9: NASA's Associate Administrator, Dr. Robert C. Seamans, Jr., moderated a panel discussion on "Reliability—The Key to Space Operations," at the 8th National Symposium on Reliability and Quality Control in Washington. Panelists John H. Rubel, Assistant Secretary of Defense; Dr. C. Stark Draper, Head of the Dept. of Aeronautics and Astronautics, MIT; Dr. Simon Ramo, Exec. VP of Thompson Ramo Wooldridge, Inc.; and Dr. Jack A. Morton, VP Device Development, Bell Telephone Laboratories discussed means of comprehensive, planned efforts to increase the Nation's percentage of success in space launchings and operations.

• Addressing SAM's Aerospace Medicine symposium, Major General C. H. Mitchell, Vice Commander of AFCS, said that the "Russian threat in space is becoming obvious . . . Clearly it is our responsibility as a nation to insure that space is used to benefit all mankind. We can insure this only through development of the ability to conduct military operations in space with maximum effectiveness . . ."

January 10: NASA announced that the Advanced Saturn launch vehicle, to be used for manned flights around the moon and for manned lunar landings with rendezvous technique, would have five-engined first and second stages. The first stage (S–IB) would be powered by five F–1 engines (total of 7.5 million pounds thrust) and the second stage (S–II) would be powered with five J–2 engines (total of 1 million pounds thrust). A third stage (S–IVB) with a single J–2 engine would be used on escape missions.

• X–15 No. 1 piloted by Cdr. Forrest Petersen (USN) made its first forced landing in 47 flights when its rocket engine failed to ignite in mid-air after two attempts. X–15 was brought down without incident at Mud Lake, Nevada.

• NASA and AEC awarded 5-yr. contract for the development of the Nerva engine to Aerojet-General Corp. At same time, Aerojet-General signed a subcontract with Westinghouse Electric Corp. for nuclear portions of the development which began in 1955.

January 11: In his State of the Union message to the Congress, President Kennedy said: "With the approval of this Congress, we have undertaken in the past year a great new effort in outer space. Our aim is not simply to be first on the moon, any more than Charles Lindbergh's real aim was to be first to Paris. His aim was to develop the techniques and the authority of this country and other countries in the field of the air and the atmosphere.

"And our objective in making this effort, which we hope will place one of our citizens on the moon, is to develop in a new frontier of science, commerce and cooperation, the position of the United States and the free world. This nation belongs among the first to explore it. And among the first, if not the first, we shall be.

"We are offering know-how and cooperation to the United Nations. Our satellites will soon be providing other nations with improved weather observations. And I shall soon send to the Congress a measure to govern the financing and operation of an international communications satellite system, in a manner consistent with the public interest and our foreign policy.

"But peace in space will help us naught once peace on earth is gone . . . ."

January 11: USAF B–52H flew nonstop and without refueling 12,519 miles from Okinawa to Madrid, breaking by 1,283.4 miles the 1946 record set by Navy P2V–1 "Truculent Turtle." Maj. C. E. Evely headed crew of eight making the 21-hour-52-minute flight.

- At SAM symposium on aerospace medicine, Lt. Col. Burt Rowen (USAF), Chief of Bioastronautics at AFFTC, presented heartbeat and breathing records of Maj. Robert White during X–15 record speed flight of November 9, 1961. "When the President's Scientific Advisory Committee first became aware of the high heart and respiration rates [of pilots in high-performance aircraft] they became concerned with the question of success of the Mercury program. . . . But now this has come to be regarded as normal." Dr. Charles Sandhous of the University of California warned that an astronaut caught in space during a solar flare might age three years or more as a result of the radiation received.

- At Eighth National Symposium on Reliability and Quality Control in Washington, W. T. Sumerlin of the Philco Corp. estimated that 3,000 engineers and others were now devoting full time to this "new field."

- E. I. DuPont de Nemours & Co. announced it had awarded grants totaling $1,693,300 to 161 American universities and colleges to strengthen the teaching of science and related subjects, to promote fundamental research, and to aid facilities for education or research in science and engineering.

January 12: John Jay Ide, European representative of NACA (1921–40, 1946–50) and U.S. representative at numerous international air law and commerce conventions, died in New York. He was a founder and fellow of the IAS, a board member of the NAA, and an honorary member of the Royal Aeronautical Society. Mr. Ide had contributed to world aviation in making known the results of NACA research and in acquiring information on European progress, as well as helping the establishment of transatlantic commercial air routes.

January 13: USAF DISCOVERER XXXVII launched from Vandenberg AFB but did not attain orbit.

- NASA launched 2-stage Aerobee sounding rocket from Wallops Station to an altitude of 130 miles to obtain planning data for future solar physics projects.

- Dr. Hans-Georg Clamann, Chief of Space Medicine at SAM, submitted that a mountain peak near the south pole of the moon may be the ideal location for a lunar base. Continuous sunlight would provide support for growing vegetation.

- Reported from New York that three-day U.S.-British discussions on U.N. problems had devoted considerable time to problems of "the law of outer space." U.S. representatives indicated that the U.N. resolution of December covering "international cooperation and the peaceful uses of outer space" had extended international law to space. In view of this, the U.N. resolution destroyed any argument that surveillance satellites violate international law, as orbiting satellites are legally no different from operating a destroyer in international waters.

January 14: National Science Foundation released comprehensive report on "Education and Professional Employment in the USSR" by Nicholas DeWitt of the Harvard Russian Research Center.

January 15: "Big Shot" suborbital inflation test of 135-foot diameter sphere in Echo II program was conducted from AMR. The 44-inch cannister containing the uninflated balloon that would be larger and more rigid than ECHO I was separated at altitude from a specially stabilized Thor booster which carried both TV and 16-mm movie cameras. These cameras monitored the separation of the cannister and inflation of the "balloon satellite," showed the too-rapid inflation that ripped the balloon apart at 250-miles altitude; capsule with movie film re-entered and was later recovered by skin divers.

• In regular press conference, President Kennedy announced that he had asked his Science Advisory Committee, in cooperation with the Federal Council for Science and Technology, to report "as quickly as possible on the specific measures that can be taken within and without the Government to develop the necessary and well-qualified scientists and engineers and technicians to meet our society's complex needs—government, educational and industrial." He prefaced his announcement with a review of the declining number of scientists and engineers educated in the United States since 1951.

• In NASA press conference in New York City, Dr. Abe Silverstein, Director of the Lewis Research Center, outlined factors in NASA's nation-wide recruitment of 2,000 scientists, engineers, and technicians: (1) NASA had already interviewed 4,000 candidates in other cities and hoped to screen another 1,000 in New York; (2) salaries in industry tended to be higher than in Federal employment but NASA offered better postgraduate experience as well as opportunity to acquire national or international reputation; (3) NASA's manpower drain on the annual pool of 40-45,000 engineering graduates was small; and (4) the new specialists trained in the space program during the next decade would be an effective argument for much of the cost of the entire space effort.

January 16: In speech to the American Astronautical Society, Dr. Edward C. Welsh, Executive Secretary of the National Aeronautics and Space Council, submitted that Russia could match the U.S. in space "only if we place short-run convenience ahead of our nation's future. . . . There are those who seem to take for granted that a country like Soviet Russia, with less than half the per capita income of the United States, can afford a major successful space program, while we can not. That is, at best, ridiculous, and, at worst, deliberate sabotage."

• Rep. George P. Miller, Chairman of the House Committee on Science and Astronautics, spoke to the American Astronautical Society on the broad benefits derived from the space program. He said: "Space exploration is of such immense importance to man's total knowledge that it will benefit and alter the course of his existence in ways no more foreseeable today than those which resulted from the invention of the wheel. . . . One of the major benefits being gleaned from this vast effort to conquer space is the stimulus which it is providing for scientific research in new and uncharted areas. . . ."

January 16: Army Pershing missile successfully fired on longest test flight to date—400 miles—from AMR.

January 17: X–15 (No. 3) flown to 3,715 mph and 133,000 ft. in test of new autopilot control system, NASA's Neil Armstrong as pilot.

• Navy F8U Crusader piloted by Cdr. George Talley (USN) made first aircraft landing on a nuclear-powered carrier, the U.S.S. *Enterprise,* off the Virginia Capes.

January 18: Administration budget for FY 63 presented to Congress by President Kennedy. NASA requests totaled $3,787,276,000, with $2,968,278,000 going to research, development, and operations, and $818,998,000 to construction. DOD requests totaled $51.64 billion, with $6.843 billion going to research, development, test, and evaluation.

• In a NASA press conference following the presentation of the Administration's budget to Congress, Mr. James Webb, NASA Administrator, commented on the general attitude toward NASA's doubled appropriation request: "I would say that the people I have talked to have felt that we ought to go forward with the effort at about the level the President has recommended. I have seen no indication as we have had advanced discussion with some of the leaders, like the Chairman of the House Committee, Congressman Miller, the Chairman of the Senate Committee, Senator Kerr, and with some of the Appropriations members; we have seen no disposition on their part to just simply throw up their hands and say, 'No, sir.' Each one of them has said that what you have makes sense and we are going to look it over very carefully, of course, but nevertheless there has been no tendency to start out with a reaction that it was simply out of line with reality. . . . Under those circumstances my judgment is that there will be a good, strong, vigorous debate, just as there was when this agency was formed, and that when the issues are clearly out on the table, and the success of the very active flight program that we are now conducting makes itself felt, that there will be support for the program and that we will end up with about the recommendations of the President."

• NASA Goddard Space Flight Center selected Rohr Aircraft Corp. to negotiate for the manufacture and erection of three 85-foot-diameter parabolic antenna systems to be located at Pisgah National Forest (Rosman, N.C.), Fairbanks, Alaska, and an undetermined location in eastern Canada. When completed, these facilities in addition to similar system completed at Gilmore Creek, Alaska, would serve as the core of NASA's wide-band satellite instrumentation network. They would receive and record telemetry from large "second generation" satellites including Nimbus and the Orbiting Astronomical Observatory (Oao).

January 19: At NASA press conference, scientists described preliminary scientific results obtained by EXPLORER XII: based on study of 10 per cent of the data, it now appears that instead of the two radiation belts (previously called the inner and outer Van Allen belts) there is one magnetosphere extending roughly from 400 miles above the earth to 30,000 to 40,000 miles out. In this magnetosphere there is only a handful of matter but a very wide range of charged particles trapped by the earth's magnetic field. Density of electrons is now considered to be $10^8/cm^2/sec$ as opposed to the 10^{11} derived from previous satellites. Radiation levels of protons in the lower portion of the magnetosphere were found to be the same as reported by previous satellites, so that the magnetosphere still poses a formidable radiation problem to extended manned flight. At about 10 earth radii the magnetosphere ends abruptly. Above it is an extended area, about 20,000 km. across, of electromagnetic turbulence and featuring very low energy particles on the order of 1 to 20 kilovolts. Beyond this is thought to be true interplanetary space.

- Ranger III lunar shot postponed because of technical difficulties with Atlas booster rocket.
- Dr. John P. Meehan of the University of Southern California reported that "Enos," 37-lb chimpanzee who orbited the earth in MA–5 on November 29, 1961, developed temporary hypertension during flight due to frustration and confusion caused by equipment malfunction.
- In speech before the Builders and Contractors Exchange, Thomas F. Dixon, NASA Deputy Associate Administrator, pointed out that one fourth of the total NASA appropriation for FY 1963 was for construction of facilities. This construction does not include that to be undertaken by other Government agencies or by the thousands of private industries, educational institutions, and others with a role in the space effort.
- NASA received patent (3,016,693) for electrothermal rocket invented by John R. Jack and Wolfgang E. Moeckel of Lewis Research Center.
- Department of Defense authorized USAF to proceed with development of a mobile mid-range ballistic missile (MMRBM).
- Unnamed Soviet scientist, called "chief designer of Soviet spaceships," reported by Tass as saying that U.S.S.R. planned to establish an "industrial undertaking" on the moon.

January 21: NASA announced the launching of two Arcas meteorological sounding rockets from Kindley AFB, Coopers Island, Bermuda, the first of a series of tests to measure the structure and density of the atmosphere in support of an atmospheric re-entry experiment utilizing a Scout rocket fired from Wallops.

January 22: Reported from AMR that MA–6 launch would be delayed until January 27 at the earliest, due to difficulties in the oxygen system of the Mercury spacecraft.

- Fuel tank of the Atlas booster of Ranger III was repaired from inside in 48 hours, first such repair while booster was upright on the pad. Effort of General Dynamics/Astronautics crew was necessitated by fuel leak.

January 23: NASA's Goddard Space Flight Center announced the selection of Motorola, Inc., Military Electronics Division, of Scottsdale, Ariz., as contractor for research and development on the Goddard range and range-rate tracking system. Intended for tracking satellites in near-space and cislunar space, the system will measure spacecraft position to within a few feet and velocity to within fractions of a foot per second by measurements of carrier and side-tone modulations.

- Institute of the Aerospace Sciences awarded: the Louis W. Hill Space Transportation Award, carrying a $5,000 honorarium, to Robert R. Gilruth, Director of NASA's Manned Spacecraft Center, for his "outstanding leadership in technical development of spacecraft for manned space flight"; the Lawrence B. Sperry Award to Douglas G. Harvey for design, development, and testing of the first two nuclear auxiliary powerplants placed in orbit; and the Sylvanus Albert Reed Award to Alfred J. Eggers, Jr., chief of Vehicle Environment Division at NASA's Ames Research Center, for his work on solving problems of space vehicle re-entry and alleviating heat problems at hypersonic, satellite, and escape speeds.

- National Academy of Sciences' Space Science Board report on the atmospheres of Mars and Venus released. Prepared by William W. Kellogg of RAND and Carl Sagen of the University of California, it reported that available evidence suggests the existence of life on Mars and that space flights during the next decade would probably resolve this question.

- Reported by Dr. Sigmund Fritz of the Weather Bureau that TIROS III had spotted fifty tropical storms during the summer of 1961.

- Construction contract for Saturn umbilical tower (Launch Complex 34, Cape Canaveral) was awarded by NASA Marshall Space Flight Center to Consolidated Steel and Ets-Hokin-Galvin. Cost of the 240-ft. steel tower was estimated at $504,900.

January 24: NASA announced Aerojet-General Corp. had been selected as the contractor for design and development of the M–1 liquid-hydrogen engine. To develop 1,200,000 pounds of thrust, the M–1 would be used in the Nova vehicle as propulsion for the 2nd (4 M–1 engines) and 3rd (1 M–1 engine) stages. Target date for the operational use of the engine was 1965, total cost of the contract expected to be $90,000,000.

- Mr. John Stack, Director of Aeronautical Research, NASA, speaking before the Institute of the Aerospace Sciences in New York, advocated a variable-wing, variable-fuselage configuration for the nation's supersonic transport aircraft. Not only were these practicable for the state of the art but the only way he saw to gain the necessary subsonic vs. sonic flight characteristics and load flexibility to make the aircraft economically usable.

- Composite I, the Navy's 5-in-1 satellite package, was launched from AMR but failed to achieve orbit when the 2nd stage of the Thor-Able-Star booster rocket misfired.

January 25: EXPLORER X detected a "shadow" on the side of the earth facing away from the sun; in this shadow there is an absence of the solar "wind", a belt of plasma moving out from the sun at about 200 miles per second but deflected around the earth by the earth's magnetic field and creating a cone-shaped "shadow" some 100,000 miles across at its larger end. The EXPLORER X findings were reported to the annual meeting of the American Physical Society in New York by Dr. Bruno Rossi of MIT.

- NASA approved Saturn C–5 development program and authorized NASA Marshall Space Flight Center to direct its development.

- "Satellite Communications Corporation" bills were introduced by Senator Robert Kerr (S. 2650) and by Rep. George Miller (H.R. 9696), which would amend the NASA Act by adding a new section which would declare that it is "the policy of the United States to provide leadership in the establishment of a worldwide communications system involving the use of space satellites." The section would create a "Satellite Communication Corporation" which would be privately owned and managed, and which would develop and operate a communications satellite system.

- Stars may devote $\frac{1}{6}$ of their lifetime energy to production of neutrinos, according to Dr. Hongyee Chiu of Yale and NASA's Institute for Space Studies, speaking before the American Physical Society in New York. Neutrinos are among the most elusive of atomic particles, having no weight and no electrical charge, passing through matter with little chance of being stopped and with minimum interaction. Dr. Chiu theorized that the universe may have once been made up entirely of neutrinos which interacted to produce other particles and elements and that eventually the universe may return to neutrinos.

- The 140-ft. radiotelescope under construction at Green Bank, W. Va., would not be completed until 1963 and its original cost of $6 million would rise to $13 million, the National Science Foundation reported. Cause of most of the delay and added cost was said to be the danger of brittle fracture in the steel used to construct the antenna, which required that the fabricated parts be scrapped and new parts made.

- Army Signal Research and Development Laboratory at Fort Monmouth announced development of a super-powered laser with a peak power of more than 3 million watts. The laser (light amplification by stimulated emission of radiation) is 300 times as powerful as lasers in general laboratory use. It has potential in the fields of communications, range-finding, space vehicle guidance and special-purpose illumination.

- Ten-year program for the study of weather, climate, and the atmosphere was submitted to the President's Special Assistant for Science and Technology, Dr. Jerome B. Wiesner. Report by Dr. Sverre Petterssen was based on six conferences among 189 meteorological specialists and called for tripled manpower and expenditures including much research in space programs.

January 26: RANGER III was launched from AMR; excessive acceleration by the Atlas 1st-stage booster caused the 727-lb. payload to
pass 22,862 miles in front of the moon on January 28 instead of
impacting as had been planned. Failure of a high-gain antenna
to home on the earth rendered signals too weak to provide usable
television photographs from the ones RANGER took of the moon
as it passed it. RANGER III went into orbit around the sun. The
flight proved out many of the systems within the payload, including the mid-flight guidance mechanism.

• During hearings of the Joint Senate and House Economic Committee, Senator Paul H. Douglas stated that "the public has
never really had a chance to consider" the space program and
that one third of the members of the American Astronautical
Society had indicated in a poll that "a man landing on the moon
was not desirable." Budget Director David E. Bell stoutly
defended the space program. Senator Sparkman questioned
Senator Douglas: "I wonder whether Congress felt the same way
when it put up money for Samuel Morse" to develop the telegraph.

January 27: With the countdown at T—29 minutes, NASA postponed
the first U.S. manned orbital flight until February 1 because of
cloud cover at the launching site that would have precluded
adequate tracking of the vital first minutes of the flight. Astronaut John H. Glenn had been in the Mercury capsule atop the
Atlas launch rocket for more than 5 hours when the launch was
postponed.

January 29: NASA announced selection of three firms to submit final
proposals on the design and development of the Rift (Reactor-
In-Flight-Test) rocket stage that would flight-test the Nerva
nuclear rocket engine. The firms were General Dynamics/
Astronautics, Lockheed Missile and Space Company, and Martin
Marietta Corp. When selected, the contractor would be responsible for designing the Rift stage, fabricating and assembling
it at the Michoud Operations Plant, and conducting tests and
checkouts related to eventual flight test. The three firms were
selected from five who submitted initial bids on January 3, following a first-phase preposals conference at Marshall Space
Flight Center on December 7 attended by 33 firms.

• The last test model of the USAF Titan I ICBM was successfully fired
from AMR. It was the 47th Titan I launching from Cape
Canaveral, of which 34 were successes, 9 partial successes, and
4 failures.

January 30: NASA announced at Cape Canaveral that manned MA–6
launch would be postponed until February 13 because of "technical difficulties with the launching booster." John A. Powers,
NASA press spokesman, quoted Astronaut John Glenn as saying:
"Sure, I'm disappointed, but this is a complicated business.
I don't think we should fly until all elements of the mission are
ready. When we have completed all our tests satisfactorily then
we'll go."

• NASA announced that Cdr. Forrest S. Petersen (USN), one of six
research pilots assigned to the X–15 flight project, returned to
Navy duty after 3½ years with the X–15.

January 31: President Kennedy submitted the 1961 "U.S. Aeronautics and Space Activities" to the Congress.   In the preface, he said: "During 1961, major attention was devoted to establishing our policy objective of space leadership and to accelerating our efforts toward achieving that objective."

- Reported by Irene Fischer of the Army Map Service that three experimental methods measured the mean distance from earth to the moon as 238,866 miles, a figure accurate within about a mile.   A fourth method gives a figure accurate within about 8½ miles.   The 238,866-mile distance is 9 miles more than previously accepted figure used for decades and based on direct observation of the parallax of the moon at its mean distance from the earth.
- USAF contracted with Boeing Co. and General Dynamics Corp. to undertake final studies on a new tactical fighter aircraft known as TFX, one suitable for both Tactical Air Command and Navy carrier operations.
- Nike-Zeus with three solid-fuel stages successfully flight-tested at White Sands Missile Range.
- EXPLORER I completed its fourth year in orbit with a life expectancy of several more years.
- First J–2 engine of Block I R&D series of Saturn C–1 was successfully test-fired by Rocketdyne Div., North American Aviation.

During January: Manned Rogallo-wing paraglider built by NASA's Flight Research Center launched by automobile tow from Rogers Dry Lake, Calif.   Tests may include launch of glider by aircraft tow, leading to development of inflatible paraglider for use with Gemini and Apollo programs.

- European Broadcasting Union in Brussels undertook detailed planning for intercontinental TV experimentation and programing via NASA's Project Relay and AT&T's Telstar satellites to be launched in late summer of 1962.   Construction of stations for participation in the tests were reported underway in England, France, West Germany, and Italy.
- Nimbus meteorological satellite underwent rigorous test program at GE's Missile and Space Vehicle Center at Valley Forge, Pa.
- NASA awarded contract to Kollsman Instrument Division for 38-inch-diameter primary mirror in the space telescope to be used in the Orbiting Astronomical Observatory (Oao).
- NASA let contract with University of Denver to study the commercial use of space technological advancements as well as possible means to increase and forecast future civil products which may be generated by current space research.
- NASA initiated a $1.2 million contract with Documentation, Inc., to operate a completely integrated technical information center for one year.   The Center will disseminate space data to NASA, its prime contractors, and designated organizations and individuals throughout the free world, using latest automation for information retrieval, communication, and data acquisition equipment.   The facility will be located in Bethesda, Md.
- Management of Centaur contracts became a direct responsibility of MSFC, with transfer of Hawthorne (Calif.) Centaur Space Vehicle Project Office to MSFC.

FEBRUARY 1962

February 1: NASA announced that MA–6 manned orbital flight would be scheduled no earlier than February 13. Repair of leak in Mercury-Atlas fuel tank would be completed by then.

- Dr. William H. Pickering, Director of JPL, predicted in a press conference at Sunnyvale, Calif., that one of the next two attempts to impact the moon with a Ranger spacecraft would be a partial success. He reported that RANGER III was now more than 500,000 miles from the earth and moving into a solar orbit at 2,600 mph.

- Ten members of NASA's Office of Advanced Research and Technology, headed by Thomas F. Dixon, Deputy Associate Administrator, visited MSFC.

February 2: NASA announced that it was negotiating with Bendix Corp. and Radio Corp. of America on a contract to operate five manned-spaceflight tracking stations at Bermuda, Grand Canary Island, Kano (Nigeria), Zanzibar, and Guaymas (Mexico), as well as test and demonstration facility at NASA Wallops Station, Va. Project Mercury network would also be used for Project Gemini.

- Rep. James G. Fulton, senior Republican on the House Committee on Science and Astronautics, said in an interview: "There's no doubt our overall space program is slipping despite the high words and fine praise coming from the White House . . . If it continues to slip, we'll be lucky to get a man on the moon before 1980."

- Reported from Cairo, Egypt, that Cosmonaut Yuri Gagarin has said that the United States "eventually" would orbit a man around the earth.

February 3: Interviewed at home, Astronaut John H. Glenn, Jr., said that scheduled MA–6 launch on February 13 "can only bode for success." Surveying the crowd of newsmen on his lawn, Colonel Glenn remarked that "it looks like Hangar–S was not such a bad place after all."

February 4: Hindu astrologers predicted disaster during the period February 4–5 when five planets, the sun, the moon, and an invisible body called Khentu moved into rough alignment. Predictions of disaster were based on astrological tables. Reaction was so widespread that it prompted Prime Minister Nehru to ridicule the predictions in an address: "I do not know the secrets of nature. We will know them gradually. But what we don't know, we should not accept blindly. At least what strikes our intelligence as completely absurd should not be accepted."

12

February 5: NASA submitted to the Congress a draft NASA authorization for FY 1963 bill. It would provide legislative authority to support the NASA budget for FY 1963, and authorize NASA to spend $2,968,278,000 for research, development, and operations and $818,998,000 for construction of facilities.

- Results of *Aviation Week* poll of the members of the House of Representatives to secure "grass-roots" opinions on the U.S. space program were published. Representatives were asked to indicate how their constituents felt about key space aspects. The majority indicated that the U.S. space program is "proceeding at the right pace"; of the minority who disagreed with this, twice as many favored "speeding up the program." Landing a man on the moon was considered "something the U.S. must do primarily to keep up with the Russians." The majority also felt that NASA rather than the military should run the space program ("emphatic support for NASA on this question by a ratio of almost 5 to 1"). Rep. Emilio Q. Daddario noted on his questionnaire that it is regrettable that the space program is linked in the public mind to the cold war contest with Russia: "I am concerned by this because the space program does not therefore stand on its own and there is the resulting danger that it will not receive the continuous support it will need over the years ahead to do what must be done . . ."
- In Voice of America broadcast, Dr. Herbert Friedman, Supt. of Atmosphere and Astrophysics Division of NRL, reported that the mechanism involved in solar flares may contain the secrets of controlled production of thermonuclear power.
- USAF Cambridge Research Lab team, headed by Dr. Richard Dunn of the Sacramento Peak Observatory, successfully observed perfect total eclipse in Lae, New Guinea, making useful observations of the chromosphere (1,000- to 10,000-mile band of energy between sun's visible surface and the corona) and the sun's corona.
- Lt. Col. John H. Glenn, Jr., visited the White House at the invitation of the President.
- Navy HSS–2 Sea King became first helicopter to exceed 200 mph in official speed trial; a claimed world record of 210.65 mph was flown by the twin-turbine helicopter over 19-km. course in Connecticut. Pilots were Lt. R. W. Crofton (USN) and Capt. L. K. Keck (USMC).
- According to Tass, Academician Blagonravov reported that Soviet radar observations of Venus last year defined the astronomical unit as 149,599,500 kilometers, plus or minus 2,000 km.

February 6: Announced at Cape Canaveral that the MA–6 launch attempt had been changed from February 13 to February 14.

- Tiros launch at AMR postponed due to technical difficulties.

February 7: In a special message to the Congress, President Kennedy proposed the creation of a "Communications Satellite Corporation" to be financed through the sale of stock to communications companies and the general public. This privately-owned corporation would have the responsibility to develop, own, and operate communications relay stations in space. The President declared that the creation of a satellite communication system would be "a measure of immense long-range importance."

February 7: In regular press conference, President Kennedy was asked for his "evaluation of our progress in space at this time" and whether the U.S. had changed its "timetable for landing a man on the moon?" He replied: "I have said from the beginning we've been behind. And we are running into the difficulties which come from starting late. We, however, are going to proceed. We're making a maximum effort, as you know, and the expenditures in our space program are enormous. And, to the best of my ability, the time schedule, I hope, has not been changed by the recent setbacks."

• President Kennedy requested $156 million supplemental appropriations for NASA during FY 1962, $85 million of which was to cover Advanced Saturn, the Centaur vehicle program, and the M–1 engine program.

• AEC and NASA announced that the Catalytic Construction Co. had been selected as construction contractor for nuclear rocket development facilities for Project Rover at the AEC Nevada Test Site near Las Vegas, Nev.

• Army launched Nike-Zeus from underground cell, at White Sands.

• Harrison E. Salisbury reported that "in the most advanced echelons of Soviet science there is emerging a tendency to seek a nonmaterialist concept of the universe." According to the former Moscow correspondent who recently toured the Soviet Union for two months, "some of the most eminent figures in the galaxy of Soviet phsyicists, astronomers and mathematicians are involved," although they are not believers in formal religion or dogma.

February 8: NASA's TIROS IV launched by a 3-stage Thor-Delta rocket from Cape Canaveral into a near-circular orbit with an apogee of 525 miles and perigee of 471. TIROS IV featured the same basic types of equipment as in previous Tiros satellites, including cameras for cloud-cover photography and infrared sensors to measure temperatures at various levels in the atmosphere. Principal innovation was a camera with new type of wide-angle lens covering an area 450 miles on a side, which was expected to provide minimum distortion. NASA press conference reported that quality of TIROS IV pictures was good.

• Navy, in conjunction with Weather Bureau and Canada, launched a major ice reconnaissance effort in the Gulf of St. Lawrence as part of a project to develop procedures and techniques for interpreting satellite readouts of ice formation. Called Project Tirec, the effort was timed to coincide with the successful launch of TIROS IV.

• Deputy NASA Administrator Dr. Hugh L. Dryden accepted invitation to become an Honorary Fellow of the British Interplanetary Society, an honor only accorded to eight persons who have rendered major services to astronautics over a number of years.

• Edmund F. Buryan was named consultant to NASA to develop plans, policies, and programs for applying technological advances to practical benefit of the U.S. economy and industry.

February 9: Weather Bureau reported that TIROS IV had taken "striking pictures, of excellent quality, of cloud cover, and of snow and ice distribution" during its first 24 hours in orbit.

February 9: NASA-wide conference reviewing NASA's Future Applications Program was held in headquarters to review best means of increasing application of technological advances derived from the space program to commercial and industrial purposes.

- NASA announced that General Electric had been selected for a major supporting role in the manned lunar Apollo project, to provide integration analysis of the total space vehicle (including booster-spacecraft interface), assuring reliability of the entire space vehicle, and developing and operating a checkout system.
- First successful captive firing of RL–10 Centaur engines on a single-tank propulsion test vehicle, at Edwards Rocket Site, Calif.

February 11: Project Oscar Association chairman, M. C. Towns, reported that signal reports on OSCAR I, placed in orbit in December 1961, had been received from radio amateurs throughout the world. No confirmation was received that OSCAR's Morse signal was heard behind the Iron Curtain. Towns announced that the 250,000 hams would have another Oscar, tentatively set for launch sometime in late spring.

- Reported in Tass that a Soviet expert on rocket propulsion, not mentioned by name, had said: "The creation of powerful rockets has been full of pitfalls. Our investigations and experiments have not always been successful—far from it. There have been difficulties . . ."

February 12: In NAS-sponsored lecture on "Geomagnetism" for Voice of America broadcast, Dr. James P. Heppner of NASA Goddard Space Flight Center said: "Each new space experiment adds to the list of unanswered questions about our electromagnetic environment." He pointed to the changes in the scientific concepts of the boundary of the earth's atmosphere and of relationships between the earth's magnetic field and solar radiation that have resulted from data already acquired by satellites and probes.

- Soviet officials indicated that newsmen may be allowed to cover the next Soviet space shot, a report later denied.

February 13: USAF Atlas-E model ICBM completed its test-flight program with a 7,000-mile strike down the Atlantic Missile Range. The Atlas-E, already operational at Vanderberg AFB, Calif.; Warren AFB, Wyo.; Fairchild AFB, Wash.; and Forbes AFB, Kan., showed an overall test record of nine successes, seven partial successes, and two failures.

- Reported from Moscow that Soviet authorities were considering an offer to purchase licenses on U.S. patents, with royalties for their use. An agreement for the visit of a Soviet trade delegation later in the year for negotiation of such an arrangement was announced by the National Patent Development Corporation in New York.

February 14: MA–6 launch postponed because of bad weather in recovery areas. At regular press conference, President Kennedy was asked about the eighth postponement of "Col. Glenn's flight." He replied as follows: "Well, it is unfortunate. I know it strains Colonel Glenn. It has delayed our program. It puts burdens on all those who must take these decisions as to whether the mission should go or not.

"I think it's been very unfortunate. But, I have taken the position that their—the judgment of those on the spot should be final in regard to this mission and I'll continue to take that judgment.

"I think they would be reluctant to have it cancelled for another three or four months because it would slow our whole space program down at a time when we're making a concentrated effort in space.

"But I'm quite aware of the strain it caused everyone and it's been a source of regret to everyone and—but I think we ought to stick with the present group who are making the judgment and they are hopeful still of having this flight take place in the next few days.

"And I'm going to follow their judgment in the matter even though we've had bad luck."

February 14: 25-ft. space environmental simulator used for the first time at Jet Propulsion Laboratory, in beginning of 3-week test of Mariner spacecraft designed for a Venus probe.

• Twenty specialists of Weather Bureau's Research Flight Facility, who make prolonged and repeated flights into hazardous weather for data for use in atmospheric research, were awarded the Department of Commerce Gold Medal.

February 15: USAF Minuteman ICBM launched from underground silo at AMR, the 5th successful silo test in a row since last November.

• Dr. Bernard Lubarsky of NASA's Lewis Research Center named one of Ten Outstanding Young Men in the Federal Government for 1961 to receive Arthur S. Flemming Awards.

February 16: Following an early morning weather briefing, Mercury Operations Director Walter C. Williams advised that weather conditions again precluded a launch attempt of the MA–6 mission. February 20 was announced as the earliest possible launch date. Notified of the decision at 12:50 AM, Astronaut Glenn said: "I guess it was to be expected. We all knew the weather was marginal."

• Dr. Robert C. Seamans, NASA Associate Administrator, outlined NASA's views of solid-propellant Nova-class motors in a letter to John H. Rubel, Deputy Director, Defense Research and Engineering:

"NASA has a strong interest in the technology of solid-propellant rockets and hopes to make use of them in NASA vehicles when the appropriate feasibility is demonstrated . . ." NASA interest was expressed for motors with thrust levels of "about 2.5 million and 5 million lbs. and burning times of approximately 115 seconds."

"The higher thrust level produced by motors about 240 inches in diameter by about 100 feet long is especially attractive from a long-range, payload growth viewpoint. In addition, the greater thrust and impulse per motor results in designs with fewer motors per stage, and thus with greater potential for reliability. . . ."

• Congressman V. Anfuso and Senator H. Humphrey introduced similar bills (H.R. 10203 and S. 2849) to provide for research into and development of practical means for the utilization of solar energy.

• Hartley A. Soulé of NASA's Langley Research Center retired after 34 years' service at Langley. Soulé was instrumental in the establishment of what is now the NASA Flight Research Center at Edwards, California, and directed the research on the X–1 (which in October 1947 first broke the speed of sound) and the X–15. More recently he was NASA Project Director for the multimillion-dollar Project Mercury worldwide tracking and ground instrumentation system.

February 16: SAC combat missile crew launched successful Atlas ICBM from Vandenberg AFB, California.

- USAF awarded contract to Martin-Marietta Corp. for study of design criteria for a Titan III standardized space launch vehicle.

February 17: 100-inch diameter, 53-foot long, solid-fuel rocket static-fired for 98 seconds and developed 600,000 pounds of thrust by Aerojet-General, the largest solid rocket fired to date.

February 18: NASA announced Project Fire, a high-speed re-entry heat research program to obtain data on materials, heating rates, and radio signal attenuation on spacecraft re-entering the atmosphere at speeds of about 24,500 mph. Information from the program would provide technology for manned and unmanned re-entry from lunar missions. Under management of Langley Research Center, Project Fire would use Atlas D boosters and the re-entry velocity package would be powered by Antares solid-fuel motor (3rd stage of the Scout).

- NASA Administrator James E. Webb, Assistant Administrator Hiden T. Cox, and other NASA speakers addressed the annual convention of the American Association of School Administrators. Mr. Webb said: "Space science and space exploration have become an integral and vital part of a great industrial and technological revolution which is now taking place in our own country and throughout the world . . . [the] rapid *rate of change* as much as the change itself is one of the dominant facts of our time." Dr. Cox outlined NASA's program of educational services to meet the needs of education in and for the Space Age.

- Four time-to-climb records claimed by USAF Northrup T–38 Talon supersonic trainer.

February 19: NASA-AEC designated the Jackass Flats area of the AEC Test Site as the Nuclear Rocket Development Station. Placed under management of the joint NASA-AEC Space Nuclear Propulsion Office (SNPO), the new station has been used since 1959 for ground tests of Kiwi reactors.

- In Voice of America broadcast, Alan H. Shapley of the National Bureau of Standards' Central Radio Propagation Laboratory, reviewed the "new dimension to research in ionospheric science" added by space vehicles. He pointed out that "scientific expeditions landing on the surface of Mars or Venus, whether manned or unmanned, will have to communicate by radio with terrestrial headquarters. We must know which frequencies will get through the planetary ionosphere, and which can be used with the radio-mirror effect to talk from one part of the planet to another. So far we have had only [the earth's] ionosphere to study, but this is like the doctor with only a single case. To understand a disease he needs many cases."

- The Air Force Space Plan, a ten-year blue-print for military space technology, was given to a House committee by Lt. Gen. James Ferguson, DCS/R&T. The Space Plan foresaw a military need for rendezvous, docking, and transfer. The USAF expected to depend on a manned rendezvous vehicle, using the two-man Gemini, built by NASA, as an initial vehicle. The Space Plan was sent to the Air Force Scientific Advisory Board for review.

February 19: Army fired Pershing ballistic missile in extended range test at Cape Canaveral, which exceeded the range of the 200-mile Redstone.

• Reported that Russian Cosmonauts Gagarin and Titov were assured membership in the Supreme Soviet. Neither was opposed in election slates and both will represent their home towns. It was also reported that the U.S.S.R. Foreign Ministry's press department had denied the rumor that Russia was preparing to let Western newsmen cover the next Soviet space shot.

February 20: Mercury spacecraft (Capsule 13), FRIENDSHIP 7, was launched into orbit by Atlas booster (109D), Lt. Col. John H. Glenn, Jr., (USMC), as astronaut. After three orbits (apogee: 162.5 mi.; perigee: 98.9 mi.), FRIENDSHIP 7 re-entered and parachuted into Atlantic some 166 miles east of Grand Turk Island in the Bahamas. Picked up by the destroyer *Noa*, Glenn remained inside capsule until on deck. Beyond being the first U.S. manned orbital space flight, MA–6 flight of FRIENDSHIP 7 provided aerospace medical data during 285 minutes of weightlessness, including consumption of solid and liquid food and disorientation exercises. Astronaut Glenn was forced to "fly by wire" (i.e., pilot the spacecraft) during 2nd and 3rd orbits due to troubles with the automatic pilot. Glenn was pronounced "hale and hearty" after his 81,000-mile flight of 4 hrs. 56 min.

• President Kennedy made a statement to the press on the lawn of the White House, expressing the "great happiness and thanksgiving of all of us that Col. Glenn has completed his trip . . . .

"I also want to say a word for all those who participated with Col. Glenn in Canaveral. They have faced many disappointments and delays—the burdens upon them were great—but they kept their heads and made a judgment and I think their judgment has been vindicated.

"We have a long way to go in this space race. We started late. But this is the new ocean, and I believe the United States must sail on it and be in a position second to none.

"Some months ago I said that I hoped every American would serve his country. Today Col. Glenn served his, and we all express our thanks to him."

• Estimated that over 60 million Americans witnessed the launch of FRIENDSHIP 7 on live TV. Voice of America carried live overseas radio coverage. The U.S. Senate recessed at 2:30 PM EST just before the landing of Mercury spacecraft. The U.S. Post Office placed Project Mercury postage stamp on sale minutes after Astronaut Glenn stepped on the deck of *Noa*.

February 21: At regular press conference, President Kennedy opened with a statement on space, saying:

"It is increasingly clear that the impact of Col. Glenn's magnificent achievement yesterday goes far beyond our own times and our own country. The success of this flight, the new knowledge it will give us, and the new steps which can now be undertaken will affect life on this planet for many years to come.

"This country has received more than 30 messages of congratulations from other heads of state all over the world . . . . I want to express my thanks to them and at the same time pay tribute to the international cooperation entailed in the successful operation of the Mercury tracking network . . . .

". . . Chairman Khrushchev of the Soviet Union [suggested] . . . that it would be beneficial to the advance of science of our countries if our countries could work together in the exploration of space. I am replying to his message today . . . .

"We believe that when men reach beyond this planet they should leave their national differences behind them . . . .

"It has been said that peace has her victories as well as war, and I think all of us can take pride and satisfaction in this history of technology and the human spirit."

February 21: Mercury officials debriefing Astronaut John Glenn at Grand Turk Island announced that he was in "excellent physical condition." Dr. Robert B. Voas, psychologist and training officer, said that it was "quite significant" that Glenn had been unable to detect any ill effects during almost five hours of weightlessness. Voas pointed out that longer flights such as the 17 orbits of Cosmonaut Titov are required to document the physical effects of prolonged weightlessness.

• In formal response to congratulatory note from Premier Khrushchev on the flight of FRIENDSHIP 7, President Kennedy said: ". . . I welcome your statement that our countries should cooperate in the exploration of space. . . . We of course believe also in strong support of the United Nations in this field and are cooperating directly with many other countries individually.

"I am instructing the appropriate officers of this Government to prepare new and concrete proposals for immediate projects of common action, and I hope at a very early date our representatives may meet to discuss our ideas and yours in a spirit of practical cooperation."

• NASA announced plans to attempt to launch 440-lb. Orbiting Solar Observatory (Oso) containing 13 experiments, within a week.

• D. Brainerd Holmes, Director of NASA's Office of Manned Space Flight, addressed the Engineers, Scientists, and Architects Day meeting, in which he said: "as a stimulus to our fervent interest in rocketry have been the realities of life in a world which is divided into two armed camps, one of which has the avowed intention of imposing its own way of life on free peoples everywhere. We are, in short, compelled to achieve and maintain leadership in space research and technology because our own fate as a free nation and, in fact, the fate of human civilization as we know it, will depend upon whether the spacecraft of the future are devoted to peaceful purposes or to the destruction of the human race . . . ."

• Thomas F. Dixon, NASA Deputy Associate Administrator, in a speech on the Apollo Program at San Jose, California, outlined NASA's progression of flight programs leading to goal of a manned lunar mission before 1970, relating the X-15, Ranger, Surveyor, Mercury, Gemini, and rocket propulsion developments to the Apollo program.

February 21: Dr. Orr E. Reynolds, formerly head of the DOD's Office of Science, named Director of NASA's Bioscience Programs in the Office of Space Sciences.   During 1949–1957, Dr. Reynolds was Director of ONR's Biological Sciences Division.

• In speech to National Conference on Application of Electrical Insulation, Gen. Bernard A. Schriever (USAF) pointed to the lessons of history:

". . . Our boldest predictions do not keep very far ahead of technical developments.

"A second lesson is that the advance of scientific and engineering knowledge is most effectively achieved through joint effort. The growth of military airpower has owed much to the scientific and industrial community . . . . I think we can predict the same mutual benefits in space exploration . . ."

• USAF announced that an unnamed satellite had been launched by a Thor-Agena B booster from Vandenberg AFB.

February 22: Before the 3rd West Coast Reliability Symposium, Dr. Landis S. Gephart, NASA's Director of Reliability and Quality Assurance Office, said that the U.S. has "reliability requirements greater than our record of achievements to date will satisfy." He pointed out that when Saturn-class vehicles will cost an estimated $20 to $30 million a launch, neither NASA or the nation can tolerate a "batting average of 500."   Dr. Gephart then discussed three publications NASA was issuing in connection with contracts to ensure tighter quality control for NASA hardware.

• In speech before the Woman's National Press Club, Adlai Stevenson, U.S. Ambassador to the United Nations, said that the U.S., U.S.S.R., and other qualified nations should pool their space efforts under the auspices of the United Nations.   "Perhaps their children," he said, "will sail to Mars together."

• First International Woman's Space Symposium held at Los Angeles, Shirley Thomas as chairman.

• The National Center for Space Studies of France, headed by Pierre Auger, released a summary of its 1962 space program: launching of sounding rockets containing a rat and later a cat to 125 miles altitude; research observations of the sun and various planets; and telecommunications experiments.   France would also cooperate in a joint satellite project with the U.S. and participate in the European Organization for Space Research.

February 23: In ceremony at Cape Canaveral, President Kennedy thanked the Project Mercury team for the successful flight of FRIENDSHIP 7:

". . . it's my great pleasure to speak on behalf of all our fellow Americans in expressing pride and satisfaction to those intimately involved in this effort.   All of us remember a few dates in this century and those of us who were very young remember Col. Lindbergh's flight, Pearl Harbor, and the end of the war, and remember the flight of Alan Shepard, Major Grissom and we remember the flight of Col. Glenn . . ."

President Kennedy then awarded the NASA Distinguished Service Medal to Robert R. Gilruth and read the citation, "for his distinguished leadership of the team of scientists and engineers that carried Project Mercury, the U.S. initial manned space flight program, from its inception to the successful accomplishment of man's flight in orbit about the earth . . ."

Lt. Col. John H. Glenn also was awarded the NASA Distinguished Service Medal by the President, the citation to which said:" . . . made an outstanding contribution to the advancement of human knowledge of space technology and a demonstration of man's capabilities in space flight. His performance was marked by his great professional skill, his skill as a test pilot, his unflinching courage. . . ."

In his remarks, Astronaut Glenn thanked fellow astronauts and the entire Mercury organization: "we all acted literally and figuratively as a team. . . . It goes across the board . . . sort of a crosscut of Americana, of industry, and military and civil service. . . . It was headed up by NASA, of course, but thousands and thousands of people have contributed as certainly much or more than I have to the project. . . ."

February 23: In later NASA press conference at Cape Canaveral, Astronaut John Glenn described his three-orbit flight. He related his observation, his pilotage during almost two orbits, and his concern when burning fragments of the retrorocket package during re-entry appeared to indicate to him that the heat shield was breaking up. He referred to "weightlessness" as a "very pleasant" sensation to which he is becoming "addicted." Administrator Webb and Director of the Manned Spacecraft Center, Robert Gilruth, spoke briefly. This conference, like all events associated with the MA–6 flight, was fully covered in all detail by the entire American press, radio, and TV.

- 12 European nations agreed on draft convention for the creation of a European Space Research Organization. Once signed by member nations (Britain, France, West Germany, Italy, Belgium, Denmark, Norway, Sweden, the Netherlands, Switzerland, Austria, and Spain), the organization would cooperate with the U.S.'s NASA and the International Committee on Space Research of which the Soviet Union is a member.

- DOD and NASA signed an agreement that neither agency would begin the development of a launch vehicle or a space booster without the written acknowledgment of the other.

- Senator Henry M. Jackson in a radio interview said there was evidence that "lives were lost" in the Russian manned space flight program. He urged the Kennedy Administration to challenge the U.S.S.R. to "lay bare" its trials and errors in achieving manned space flight.

- Drew Pearson repeated rumors previously published in his column that 5 Soviet cosmonauts may have been killed in manned space flight attempts.

- USAF Titan I ICBM launched by military crew at Vandenberg AFB, Calif.

- Reported that two radio hams of Ann Arbor, Mich., J. Schmidt and E. Nuttle, recorded the voice transmissions of Astronaut Glenn during the three-orbit flight of FRIENDSHIP 7.

February 24: Robert B. Leighton of Cal Tech reported on first direct detection of elastic waves 40 to 50 miles high that carry heat from the sun to its corona. Photographs made through Mount Wilson Observatory detected the energy transfer mechanism of waves which rose and fell at the rate of 1,000 mph and in five-minute periods.

- Department of Defense Directive (No. 5030.18) issued which specified responsibilities and procedures for the support of NASA "in order to employ effectively the Nation's total resources for the achievement of common civil and military space objectives." It assigned responsibilities within DOD to cover existing DOD-NASA Agreements on program content, funding, scheduling, and assignment of respective responsibilities.

- Studies conducted by American Airlines revealed that 78% of Americans have never traveled on commercial airliner, while 8 million use the airlines regularly (15% of these accounting for 64% of commercial air passengers).

February 25: Soviet scientists claimed to have discovered the third radiation belt around the earth and published such findings two years before the findings of EXPLORER XII were made public by NASA on January 19, 1962. Academician Blagonravov, Vice President of the International Committee on Space Research, said in an *Izvestia* interview that the existence of a dense belt with energies of 200 to 20,000 electron volts at a distance of 25,000 to 50,000 miles was recorded by Soviet space launchings in 1958. Such findings, he said, were published by Dr. K. Gringauz and associates in the February and April issues of the *Soviet Academy of Sciences Proceedings* in 1960, and in later publications. Commenting on the view that the three radiation belts really formed a single large pulsating band that might be called a "magnetosphere," Blagonravov agreed that the boundaries might be arbitrary but that the charged particles in each belt had distinctive characteristics and that it would be "inexpedient" to reject the theory of three belts.

February 26: John Glenn Day in Washington, D.C. An estimated 250,000 persons lined the rain-drenched parade route from the White House to Capitol Hill. In 20-minute address to the joint session of the Congress, Astronaut Glenn paid tribute to the Project Mercury team and pointed to the unbounded future of space exploration:

". . . I feel that we are on the brink of an area of expansion of knowledge about ourselves and our surroundings that is beyond description or comprehension at this time.

"Our efforts today and what we've done so far are but small building blocks on a very high pyramid to come . . .

"We're just probing the surface of the greatest advancement in man's knowledge of his surroundings that has ever been made . . . There are benefits to science across the board. Any major effort such as this results in research by so many different specialties that it's hard to even envision the benefits that will accrue in many fields.

"Knowledge begets knowledge. The more I see, the more impressed I am not with how much we know but with how tremendous the areas are that are yet unexplored . . ."

February 26: In NAS lecture for Voice of America broadcast, Dr. Joseph W. Chamberlain of the University of Chicago's Yerkes Observatory predicted that space-age research techniques may soon provide final answers to questions about the aurora that have eluded scientists for centuries. "Better understanding of physical processes producing the aurora," he said, "should allow the spectroscopist to derive even more information from his observations of the auroral spectrum about the physics and chemical processes in the high atmosphere." He suggested that space probes may detect auroras on other planets which will provide an entire new set of data different from terrestrial phenomena with which to test hypotheses for auroral bombardment mechanisms.

- Technical difficulties with third stage of Thor-Delta booster postponed launching of Oso satellite at AMR.

- Senators Estes Kefauver and Wayne Morse proposed legislation calling for a government-owned corporation to operate space satellite communications network.

- Soviet representative at the United Nations, Valerian A. Zorin, told a news conference that the U.S.S.R. would cooperate with the U.S. and other member nations of the U.N. Committee on the Peaceful Uses of Outer Space meeting on March 19. Zorin said that he hoped that the committee would be able to agree on a work program for "mutually beneficial and advantageous cooperation."

February 27: NASA Administrator Webb introduced Robert R. Gilruth and Astronauts Shepard, Grissom, and Glenn to NASA Headquarters' personnel assembled in FOB No. 6, Washington.

- House Committee on Science and Astronautics began hearings on $3.7-billion NASA authorization request. Administrator Webb, Robert R. Gilruth (Director MSC), and Astronauts Glenn, Grissom, and Shepard appeared as witnesses.

    Mr. Webb outlined NASA's projected program to be covered in detail in subsequent testimony: (1) total ten year program ahead may cost over $35 billion; (2) new astronaut training program to include scientists will be announced shortly; (3) future manned spacecraft will be capable of landing on land or water; (4) NASA is opposed to using astronauts as goodwill ambassadors, as they are needed on the current program; (5) 92% of NASA's budget will be spent through industry and universities.

    Astronaut Glenn pointed out that the Nation "must be prepared" for failures in future manned space flights. "We don't envision every flight," he said, "coming back as successfully as the three we have had so far . . . I hope we will always have the confidence in the program that we now have despite the fact there will be times when we are not riding a crest of happiness and enthusiasm as we now are."

- NASA witnesses appeared before the Senate Committee on Aeronautical and Space Sciences hearings on communication satellites system. Deputy NASA Administrator Dryden stated that the U.S. would have operational satellite communications within five years.

- USAF DISCOVERER XXXVIII launched into orbit with undisclosed payload.

February 27: Second Saturn flight vehicle (SA–2) arrived at Cape Canaveral aboard the barge *Promise.*

- In testimony before closed session of the House Appropriations Committee, Prof. James A. Van Allen stated that man-in-space programs were not necessary for scientific space exploration. "For the same investment of effort we learn much more without the man. . . . A monkey made the first orbital trip and he made out alright." In testifying on behalf of the National Science Foundation appropriation, Van Allen said that the U.S. should not rush into a cooperative space program with Russia; failures would worsen relations since Russia has more to gain from such a program. The U.S., he said, had surpassed Russia "in nearly all areas of purely scientific investigation of outer space" but some American scientists have "an uneasy, creeping feeling that Russia is about to launch a manned flight around the moon." Testimony was released on May 14.

- USAF witnesses before the Senate Military Appropriations Subcommittee testified under questioning that DOD budget was not adequate in all areas. Gen. Curtis E. LeMay, USAF Chief of Staff, said that funds should be provided for "at least 100 more Minuteman ICBM's" than in present budget, and that he favored a "little more emphasis" in the military space program on early-warning systems. Secretary Eugene Zuckert indicated that he favored development of the RS–70 as a full-fledged weapon system.

- Dr. Nancy Roman, Chief of Astronomy and Solar Physics in NASA's Office of Space Sciences, was one of six receiving the Federal Women's Award.

February 28: Mercury astronauts Shepard, Grissom, and Glenn appeared before both the House Science and Astronautics Committee and the Senate Aeronautics and Space Committee, answering a variety of questions on future manned space flight. Dr. Robert Seamans, NASA Associate Administrator, summarized specific NASA program, manpower, and money requirements before the House Committee.

- USAF Atlas–E ICBM launched from Vandenberg AFB, the first such launch from this base.

- NASA announced that it would acquire launch facility at the Pacific Missile Range to be used for all NASA Thor-Agena B launches requiring a polar orbit (e.g. Echo II). The basic Thor pad structure is being provided by the USAF in addition to management for certain pad modifications, while the Army provided a surplus Redstone Service Tower and the Navy operates the Pacific Missile Range. NASA field responsibility on the West Coast is provided by the NASA Launch Operations Directorate Test Support Office, Point Mugu, headed by Simon J. Burttschell.

- H. Douglas Garner and Henry J. E. Reid, Jr., aerospace technologists at NASA Langley Research Center, awarded $1,000 for invention of Horizon Scanning Active Attitude Orientation of Stabilized Space Vehicles. James S. Albus, engineer at NASA Goddard Space Flight Center, awarded $1,000 for his invention of a Digital Solar Aspect Sensor. Invention awards of less than $1,000 were made to following NASA employees: William H. Kinard and

Sidney A. Batterson, Langley Research Center; Estin N. Baker, Jess S. W. Davidsen, Ralph L. Mossino, and Gilbert G. Robinson, Ames Research Center; Wilhelm Angele, Marshall Space Flight Center.

February 28: Under Secretary of State George C. McGhee testified before the Senate Aeronautics and Space Committee in support of the Administration bill to create a private corporation for the U.S. portion of a global satellite communications system. He said that it "would lead to opportunities for all members of the family of nations . . . to participate in a truly joint venture which will be clearly to the benefit of all mankind . . . ."

- USAF released unclassified "U.S. Air Force Space Policy" Report to the House Armed Services Committee in executive session. It was reported as saying that the U.S. must "recognize the peril" of Soviet space supremacy, take the lead in military space technology, and called for continued close cooperation with NASA.

- Ejection capsule test at Edwards AFB, Chief W/O Edward J. Murray (USAF) was ejected from B-58 flying at 565 mph and capsule parachuted to earth successfully.

During February: Kenneth C. Sanderson of NASA's Flight Research Center was named Chairman of the Dyna-Soar Instrument Development Team, a joint USAF-NASA-Industry group formed to provide technical direction over the development of sensors, communications systems, signal conditioners, airborne recorders and telemetry, ground recovery devices, and data processing equipment. Sanderson succeeded Frank Smith of NASA's Langley Research Center.

- Pilot course in the practical engineering aspects of space satellites began at Texas A&M College under an agreement with NASA.

- Major problem under study by Air Force Cambridge Research Laboratories' Sacramento Peak Observatory was the prediction of safe intervals for manned space flight. Since March 1961, intensive study has been made of proton showers originating from solar flares and arriving at the earth from 30 minutes to 6 hours after the flare has peaked. One major flare out of every four produces proton showers which follow the 11-year cycle of activity. Sunspot maximum occurred in 1957-1958 and the minimum is expected in 1964-1965.

- Army announced establishment of Space Construction Office in the Office of the Chief of Engineers to coordinate all activities carried out for NASA, to be headed by Brig. Gen. T. J. Hayes, III.

- Dr. Alla G. Masevich, Deputy Chairman of the Soviet Astronomical Council, published first serious mention of ECHO I in a leading non-technical Russian magazine (*Ogonyok*). Virtually nothing has been said about ECHO I in the Soviet press since its launch in August 1960, although it has remained in orbit longer than any Soviet satellite and was visible to Soviet citizens.

- Draft convention presented to Australia covering joint use of the Woomera Rocket Range by space research program of the western European nations.

- NASA's Marshall Space Flight Center was presented an Award of Honor by the National Safety Council for having logged nearly 4 million working hours without a disabling injury or fatality.

During February: Astronaut John Glenn's orbital space flight produced a great propaganda dividend for the United States, Mr. Edward R. Murrow, Director of the United States Information Agency, said. "The contrast was immediately drawn around the world between the openness, the complete coverage of the flight, compared to the Russian efforts in this area . . . More newspapers normally hostile to the United States had to admit not only the scientific achievement but the fact that we were prepared to let our own people and the rest of the world see the entire operation from beginning to end."

MARCH 1962

March 1: John Glenn Day in New York City. An estimated four million people lined the streets to cheer the Mercury astronauts. Mayor Robert Wagner presented the city's highest award. the Medal of Honor, to Col. Glenn and Robert R. Gilruth.

- Third successful static-test firing of "Old Salty"—an H–1 engine that had been submerged in salt water for four hours before each of its test runs. Purpose of tests was to determine feasibility and costs of recovering, rebuilding, and reusing booster engines as opposed to cost of new engines. Early results indicated that engines could be recovered and rebuilt at a cost of about 10% of new engines.
- NASA fired a Scout rocket from Wallops Station, Va. The rocket flew 135 miles high and about 800 miles downrange in a re-entry test reaching speeds of 19,000 mph.
- Rift (Reactor-in-Flight-Test) bidders conference held at Marshall Space Flight Center.
- Morton J. Stoller was named Director of the Office of Applications in NASA Hq., a post he had held in an acting capacity since January 1962. He was made Deputy Director in November 1961.
- Reported in the press that a 3-ft. by 2-ft. metal fragment found on a farm in South Africa on February 21 was identified by numbers stamped on it as a part of the Atlas booster that placed Astronaut John Glenn into orbit on February 20. Local reports of an explosion about 1 AM on February 21 indicated that the fragment came to earth after about 8 hours in orbit. Fragment was reported as 4 ft. by 2-ft. "stainless steel" in Capetown telegram to NASA Feb. 26, 1962.
- NASA selected Chance-Vought Corp. as contractor to study rendezvous of space vehicles. A primary part of the contract would be a flight simulation study exploring the capability of an astronaut to control an Apollo-type spacecraft.
- American Airlines 707 jet transport crashed in Jamaica Bay, N.Y.. seconds after takeoff from New York International Airport, and all 95 persons on board were killed, the highest toll in U.S. history in the crash of a single airplane. The crash occurred about one hour before the start of the giant parade in honor of Lt. Col. John H. Glenn, Jr.

March 2: Mercury astronauts visited the United Nations, and John Glenn, during an informal reception given by Acting Secretary-General U Thant, said: "To be here at the United Nations this morning and have all these tributes to our project and all the people that are working on it from people of this calibre, is indeed overwhelming all over again after yesterday. . . .

"As space science and space technology grows still further and our projects become more and more ambitious, we will be relying more and more on international teamwork.

"And the natural center for this teamwork is the United Nations . . . ." Colonel Glenn also expressed the gratitude of Project Mercury to those countries who had cooperated in the flight operations.

March 2: In Nation-wide radio-TV address, President Kennedy announced that the U.S. would begin atmospheric nuclear tests from Johnston and Christmas Islands in the Pacific during April unless such tests were rendered unnecessary by the clear willingness of the U.S.S.R. to take mutual steps toward "general and complete disarmament."

- Five Nike-Cajun rockets were fired from Wallops Station, Va., in a series of tests to measure atmospheric conditions at high altitudes. At 5:47 AM NASA scientists fired the first rocket to spread a trail of water vapor up to 89 miles altitude. At 5:54 AM the second rocket was fired. It loosed a cloud of sodium vapor starting at 26 miles and rising with the rocket to 84 miles. The cloud was spectacular in the sunrise and visible for hundreds of miles along the Atlantic coast. At 6:15 AM the third rocket jettisoned and detonated grenades one at a time at altitudes from 24 to 55 miles. Two additional Nike-Cajun rockets were fired in the evening with sodium and grenade experiments.

- USAF reported that "an Air Force satellite had provided the first continuous data on the actual size and intensity of the inner belt" of proton radiation over the Equator at altitudes of 600 to 3,000 miles. Stating only that the data was acquired from a 23-lb. experiment aboard an Agena, the report said radiation intensity at the center of the belt was 600,000 proton particles passing through one sq. inch per second, previously estimated at 100,000.

- A simple satellite plotter costing only a few hundred dollars and accurate to within 10 miles on any orbit was devised by Major William Gamble of the Canadian Defence Research Board. The Defence Research Board would use the device to give quick fixes on satellite orbits to radar stations.

March 3: Astronaut John H. Glenn, Jr., was welcomed in his home town of New Concord, Ohio, by a crowd estimated at 50,000 people (normal population of New Concord, 2,127).

- Capsule from DISCOVERER XXXVIII recovered in midair over the Pacific by a USAF C-130 piloted by Capt. Jack R. Wilson. It was his second recovery, the 8th midair recovery, and the 12th recovery air or sea during the Discoverer series. The capsule had been in polar orbit 4 days since its launch on February 27.

- NASA selected Sverdrup and Parcel of St. Louis as contractor for the design criteria and initial master planning for the Mississippi Test Facility, the 13,500-acre site that would eventually contain 6 or more static test stands for testing booster stages of Saturn and Nova vehicles.

- U.S. NAVY filed claims with Fédération Aeronautique Internationale for new time-to-climb records for the McDonnell F4H Phantom II fighter (which is also slated for USAF service as the F-110), superseding records held by the USAF's F-104 since 1958 and apparently bettering new USAF claims for the same marks based on flights made by the T-38 Talon trainer on Feb. 19, 1962. The Phantom claims: 3,000 meters: 34.523 sec, LCdr. J. W. Young (USN); 6,000 meters: 48.787 sec, Cdr. D. M. Longton

(USN); 9,000 meters: 61.629 sec, Lt. Col. W. C. McGraw, Jr. (USMC); 12,000 meters: 77.156 sec, Lt. Col. W. C. McGraw, Jr. (USMC); 15,000 meters: 114.548 sec, LCdr. D. W. Nordberg (USN).

March 4: USAF's first balloon launching in Project Stargazer ended when the large helium balloon exploded at 42,000 feet two hours after launch. It dropped its gondola loaded with telescopic camera equipment into deep Sierra Mountain snow. Launched from Chico, Calif., the balloon was to have ascended to the upper atmosphere where its telescopic cameras would have photographed the stars and planets free from the filtering action of the atmosphere.

March 5: USAF B–58 Hustler bomber claimed 3 transcontinental speed records after a Los Angeles-to-New York-to-Los Angeles nonstop flight: round trip, 4 hrs. 42 min. 12 sec.; west-to-east, 2 hrs. 1 min. 39 sec.; east-to-west, 2 hrs. 15 min. 12 sec. Flying most of the way at 50,000 ft. and at an average speed of 1,044.3 mph, the B–58 trailed a 40-mile-wide sonic boom along the ground in its wake. Crew was: Capt. Robert Sowers, pilot; Capt. Robert MacDonald, navigator; and Capt. John Walton, defense systems operator.

- U.S. reported to the United Nations that a total of 72 U.S. space vehicles and associated objects were in orbit around the earth as of February 15.

- Department of Defense has issued a directive providing for closer and more effective cooperation between DOD and NASA, DOD's Deputy Director of Research and Engineering John H. Rubel told the Senate Aeronautical and Space Committee. The directive gave the USAF authority to work with NASA on all levels for purposes of coordinating efforts and exchanging technical information but reserved to DOD the responsibility for establishing new programs and assigning military support to NASA.

- Tiros weather satellites have been of major assistance in improving forecasts for airline operations. Mr. Silvio Simplicio, supervising forecaster at New York International Airport, said: "We can do a job nowadays we never thought we could do years ago . . . With the aid of the Tiros satellites we never had it so good, and with the Nimbus meteorological satellite now undergoing tests, we expect even better results."

- NASA selected American Telephone and Telegraph Co. as contractor for system engineering support for the manned space flight program. It was estimated that some 200 specialists would be available under the contract for quick fact-finding in regard to critical decisions on the complex program.

- Caledonian Airlines DC–7 with 111 persons aboard crashed near Douala, Cameroon, the worst single aircraft disaster in civil aviation history.

March 5–7: NASA Wallops Station, Va., suffered storm damage first estimated in excess of $1 million as a result of high seas accompanying series of severe storms. Facilities of the main base on Wallops Island unused by NASA were subsequently opened to house evacuees from Chincoteague and surrounding areas.

March 6: Each of the series of weather satellites, Tiros, Nimbus, and Aeros, would not only provide an advance in capability but would extend observations into vital new areas, Morton J. Stoller, NASA's Director of Applications, testified before a subcommittee of the House Committee on Science and Astronautics. Speaking of Nimbus and Aeros, he said: "With one Nimbus in orbit, about half a day passes before we again see the same area. This will be a reasonably satisfactory time for the observation of such systems as cyclonic storms and hurricanes whose normal life is considerably greater than 12 hours. But, if we attempt to observe thunderstorm cells or tornadoes, systems whose life is appreciably less than 12 hours and often less than 6 hours will usually form, move and die without ever being detected. It is for this reason that the Aeros satellite with its capability for focusing on small short-lived storm systems and tracking them continuously, is considered to be an integral component of the eventual operational system."

- Secretary of the Air Force Eugene M. Zuckert, speaking before the Los Angeles Chamber of Commerce, cited Dr. Hugh L. Dryden's summation of the purposes of the national space program—"(1) insurance of the nation against scientific obsolescence in a time of explosive advances in science and technology; and (2) insurance against the hazards of military surprise in space." The Secretary added: "He defined in these two purposes the peacetime role of NASA in the national space program, and the defense role of the Air Force. The two must advance in harness, and they do. They are interdependent. One cannot move without the other. Furthermore, the nation cannot stand for any diversion of the efforts of either toward jurisdictional conflicts which which can rob us of the most precious ingredient—time."

- Dr. Arthur E. Raymond was appointed a special consultant to the NASA Administrator. He would be concerned with organization and management of research and development programs, especially those involving advanced research and over-all systems planning. Dr. Raymond retired in 1960 as senior vice president of engineering for Douglas Aircraft Co., having spent 35 years with the company. He served as a member of NACA from 1946–1956.

- AEC selected the Martin-Marietta Corp. as contractor to design, build, and ground test Snap 11, the thermoelectric nuclear generator for use on NASA's Surveyor spacecraft. Snap 11 would weigh about 30 lbs., including shielding, and would provide a minimum of 18.6 watts of power continuously for 90-day lunar missions.

- Army fired a Nike-Zeus antimissile missile from Kwajalein Island and successfully intercepted an electronically simulated ICBM warhead.

March 7: OSO I (Orbiting Solar Observatory) was successfully launched into orbit from Cape Canaveral, marking the seventh straight success for the Thor-Delta booster. The 458-lb. satellite, with an apogee of 370 mi. and a perigee of 340 mi., immediately began sending back signals on the sun's radiation in the ultraviolet, x-ray, and gamma ray regions from its position above the filtering action of the earth's atmosphere. By an intricate positioning apparatus, oso's 13 instruments were focused constantly on the sun with a pointing accuracy of 1 minute of arc. This was the first of a series of Osos to be launched by NASA in the next 11-year sun cycle.

March 7: NASA established the NASA Launch Operations Center at Cape Canaveral, with Dr. Kurt H. Debus as Director. Reporting to the Director of Manned Space Flight at NASA Hq., the new Center would serve all NASA projects launched from Cape Canaveral, absorbing Marshall Space Flight Center's Launch Operations Directorate. Similarly at Point Mugu, Calif., the NASA Test Support Office was redesignated the NASA Pacific Launch Operations Office, with Cdr. Simon J. Burttschell as Acting Director.

- D. Brainerd Holmes, NASA's Director of Manned Space Flight, testified before a subcommittee of the House Science and Astronautics Committee on the very large and complex manned space flight program. The most important step in space capability in 1961, he said, "was not that of awarding particular contracts or making technical decisions, but rather Mr. Webb's carefully planned reorganization of the National Aeronautics and Space Administration." Speaking of future launch sites for the Advanced Saturn vehicle as the workhorse of the manned lunar program, he said: ". . . it shows our plans for an entirely new concept of launching our large space vehicles. Instead of literally assembling or building the launch vehicles on the pads as we do at present, this arrangement would permit assembly and test at a location remote from the launch pads and under cover, protected from weather. It will be possible to have vertical transfer of the entirely assembled vehicle to the launch pad in a tested condition. This arrangement should offer many advantages in order to permit better and more thorough checkout, and to assure more rapid launching and efficient utilization of each pad. For the extremely tight schedules that will be required for launch operations in support of orbital rendezvous, this rapid launch capability is highly desirable."

- Escape-velocity payloads with the nuclear-engine Rift as the 3rd stage on an Advanced Saturn booster would be more than double that of a 3-stage, all-chemical Advanced Saturn and higher than that of the Nova 12-million-pound-thrust vehicle, according to testimony given to a subcommittee of the House Committee on Science and Astronautics by Harold B. Finger, NASA's Director of Nuclear Systems.

- Dr. Homer E. Newell, NASA's Director of Space Sciences, testifying before a subcommittee of the House Committee on Science and Astronautics, said: "Because of this great breadth and scope of space science, and because of the many basic scientific problems that it encompasses, our country must have a sound and vigorous space science program if we intend to maintain our position of leadership in world science . . .

"The space science of today is needed to sow the seeds for the harvest of future applications of space knowledge and technology. The weather, communications, and navigation satellites of today grew out of the scientific engineering and research of the past decades. Their perfection, and the development of new applications, will rest upon the space science and engineering of today and the years to come."

March 7: USAF launched an unidentified satellite with an Atlas-Agena B from Point Arguello, Calif.

March 8: John L. Sloop, NASA's Director of Propulsion and Power Generation, testifying before a subcommittee of the House Committee on Science and Astronautics, said: ". . . some of the new solid propellants promise an increase in performance of over 18 percent. . . . If applied to our new Surveyor spacecraft designed to land instruments on the moon, this 18 percent increase in solid propellant specific impulse could make possible an increase in the weight of instruments aboard by more than 50 percent."

- Tracking network that operated during Glenn orbital flight would for the most part be sufficient to handle the 18-orbit flights to follow, according to Edmond C. Buckley, NASA's Director of Tracking and Data Acquisition, in testimony before a subcommittee of the House Committee on Science and Astronautics. For the Glenn flight, the ground stations had to provide five minutes of contact and communication for every 15 minutes of flight during the 4½-hour mission. For 18 orbits the current requirement was one contact per orbit. This would require repositioning of the two ship stations and adding more telemetry-receiving equipment and command systems to some of the existing sites.

- Dr. Robert C. Seamans, Jr., NASA Associate Administrator, speaking before the IAS National Propulsion Meeting in Cleveland, said: "As we move on toward the moon, with Saturn, Advanced Saturn, and Nova as our launch vehicles, the reliability problem takes on a new dimension.

  "These big space vehicles will be the first developed especially for manned space exploration. The unit cost will be high. Only a few of each version will be produced. The systems and the missions are extremely complex, involving long periods of operation in space and a return launch from the moon without the help, needless to say, of a lunar Cape Canaveral."

- Army launched a 3-stage Nike-Zeus anti-ICBM missile in its first full-missile test from Point Mugu, but the missile was destroyed in midair when it seemed to be going off its planned trajectory.

- USAF Minuteman ICBM fired from a silo at Cape Canaveral and flew more than 3,000 mi. down the Atlantic Missile Range, the sixth straight success in the underground launchings.

March 9: In a Pentagon ceremony, Lt. Col. John H. Glenn, Jr., was awarded the Navy's astronaut wings and a new Marine Corps device—a winged rocket superimposed on a globe—that would also be given to all future Marines who orbited the earth in a space capsule.

- Titan II underwent a successful 20-second full-power captive firing at Cape Canaveral and was considered ready for its first flight test in about a week. USAF-developed Titan II produced 430,000 lbs. thrust in its 1st stage; the 2nd stage would add another 100,000 lbs. thrust; compared with 400,000 lbs. for the Atlas E and 380,000 for Titan I. Titan II, already selected as booster for NASA's 2-man Gemini capsule, burns storable fuels that emit a stream of hot, clear gases rather than flame.

March 10: In address at fund-raising dinner at Miami Beach, President Kennedy pointed out that NASA would spend five times as much in Florida during next fiscal year as last, and that "research, development and production expenditures will also increase at the Cape, in this state's private industry, and in your great universities."

• The Astronautics Committee of the 51-nation Fédération Aeronautique Internationale, meeting in Paris, certified Soviet Cosmonaut Gherman Titov as holder of two new world records for space flight: duration of flight, 25 hrs. 11 min.; and length of flight, 436,911 mi. In certifying Titov's records, the committee relaxed its rule that the pilot must land with his vehicle (Titov parachuted to earth). The heaviest payload in orbital flight still belongs to the other Soviet cosmonaut, Yuri Gagarin (10,419 lbs.).

March 11: Second anniversary of the launching of the unsurpassed PIONEER V space probe. PIONEER V produced first data on the nature of interplanetary space, including solar flare effects in interplanetary space which were compared with earth-orbiting-satellite readings, and sent back telemetry 22.5 million miles from earth on June 26, 1960, a communications record unmatched until the flight of MARINER II.

• Space Science Board of the National Academy of Sciences announced that improvements had been made in the second Project West Ford package soon to be launched into orbit: (1) weight had been reduced from 75 to 50 lbs., thus reducing the number of dipoles in the package from 350 million to 250 million; (2) the container had been redesigned to ensure that it would be imparted with the proper spin by the rifle-like barrel from which it is ejected (this caused failure of the first experiment to disperse); (3) VHF beacon and telemetry were included to ensure tracking capability; and (4) a fail-safe device would prevent the capsule from ejecting at any other than the proper orbit. Report sent to International Astronomical Union to assure astronomers that every care was being taken that the filaments not interfere with astronomical observations.

• NASA announced that ECHO I, the 100-ft. balloon-type passive communications satellite launched on August 12, 1960, had recently become increasingly difficult for observers to see. The sphere now presented only ½ to ¼ its original size, due either to shrinkage or distortion during its year and a half in orbit.

• Reported in London that U.S.S.R. has nearly completed negotiations with the United Arab Republic and Iraq for supply of Soviet guided missiles.

March 12: U.S.S.R. had shown a "change in attitude in recent weeks" on cooperating with the U.S. on development and use of a weather satellite system, Dr. Francis W. Reichelderfer, Chief of the U.S. Weather Bureau, testified before a subcommittee of the House Committee on Science and Astronautics.

• Orbital data from satellites have suggested that scientists do not really know the shape of the earth, Dr. George P. Woollard, Director of the Geophysics and Polar Research Center of the University of Wisconsin, said in a Voice of America broadcast. Such studies have indicated that the earth may have not only the "flattened 'tomato shape' that has been assumed since the time of Isaac Newton but also a superimposed 'pear shape.' " Citing the number of forces that can operate on a satellite, thus requiring extremely careful analysis of its data, Dr. Woollard pointed out that the Committee on Geodesy of the Space Science Board of the National Academy of Sciences had recommended the launching of a specially instrumented geodetic-research satellite to resolve some of the questionable areas of influence on magnetic readings.

March 12: *U.S. News and World Report* carried article on "Red Space Failures," pointing to official U.S. policy with regard to Soviet launchings: "No Soviet feat [in space] has ever been publicly challenged in Washington and no failure publicly announced." It reported that while Premier Khrushchev was at the United Nations in September 1960, "a Soviet cosmonaut was sent more than a hundred miles into space and killed"; that in a five-month period in 1960–61, four Soviet space probes were fired toward Mars and Venus, but failed; and that during the same period "a dozen other Soviet space shots went awry . . ."

March 13: Dr. Abe Silverstein, Director of NASA's Lewis Research Center, received the National Civil Service League's Career Service Award, one of 10 Government leaders so recognized.

• Sharp rise in heating of re-entry vehicles returning from manned lunar missions at speeds of 25,000 mph was described to the subcommittee of the House Committee on Science and Astronautics by Milton B. Ames, Jr., Director of Space Vehicle Research and Technology, NASA. Mr. Ames stated that a simulation study at NASA's Ames Research Center had "demonstrated great promise for pilot-controlled re-entries during orbital and lunar missions, and we plan to extend our work in this area."

• USAF launched its second Stargazer balloon from Chico, Calif., and achieved an altitude of some 88,000 ft. Balloon performed well in this test preliminary to manned flights that would send two men and a stabilized telescope to 16 miles. The first Stargazer burst at 40,000 ft. on March 4.

March 14: NASA's Director of Aeronautical Research, John Stack, testifying before a subcommittee of the House Committee on Science and Astronautics, spoke of NASA support of DOD missile programs: "Included in this work are such as Pershing, Atlas, Nike, Nike-Zeus, Eagle, GAR–9 and Skybolt. The NASA contributions have been in two areas, one providing critical aerodynamic information for performance determination and control system characteristics and the other in suggesting configuration improvements to improve the over-all characteristics."

• NASA announced that sea damage to Wallops Station, Va., occurring earlier in the month was not extensive, and that rocket firings would be resumed in a few days. Although launch pads and some tracking gear on the island itself had been damaged by water, the more important tracking and data acquisition facilities on the mainland were undamaged.

• EXPLORER IX, launched on February 16, 1961, provided new and refined information on the density of the upper atmosphere, a press conference of NASA and Smithsonian Astrophysical Observatory scientists reported. They confirmed a figure previously released—that the density of the atmosphere at 420 mi. was 3 x 10^{-17} grams per cubic centimeter, or 1/40 million millionth of that at sea level. EXPLORER IX density values were about 10 times lower than those computed in 1959 from earlier satellites. Most of this decrease is attributed to the decrease in solar activity since the peak of the solar cycle in 1958 and 1959. Changes were also clearly related to the 27-day rotational period of the sun and to the occasional violent solar storms that affect the earth's atmosphere. EXPLORER IX, a 12-ft. aluminum-foil sphere painted with white "polka dots," was expected to have an orbital life of two more years. As it spiraled down into denser atmosphere, it was expected to provide much more information on density at altitudes down to 100 mi.

March 14: United Nations opened a public register on satellites in orbit. At the time of opening, it contained only the U.S. report submitted on March 5, which reported "72 U.S. space vehicles and associated objects in sustained orbit or space transit" as of February 15, 1962.

- NASA and the Jet Propulsion Laboratory announced the selection of Military Electronics Division of Motorola, Inc., as contractor to manufacture and test radio equipment in the first two phases of a program to augment the Deep Space Instrumentation Facility by providing "S" band capability for stations at Goldstone, Calif.; Woomera, Australia; and Johannesburg, South Africa. With these stations located some 120° apart around the earth, DSIF would have a high-gain, narrow-beam-width, high-frequency system, with very little interference from cosmic noise and would provide much improved telemetering and tracking of satellites as far out as the moon and nearby planets.

- Walter C. Scott, Chief of NASA's Space Power Technology Program, testifying before the House Committee on Science and Astronautics, said one of the promising developments in solar cells was the possibility of producing thin strips of silicon solar cells on some kind of substrata. If this can be done, the "much lower cost, lower weight and improved mechanical properties will be attractive to the space industry and for that matter to the civilian economy. Thin films of cadmium sulfide have been produced in small areas with conversion efficiencies of 3%. It has been stated that if this efficiency could be increased to 5% and produced in large areas, solar power would be economically competitive with conventional sources at rates of 3 to 4 mils per kilowatt hour."

- Navy launched solid-propellant Terrier-Asp IV sounding rocket from Point Arguello in successful first flight test.

- Theodore Shabad, Moscow correspondent for the *New York Times*, described the Clubs of Young Cosmonauts in U.S.S.R. The Moscow club was organized in January 1962 under joint sponsorship of *Moskovsky Komsomolets*, newspaper of the Young Communist League, and the cosmonautics section of the Soviet Federation of Aviation Sports. Its purpose was to interest youngsters in space and also to publicize Soviet achievements in space.

- USAF announced it would extend space surveillance to the moon through a system developed by the Space Track Research & Development Facility at Laurence G. Hanscom Field, Mass.

There will be surveillance of orbiting man-made or natural objects beyond the earth's nominal, low-altitude orbital areas out toward the moon.

March 14: Six candidates were selected by the USAF for Dyna-Soar, including four Air Force and two NASA test pilots. Dyna-Soar is designed to be boosted into orbit by a Titan III.

March 15: NASA announced that LCdr. M. Scott Carpenter (USN) would be the pilot on the next Mercury orbital space flight (MA–7). Major Donald K. Slayton (USAF), the astronaut originally scheduled for the flight, was disqualified because of an "erratic heart rate," after the medical findings had been reviewed by an Air Force medical board and a board of civilian cardiologists. Astronaut Walter M. Schirra was named as Carpenter's backup pilot.

• Editorial in the Royal Aero Club magazine *Flight* claimed that because the U.S. had made "six major launches on which basic orbital information had been withheld from world scientists," the U.S. was practicing "brinksmanship in space."

• National Rocket Club announced the establishment of the Dr. Robert H. Goddard Historical Essay Award, an annual competition in the history of rocketry and astronautics, the first such competition in the field of history.

March 16: USAF Titan II ICBM successfully launched on first flight, from Cape Canaveral. Initiated less than 2 years ago, Titan II is powered by storable propellants (430,000-lbs. thrust in 1st stage, 100,000-lbs. in second stage), and will also be used as the boosters for NASA's Gemini two-man spacecraft.

• Premier Khrushchev claimed that the U.S.S.R. had a new "invulnerable global rocket," announced the orbiting of COSMOS I (the 16th Russian satellite), and stated that the U.S.S.R. would carry out new atmospheric nuclear tests if the U.S. resumes its atmospheric tests. Tass released a news story on the new satellite, giving its orbital data as apogee, 609 mi.; perigee, 135 mi.; period, 96 min.; and inclination, 49° to the equator. A scientific payload included measurements of meteoric impacts, low-energy solar radiation, earth's radiation belts, cosmic rays, earth's magnetic field, short-wave radiation from sun and other celestial sources, and atmospheric cloud patterns.

• First anniversary of the dedication of the Goddard Space Flight Center, NASA. During that year, seven Goddard satellites were orbited, the Center successfully operated the new 18-station world tracking network for the Glenn flight, began expansion of the 13-station scientific satellite tracking and data network, saw some 70 of its sounding rocket payloads launched from Wallops Station, established the Institute for Space Studies in New York, and added three buildings and 700 persons to the Goddard staff.

• At Robert H. Goddard Memorial Dinner, Vice President Lyndon B. Johnson pointed out areas in which international cooperation should begin in outer space. "As we attempt to look 25 years into the future, we catch visions of breathtaking journeys in large man-made planets around the sun to Mars and Venus—of a new freedom of movement of man across millions of miles of space—of a permanent colony on the moon and of large space stations or space forts at key locations for the conduct of space research, for aid to space navigation, and for rescue operations."

Major awards of the National Rocket Club were presented: the premier Dr. Robert H. Goddard Memorial Trophy was presented by Mrs. Esther C. Goddard and the Vice President to Robert R. Gilruth, director of Project Mercury; the National Rocket Club Award was given to the *New York Times*; the Nelson P. Jackson Aerospace Memorial Award was given to the Radio Corp. of America for its design and construction of the Tiros meteorological satellites; and the Astronautics Engineer Award was given to William G. Stroud of NASA's Goddard Space Flight Center for his contribution to the technology of meteorological satellites.

March 17: VANGUARD I began fifth year in orbit, having traveled 543,195,264 miles in 15,712 orbits and still transmitting. Its rotation had slowed from three revolutions per second to one revolution in 23 seconds due to dampening effect of the earth's magnetic field.

- President Kennedy wrote Premier Khrushchev proposing specific areas of peaceful cooperation in space to be undertaken by the U.S. and U.S.S.R.: (1) in furtherance of an operational weather satellite system, proposed each country launch a weather satellite in near-polar orbit, disseminate data to all nations; (2) in space tracking, proposed each country build a tracking station in the other's territory and train personnel of the host country to maintain and operate the station; (3) in mapping the earth's magnetic field, proposed each country launch a magnetic-measurement satellite, one in near-earth orbit and one farther out, and disseminate data to all nations; (4) in communications satellites, proposed that U.S.S.R. join in cooperative program to establish an operational communications satellite system, to be made available to all nations; (5) in manned space flight, proposed that the two countries pool their knowledge of space medicine; (6) proposed general continued future cooperation, as in joint exploration of the lunar surface or scientific investigation of Mars and Venus. Further proposed that any agreements reached be reported to the U.N. Committee on the Peaceful Uses of Outer Space.

- Soviet News Agency Tass reported that COSMOS I satellite was transmitting a steady stream of technical data.

March 19: U.N.'s reorganized 28-nation Committee on the Peaceful Uses of Outer Space opened its meetings with the reelection of Dr. Franz Matsch of Austria as chairman and his subsequent ruling that the committee would make decisions by "consensus" rather than by vote. U.S. delegate was Francis T. P. Plimpton; his deputy was Richard N. Gardner, Deputy Assistant Secretary of State for International Organization Affairs. Dr. Hugh L. Dryden, Deputy Administrator of NASA, was Plimpton's technical adviser. Dr. Homer Newell of NASA and Leonard C. Meeker of the State Department were alternate representatives. Congressional advisers were Senators Howard W. Cannon and Margaret Chase Smith, and Representatives James G. Fulton and George P. Miller. U.S.S.R. representative was Platon D. Morozov. V. Dobronravov, member of the Soviet Academy of Sciences and space expert, was his scientific adviser.

- Space science might provide an answer within five years to the question of whether gravitation is growing weaker with time,

Dr. Robert H. Dicke, Professor of Physics at Princeton University's Palmer Physical Laboratory, said in a lecture for the Voice of America. This new theory of gravitation, not in accord with Einstein's theory of gravitation, could be tested with a satellite that was instrumented to act as a "gravitational clock" and checked against an atomic clock.

March 20: Premier Khrushchev replied to President's letter of March 17 proposing U.S.—U.S.S.R. cooperation on peaceful uses of outer space by accepting in principle and specifically approving of cooperation in certain areas (letter released on March 21): (1) communications satellites; (2) weather satellites; (3) coordination of tracking and data reception on scientific satellites; (4) international search and rescue operations to recover space crews who have returned to earth outside the planned recovery areas; (5) mapping the earth's magnetic field and exchange of information on space medicine; (6) joint agreement on problems of international law as related to space exploration, beyond those points already agreed to in the U.N., especially the prohibition of space experiments that would inhibit the space experiments of other nations. Premier Khrushchev warned, however, that space cooperation would "depend to some extent on the settlement of the disarmament problem," since much military hardware and techniques were used in peaceful space exploration vehicles.

- U.S.S.R. representative to U.N.'s Committee on Peaceful Uses of Outer Space pledged that U.S.S.R. would cooperate "by deeds" in international programs on space. Promised reports to U.N. on Soviet satellite and rocket launchings, cooperation on communications satellites but on the basis of "international regional agreements."

- USAF awarded contract to North American Aviation, Inc., for development of RS–70 prototypes, including fabrication of three aircraft.

March 21: President Kennedy opened his news conference with the announcement that he had received Premier Khrushchev's reply to his letter on cooperation in the peaceful uses of space and stated: "I'm gratified that this reply indicates that there are a number of areas of common interest." He further announced that Dr. Hugh L. Dryden, Deputy Administrator of NASA, had been designated to lead the technical discussions with Soviet representatives beginning in New York on March 27.

- House of Representatives passed by a vote of 430 to 0 a bill authorizing $13 billion to DOD for procurement of ships, aircraft, and missiles. Included was $491 million for development of the controversial RS–70 but without the previous Congressional language directing DOD to use that amount for the RS–70 in FY 1963. Rep. Carl Vinson, Chairman of the House Armed Services Committee, read letters from the President and Secretary of Defense in which they asserted their Constitutional obligation to make the final determination in such matters but promised, in view of the Committee's concern over continued bomber development, to re-examine the RS–70 program.

- House Science and Astronautics Committee's Panel on Science and Technology recommended that Project Anna, DOD's geodetic satellite, be declared an international scientific satellite.

March 21: NASA announced selection of Westinghouse Electric Corp. as contractor for constructing and testing prototypes and flight models of the S–52 U.S.-U.K. scientific satellite. Second of the three satellites in this program, S–52 was scheduled for launch from Wallops Station, Va., in 1963, would contain 3 major experiments for measurement of galactic radio noise, of vertical distribution of ozone, and of micrometeorite flux.

March 21–22: Fourth meeting of the Panel on Science and Technology of the House Committee on Science and Astronautics held. Fifteen leading American scientists and engineers plus Dr. George B. Kistiakowsky, Dr. Harrison S. Brown, and Sir Bernard Lovell discussed mapping and geodetic satellites, propulsion problems, use of boron as a fuel element in space flight, and the need to strengthen universities in the training of scientists and engineers. Dr. James Van Allen commended NASA for "marked improvement" during the past two years in its fostering of university scientific research and training programs and suggested that five per cent of the national space budget be so invested.

March 22: Experimental ejection-seat capsule for use in supersonic aircraft was ejected from USAF B–58 at 870 mph and 35,000 ft. Bear and capsule landed safely by parachute near Edwards AFB. Calif., 7 min. 49 sec. later.

- USAF launched a Minuteman from a silo at AMR and sent it 4,000 miles downrange, the seventh straight success in the Minuteman silo-launch test series.

- Sir Bernard Lovell, Director of the Jodrell Bank Experimental Station in England, testified before the House Science and Astronautics Committee that life on other planets was a real possibility, estimated that four per cent of the billions of stars in the universe must have planets capable of sustaining life. Listening for messages from other planets would not be worthwhile on a random basis; if disarmament came, military radar might be devoted to this purpose, he said.

- NAA announced that Mrs. Constance Wolf would receive the 1961 Montgolfier Award from the FAI for her balloon flight from Big Springs, Tex., to Boley, Okla., on November 19–20, 1961. Claiming 15 world records, Mrs. Wolf would be the first woman to receive this international award.

March 23: President Kennedy, speaking at the University of California in Berkeley on the occasion of the University's Charter Day and the awarding to the President of an honorary degree of Doctor of Laws, said: ". . . history may well remember this week for . . . the decision by the United States and the Soviet Union to seek concrete agreements on the joint exploration of space . . . the scientific gains the joint effort would offer might be small compared to the gains for world peace. For a cooperative Soviet-American effort in space science and exploration would emphasize the interests that unite us instead of the conflicts that divide us. It offers us an area in which the stale, sterile dogmas of the cold war can be left literally a quarter of a million miles behind. And it would remind us on both sides that knowledge, not hate, is the passkey to the future—that knowledge transcends national antagonisms—that it speaks a universal language—that it is the possession, not of a single class, a single nation or a single ideology, but of all mankind."

March 23: USAF combat missile crew launched an Atlas D ICBM from Vandenberg AFB, Calif., during President Kennedy's visit to the missile base, the first time an American President had witnessed a live launching of an ICBM.

- NASA launched a Nike-Cajun rocket from Wallops Station, Va., which released an 80-mile-long sodium-vapor cloud to measure air density and wind direction in the upper atmosphere.
- Early tests of pressure and dynamic stability models of the Apollo spacecraft were completed in wind tunnels at JPL and Langley Research Center.
- D. Brainerd Holmes, NASA's Director of Manned Space Flight, speaking before the Explorers Club in New York, listed one of the NASA organizational accomplishments of recent months as "the establishment of a liaison office with the Department of Defense, particularly with the Air Force Systems Command."
- Bell Telephone Laboratories announced the formation of a new corporation, Bellcom, Inc., to supply system engineering support to NASA's space program. The new corporation would be owned jointly by AT&T and Western Electric Corp.
- First U.S. patent granted to a citizen of U.S.S.R. in 10 years was granted to Nikolai V. Soodnizin of Moscow on a powered device for coupling and uncoupling lengths of pipe in an oil well. U.S.S.R. recently began encouraging its citizens to apply for foreign patents.

March 24: Two-week meeting of Study Group IV of the International Radio Consultative Committee (CCIR) of the International Telecommunication Union (ITU) concluded in Washington, D.C. Study Group IV reviewed technical phases of space communications, including the selection of frequencies for telecommunications with and between space vehicles. Over 200 representatives from 30 countries participated in these sessions, Dr. John P. Hagen serving as head of the U.S. delegation.

- Prof. Auguste Piccard, pioneer Swiss explorer of the stratosphere and ocean depths, died in Geneva at the age of 78. In 1931 Prof. Piccard ascended to 51,777 ft. in the gondola of a balloon of his own design. In numerous balloon ascents in the 1920's and 1930's, he made many studies of radioactivity, atmospheric electricity, cosmic rays, and other scientific phenomena. Turning from the air to the oceans, Prof. Piccard designed his bathyscaphe and in 1953 he and his son reached a depth of two miles. In 1960 the U.S. Navy took an improved version of his craft down to 37,800 ft.
- Soviet Marshal Mikhail Tukhachevsky, executed in 1937 on Stalin's orders and exonerated in 1956, was honored in the official military newspaper, *Red Star.* It was pointed out that in 1932 Tukhachevsky insisted on the development of rocket engines and that he persisted in this effort until purged.

March 25: National Academy of Sciences announced that the John J. Carty Medal would be awarded Provost Charles H. Townes of MIT for his pioneering work in the development of the maser (microwave amplification by stimulated emission of radiation). Dr. Townes was credited with the conception of the idea for the maser in 1951, which was subsequently used in radio astronomy in recording and charting minute radio signals emanating from celestial bodies and to receive echoes of a radar signal reflected from the planet Venus.

March 26: U.S.S.R. submitted information on 16 Soviet space flights for inclusion in the U.N. public registry on space launchings. Included were the manned orbital flights by Majs. Yuri A. Gagarin and Gherman S. Titov. U.S. submitted a similar list to the U.N. on March 5, covering U.S. space launchings to Feb. 15, 1962, that were still in orbit.

- Coordination of international arrangements for exchange and dissemination of weather data, including weather satellite information, was on the agenda of the third session of the U.N.'s Commission for Synoptic Meteorology as it opened a 26-day session in Washington. Technical experts from more than 100 nations had been invited.

- Dr. James A. Van Allen, Chairman of the Department of Physics and Astronomy at the State University of Iowa, summarized the satellite and space probe findings about the structure of the geomagnetic field surrounding the earth. Describing the inner zone as relatively stable in intensity of charged particles and caused by a combination of internal and external forces, he said the outer zone showed fluctuations in intensity "by factors of 100 to as much as 1,000," said these were complexly related to solar flares and geomagnetic storms. The only convincing explanation for the origin of the outer zone yet made is "the capture of ionized solar gas which sweeps by the earth in great clouds from time to time. . . . But the detailed mechanism for producing the observed energy spectrum remains obscure and the nature of this mechanism is perhaps the most interesting unresolved problem in the subject of trapped radiation."

- John A. Johnson, General Counsel of NASA, testifying before the Senate Subcommittee on Monopoly, stated that as of Feb. 28, 1962, a total of 297 inventions had been reported by NASA contractors under the patent provisions of the National Aeronautics and Space Act of 1958. Of these, 268 had been determined to be Government property, 29 had been or are being considered for waivers.

March 27: U.S. and U.S.S.R. technical representatives held the first of a series of talks on the possibility of joint cooperation in space research and exploration. Dr. Hugh L. Dryden, Deputy Administrator of NASA, represented the U.S.; Dr. Anatoli A. Blagonravov, of the Soviet Academy of Sciences, represented the U.S.S.R.

- NASA fired a Nike-Cajun rocket from Wallops Station, Va., which released a sodium vapor cloud between 25 and 74 miles altitude. Rays of the setting sun colored the sodium cloud red, instead of sodium vapor's normal yellow.

- In address to the Institute of Radio Engineers in New York, NASA Deputy Administrator Hugh L. Dryden stated: "The costs of satellites and space probes are so great that every possible step must be taken to assure success. In our scientific experiments we insist that the bench prototype developed by the physicist be redesigned by engineers to meet the environmental requirements, that the engineered prototype be tested on vibrators, in vacuum and temperature chambers and other ground equipment simulating the space environment beyond the required conditions, and that, if possible, the equipment be tested on the less expensive sounding rockets prior to use on the more expensive satellites. . . .

"To state the matter another way, we desire assured and demonstrated performances and reliability. The cost of the necessary engineering and tests is small compared to the cost of a single failure. We do not wish to be the first to use the newest and most advanced device nor the last to give up the obsolete. A moderately conservative engineering approach with adequate analysis and test is indicated as the best design philosophy."

March 27: Dr. Edward Teller, testifying before the House Committee on Science and Astronautics, recommended a program for establishing a large and independent colony on the moon, as a means of having a working base in space and control of near space from a standpoint of national security. A nuclear reactor should be developed to operate on the moon, eventually to furnish the power to extract water from the moon's rocks and soil, he said.

March 28: U.S. submitted to the U.N. a supplemental list of U.S. space launchings, covering the period of Feb. 15 to March 15, 1962, updating the coverage of the first U.S. list submitted on March 5, 1962. This second official list did not include Astronaut John Glenn's 3-orbit flight, since the U.S. contended that the U.N. roster was supposed to contain only those space objects still in orbit, not those that had already re-entered. U.S.S.R. listed all its space flights in its report to the U.N. on March 26. Although MA–6 flight was not registered, the U.S. submitted information on the Glenn flight to the U.N. on April 3.

• U.S. and Soviet space officials ended two days of technical discussions in New York on possible cooperation in outer space. No public announcements were made.

• Senate Aeronautical and Space Sciences Committee unanimously approved a bill for ownership and operation of the Nation's commercial communications satellites.

• In speech at Worcester, Mass., Gen. Curtis E. LeMay stated that the U.S.S.R. is "moving at full speed for a decisive capability in space" and that the U.S. could not afford a "fatal technological surprise in the 1970's." Gen. LeMay pointed out that many people looked at military space operations as "merely an extension of nuclear weapons. . . . This may not be the case at all. Our national security in the future may depend on armaments vastly different from any we know today, and believe me they won't be ultimate weapons either."

• David Sarnoff, Chairman of the Board of Radio Corp. of America, speaking to the Institute of Radio Engineers in New York, called for the free world to organize an international community of science, staffed by its best scientific brains, to expand knowledge of basic science and to attack pressing world problems such as global communications and weather and adequacy of world's food, water, and power supplies.

March 29: In testimony before a subcommittee of the House Committee on Science and Astronautics, Gen. Bernard Schriever, Commander of AFSC, discussed the "reciprocal support" between NASA and the USAF, and the national requirement for military space weapons systems. He pointed out that: (1) 102 USAF officers were assigned full-time with NASA as well as 23 Navy and 33 Army; (2) $49 millions of NASA support was funded by the USAF through FY 61 and over $200 millions in FY 62; (3) $270 millions of NASA funds were processed under cognizance of AFSC during FY 61; and (4) NASA use of USAF booster, launch, and tracking capabilities and the joint X-15 program further demonstrated "the work NASA and the Air Force are doing together as a team." Gen. Schriever pointed to the fundamental capabilities of future space weapon systems and the assignment by the Secretary of Defense of major responsibility for military space systems development to the Air Force.

- President Kennedy sent to Congress a reorganization plan that would establish an Office of Science and Technology within the White House staff. Its director would advise and assist the President on (1) major science policies and programs in the Federal government; (2) assessment of scientific and technical developments for effect on national policies; (3) review, integration, and coordination of major Federal science activities; (4) ensuring "good and close relations" with Nation's scientists and engineers. President said this would leave the National Science Foundation free to concentrate on fostering basic research and science education.

- NASA announced selection of Republic Aviation Corp. as contractor for Project Fire, the flight re-entry research program. Contract would call for construction of two re-entry spacecraft to be flown in the second half of 1963 and re-enter the earth's atmosphere at a speed of 25,000 mph.

- A four-stage NASA Scout rocket carried P-21A probe payload 3,910 miles into space and 4,370 miles downrange from Wallops Station, Va. In 97-minute flight measurements were taken of the ionospheric electron density profiles, ion density, and types of ions, data needed to improve communication between earth and space.

- European Launcher Development Organization (ELDO) formally came into being when a convention was signed in London between Great Britain, France, West Germany, and Italy to develop the rocket vehicle to launch a satellite from Woomera, Australia, in 1965. First stage would be the British Blue Streak rocket, second stage the French Veronique, third stage would be developed by West Germany. The Netherlands, Belgium, and Australia were expected to join the organization soon.

- Second altitude record attempt in two days by X-15 was scrubbed just before drop from B-52, caused by failure of inertial guidance system.

- McDonnell Aircraft Corp. first showed Gemini capsule mock-up, at St. Louis plant.

March 29: Subcommittee of the Senate Antitrust and Monopoly Committee headed by Sen. Estes Kefauver opened hearings on propriety of establishing a privately-owned communications satellite system.

• Ambassador Francis T. P. Plimpton, chief of the U.S. delegation to the U.N. Committee on the Peaceful Uses of Outer Space, invited the 28-nation committee to inspect launching facilities at Cape Canaveral in April. Chairman Franz Matsch of Austria announced that technical and legal subcommittees would meet in Geneva on May 28 to start detailed work on space control programs.

March 30: Dr. Hugh L. Dryden, U.S. negotiator on outer space cooperation, read joint statement covering three conferences held with Anatoli Blagonravov, Soviet negotiator. Statement said that informal, exploratory discussions would be resumed either at the International Committee on Space Research sessions April 30–May 10 in Washington, or at Geneva on May 28 when technical and legal subcommittees of the U.N. Committee on the Peaceful Uses of Outer Space began discussions.

• House Appropriations Committee cut a $65 million NASA item for purchase of land at Cape Canaveral and in Mississippi from a supplemental appropriations bill for the balance of FY 1962. The action was taken because the House Science and Astronautics Committee had not authorized the expenditure. Approved was $80 million for development, $5 million less than NASA asked for.

• Army announced that Nike-Zeus antimissile missile had intercepted 3,000-mph Nike-Hercules missile in test over White Sands.

• France carried out six successful missile tests in the Sahara during the previous week. Two of the launchings were of Agate missiles, with which France has planned to develop a missile striking force and eventually to launch a satellite.

• RAF claimed an international speed record for its Vulcan bomber in a flight from London to Aden, 3,895 miles in 6 hrs. 13 min., for an average speed of 627 mph.

• LCdr. F. T. Brown piloted the F4H Phantom II at NAS Point Mugu on record flight to 20,000 meters, claiming new time-to-climb record of 178.5 seconds.

• Navy Polaris missile destroyed over Cape Canaveral after launch.

March 31: Dr. Ernst Stuhlinger, Director of Research Projects Division at NASA's Marshall Space Flight Center, was awarded the $1,000 Galabert Prize in Paris for his studies on nonchemical booster rockets.

During March: NASA completed work on its first major launching facility on the West Coast, a Thor-Agena pad at Vandenberg AFB, Calif. A used gantry was shipped from MSFC and installed at a $1 million saving over cost of new construction. Pad would be used for NASA polar-orbit launches, such as Echo II, Nimbus, and Pogo (Polar Orbiting Geophysical Observatory).

• USAF announced selection of four USAF and two NASA pilots as "pilot-engineering consultants" for the Dyna-Soar program. A decision would be made later whether the men would actually fly the spacecraft. USAF officers selected were: Maj. James W. Wood, Capt. Henry C. Gordon, Capt. Russell L. Rogers, and Capt. William J. Knight. NASA pilots selected were: Neil A. Armstrong and Milton O. Thompson.

During March: White House panel headed by Dr. Paul Beeson of Yale University, assisted by Dr. James Hartgering, special assistant in the President's Scientific Adviser's office, recommended that NASA make greater use of USAF capabilities in bioastronautics.

• Specific use of the fuel cell as source of auxiliary power was confirmed when North American Aviation, Inc., prime contractor for NASA's Apollo vehicle, awarded two fuel-cell development contracts to Pratt & Whitney Aircraft Div. of United Aircraft Corp. and Tapco Div. of Thompson Ramo Wooldridge.

APRIL 1962

April 1: Beginning of third year of successful weather satellite operation by the U.S. NASA's TIROS I, launched on April 1, 1960, performed beyond all expectations, operated for 78 days, transmitted almost 23,000 cloud photos, of which some 19,000 were useful to meteorologists. TIROS II, launched November 23, 1960, transmitted more than 33,000 photos and one year after launch was still occasionally taking useful photos. TIROS III, launched July 12, 1961, took 24,000 cloud photos and was most spectacular as a "hurricane hunter." TIROS IV, launched February 8, 1962, has averaged 250 operationally-useful photos per day.

• Senator Margaret Chase Smith, in a radio interview with Senator Kenneth B. Keating, expressed doubt on the value of space cooperation with the U.S.S.R. A member of the Senate Aeronautical and Space Sciences Committee, Sen. Smith said the U.S. had "decided superiority over Russia on really important space development. Russia has the edge over us on thrust power but when it comes to finer scientific space developments and information, we have great superiority over Russia. Perhaps one of our greatest areas of superiority is in the development of miniaturization. I think we could learn something from Russia but there is a question as to just how far we can trust her and what we get in return."

April 2: OSO I, launched March 7, 1962, was reported by NASA to be performing well. As of this date, 360 telemetry data tapes had been recorded from 403 orbits. Tapes would require about one year for complete analysis of data.

• Dr. John A. Simpson, of the University of Chicago's Enrico Fermi Institute for Nuclear Studies, said in a Voice of America broadcast that space probes and satellites have offered the scientist the first tools for direct measurement and study of electrodynamics in the interplanetary medium. EXPLORER X, for example, had recorded a density of protons in space substantially below predicted values and "almost one million billion times less than the best vacuum that man creates in the laboratory.

"Such experiments in space emphasize the fact that there occur phonemena in nature which cannot be scaled down without losing the essence of their physical properties."

• USAF announced it would build a new phased-array radar system that would have a multiple-track capability and quicker reflexes than present radar for the detection, tracking, identification, and cataloging of satellites. Contractor woud be Bendix Aviation Corp.

April 3: Senate Permanent Investigations Subcommittee opened hearings to determine whether the U.S. has paid "unnecessary and excessive profits" to companies working on DOD contracts. Sen. John L. McClellan, chairman of the committee, announced that investigation would begin with the Army's family of Nike missiles, to be followed by the USAF's Atlas.

April 3: USAF said it would procure twelve additional Thor space boosters to meet future requirements. Since January 1957 there have been 142 launches of Thor with 111 successes, thirteen partial successes, and eighteen failures.

- LCdr J. W. Young piloted F4H–1 to its seventh world time-to-climb mark, reaching 25,000 meters in 230.44 seconds.

April 4: NASA Administrator James E. Webb, speaking before the National Association of Broadcasters in Chicago, cited NASA's space sciences program as "a quest for fundamental knowledge" without which the applied science and technology "would soon run dry. . . . Suppose we had laid out a program five or six years ago, directly tied to manned space flight. We would never have discovered the Great Radiation Belt, which . . . constitutes one of the severest manned-space flight problems we face."

- Soviet cosmonauts have had changes introduced into their training, including special gymnastics, in an attempt to offset nausea induced by prolonged weightlessness, according to *Trud*, newspaper of the Central Labor Union.
- Army fired Pershing missile 200 miles at AMR, 29th success out of 33 test firings.
- AFSC formed the Research and Technology Division, Provisional, to "plan and manage AFSC's basic research, applied research, and advanced technology and . . . create a broad base of research and technology for rapid use in the development of Air Force aerospace systems."
- AEC and DOD in a joint announcement designated an 800-by-600 mile area around Christmas Island in the Pacific as a U.S. nuclear test area, effective April 15.
- U.S.S.R. was converting a 1300-mph bomber into a transport so it can claim the first supersonic commercial aircraft, Najeeb E. Halaby, FAA Administrator, stated.
- James T. Koppenhaver was appointed NASA's Director of Reliability and Quality Assurance, succeeding Dr. Landis S. Gephart who resigned to take a position with private industry.

April 5: X–15 No. 3 flown to speed of 2,830 mph (mach 4.06) and to altitude of 179,000 ft. in a test of new adaptive control system to be used in Dyna-Soar and Apollo vehicles. NASA's Neil A. Armstrong was pilot. Whereas the previous control system was automatic only while the X–15 was within the atmosphere and the pilot had to control flight with reaction jets while in space, the new system would be automatic in both regimes.

- Two outstanding challenges to science today were the conquest of space and the achieving of controlled fusion power, Dr. Peter L. Kapitza, one of Russia's leading physicists, wrote in the April issue of the *Bulletin of the Atomic Scientists*. Very-long-range rockets would have little practical use except as a means of cheap, safe disposal of radioactive waste in outer space, he said.
- Army launched Nike-Zeus missile in what was described as a successful intercept of a simulated ICBM nose cone.

April 6: NASA sponsored a day-long technical symposium in Washington on results of the MA–6 three-orbit space flight. Astronaut John H. Glenn, Jr., and officials of Project Mercury reviewed the findings of the Feb. 20 flight and stressed the conclusion that the presence of the astronaut had been indispensable to successful completion of the three-orbit mission. On the Glenn effect—the firefly-like particles Glenn reported seeing outside his capsule during each of the three sunrises—Dr. John A. O'Keefe, Assistant Chief, Theoretical Division, Goddard Space Flight Center, reported that study had shown them to be flakes of paint from the spacecraft.

- NASA research pilot Glenn W. Stinnett and Stanford physiologist Dr. Terence A. Rogers emerged from 7-day confinement in a simulated lunar spacecraft. Unique study by Ames Research Center involved performance of continuous and realistic professional tasks while under confined space flight conditions.

- Soviet Union announced launching of COSMOS II satellite (975-mile apogee, 133-mile perigee, 49° inclination, 102.5-minute period). COSMOS II reportedly had the same instrumentation as COSMOS I launched on March 16: investigation of radio transmission, radiation belts, magnetic field of the earth, distribution and formation of cloud cover, and to test "elements of space vehicle construction."

- Test flight of first Atlas-Centaur rocket canceled because of heavy cloud-cover at Cape Canaveral.

- In ejection capsule test, brown bear "Big John" rode in capsule ejected from B–58 at 1,060 mph and 45,000 ft. and parachuted to earth near Edwards AFB, Calif.

- In testimony before Senate subcommittee, Federal Communications Commissioner T.A.M. Craven reported that talks with Russian and other representatives had been encouraging on establishment of a global communications satellite system. FCC Chairman Newton N. Minow said that FCC "could live with" the proposed U.S. corporation with stock to be owned by the public and by communications companies. The FCC was, Minow stated, ill equipped to handle a Government-owned system.

- Dr. Homer E. Newell, NASA's Director of Space Sciences, speaking to a sectional meeting of the American Society for Engineering Education at Texas Technological College, Lubbock, Texas, said: ". . . accomplishing the space mission absolutely requires the strong and vigorous participation of our colleges and universities, which in turn requires that NASA bear its fair share of the support required to make it possible for the universities and colleges to participate. The universities must bear their share of responsibility to the space program and allocate an appropriate fraction of their material, as well as human, resources to the effort. In such a partnership, NASA stands ready to invest an appreciable fraction of its resources."

- Soviet Academician Leonid Sedov denied in interview with *Trud* that U.S.S.R. had launched other men into orbit besides Majs. Gagarin and Titov. U.S. press had speculated that as many as five Russian cosmonauts had been killed in unsuccessful flights.

- Space-age religion must be an "open undogmatic religion which itself reaches out for undiscovered truth," Rabbi Sheldon H.

Blank, professor at the Hebrew Union College-Jewish Institute of Religion, said in Williamsburg, Va. If space exploration discovers rational people living on other planets, religious people on earth "will have to abandon parochialism and reach for a God who is less comfortably near than we have wanted to believe."

April 7: Atlas-Centaur launching from AMR was again postponed, this time because of high-altitude winds.

• A censored transcript of testimony before a closed hearing of the House Defense Appropriations Subcommittee by Dr. Harold Brown, DOD's Director of Research and Engineering, on March 19, was released, disclosing questions by Chairman George H. Mahon on security aspects of the prepared statement which outlined DOD's R&D projects. Dr. Brown replied that it was necessary under the U.S. system to inform the public, that no numbers of weapons or specific capabilities of weapons were given, and that "having the other side know what we are thinking can in some cases be useful."

• Sir Bernard Lovell, Director of England's Jodrell Bank radiotelescope, said in an interview in Los Angeles that Soviet secrecy with regard to their space launches was justified. "There are two ways of looking at this thing, you know. In scientific circles it is customary not to announce an achievement until it has actually proven a success."

• Educational Testing Service received a $30,300 grant from the National ·Science Foundation for the showing of science films over stations of the National Educational TV Network of 58 stations.

• State Board of Education in Minnesota voted to stop requiring any science or mathematics as prerequisite to high-school graduation. Members argued that some students were not suited to courses in science and mathematics.

April 8: Soviet Cosmonaut Yuri Gagarin, in a Tass interview, said: "It is now necessary to determine how long a man can operate in outer space without prejudice to his health—five days, ten days, or more. This is one of the main tasks before proceeding to more complex flights to the moon, for example, or to the establishment of orbital or interplanetary stations." He criticized the U.S. for being "reluctant" to file details of the Glenn flight with the U.N. "The best information about outer space has been obtained in the U.S.S.R. The Americans can benefit more by our experience than we by theirs. Some people in America are not interested in exchanges of scientific information." The U.S. had submitted information concerning the Glenn flight to the U.N. on April 3, 1962.

• Surveyed by the *Washington Post* on the proposal to fire rockets loaded with radioactive waste into outer space, Washington-area clergymen responded variously. Comments on the proposal ranged from a "dreadful desecration of the Heavens" to a "fulfillment of the divine commandment to protect man and the earth on which he dwells." Soviet nuclear scientist, Dr. Peter Kapitza, had made the proposal in the April issue of the *Bulletin of the Atomic Scientists*.

April 9: Astronaut John H. Glenn was awarded the Hubbard Medal of the National Geographic Society "for extraordinary contributions to scientific knowledge of the world and beyond as a pioneer in exploring the ocean of space." Awarded only 20 times since it was struck in 1906, the Hubbard Medal honorees have included Adm. Robert E. Peary, Charles A. Lindbergh, Roald Amundsen, and Adm. Richard E. Byrd.

- Aerospace Medical Association, meeting in Atlantic City, heard research papers indicating new findings about space radiation. Dr. Sol M. Michaelson, of the University of Rochester School of Medicine and Dentistry, said preliminary evidence from animal tests indicates the possibility that one type of radiation exposure—mild amounts of microwave—might enhance recovery from x-ray damage because it generated increased activity in the bone marrow. Maj. Robert W. Zellmer (USAF), of the School of Aerospace Medicine, Brooks AFB, Texas, said tests on rats indicated the possibility that seriously excessive G forces, such as experienced during launch and re-entry of a spacecraft, might also give some protection against radiation.

- USAF launched its fifth unidentified satellite employing an Atlas-Agena B booster, from Point Arguello, Calif. In Washington, the State Department said the satellite would be registered with the U.N. if it went into orbit and stayed in sustained orbit.

- President Kennedy nominated Dr. J. Herbert Holloman to the new post of Assistant Secretary of Commerce for Science and Technology. Dr. Holloman was general manager of the General Engineering Laboratory of GE in Schenectady, N.Y.

- An Atlas ICBM blew up on the pad during launching at AMR. The missile had been heavily instrumented in an effort to determine causes of difficulties on recent flights.

- Joint AEC-DOD announcement designated a second nuclear test area in the Pacific, this one centering on Johnston Island and enclosing an area with a radius of 470 mi. on the surface, expanding to an area with a radius of 700 mi. at 30,000 ft. and above. Ban on entering the area would become effective April 30.

- U.S. space experiments with regard to the moon may well give scientists new information on "how the Earth and the planets and the Sun were formed," Dr. Gordon J. F. MacDonald, Associate Director of the Institute of Geophysics and Planetary Physics of the University of California, said in a Voice of America broadcast. The seismometer to be landed on the moon by NASA's Ranger, for example, would help determine whether there are "moonquakes" some eight times more frequent than earthquakes, as there should be if the moon contains the same concentration of radioactive elements as the earth.

- European Broadcasting Union ended a 20-nation conference in Seville, Spain, with an announcement that preliminary arrangements had been made for an international television service by use of communications satellites. Three U.S. television networks—CBS, NBC, and ABC—would join with USIA to operate the system from the U.S., transmitting from Andover, Maine, and the European receiving station would be the British General Post Office Facilities in Cornwall, which would relay the signals to other members of the European Broadcasting Union.

April 9: U.S.S.R. Presidium of the Supreme Soviet declared April 12, anniversary of the first Soviet space flight, to be a national annual holiday in Russia. In announcing ceremonies for Cosmonautics Day in the Soviet Union, Leonid Korneyev reviewed the U.S.S.R.'s space program from its beginning in April 1932, with the formation of a group to study jet propulsion. Between 1932 and 1941, Korneyev pointed out, the U.S.S.R. had developed 118 different liquid-fuel rocket engines.

April 10: Fifty-five delegates from the U.N. toured the Atlantic Missile Range at the invitation of the State Department. Six Communist countries were represented, but delegates from the U.S.S.R. had declined at the last minute. Among the sights were the U.S.-U.K. satellite scheduled for launch that day but postponed because of trouble in the second stage of the Thor-Delta booster, the Saturn rocket being prepared for launching, and the Project Mercury site and control room.

• Assistant Attorney General Nicholas Katzenbach testified before the Senate Commerce Committee that the compromise communications satellite bill, while generally acceptable to the Administration, would place control of the system in the communications industry. "There is real danger," he said, "that ground stations, if separately owned by the carriers, may because of their high cost represent an obstacle to technical growth so as prematurely to freeze the type of system." The compromise bill, approved by the Senate Space Committee two weeks ago, provided for a 50–50 split of the stock in the satellite corporation between the communications industry and other companies and investors.

• Gen. Bernard A. Schriever, Commander of AFSC, said in a speech at Mississippi State College: "The Materials Laboratories at Wright-Patterson Air Force Base developed the basic material and structures used in 'sandwich construction' of aircraft; titanium alloy as a structural material; plastic, ceramics, and metallic alloys to withstand the high temperatures encountered during re-entry into the earth's atmosphere."

• The Boys Clubs of America honored Lt. Col. John H. Glenn, Jr., for his example of good citizenship to youth, presenting him with the "Man and Boy Award" at Alexandria, Va.

• First stage (S–I) for Saturn vehicle SA–3 had its first flight qualification test at MSFC, all eight engines firing for 30 sec.

April 11: NASA Administrator James E. Webb, testifying before the Senate Committee on Commerce, supported the President's bill setting up a communications satellite corporation and approved of the Senate amendments, except for a caution on the one that would direct the FCC to encourage communications common carriers to build and own their own ground stations. Mr. Webb summarized NASA's responsibilities under the bill: (1) to advise FCC on the technical characteristics of the system; (2) to advise FCC on the technical feasibility of attaining the desired technical characteristics; (3) to coordinate its space communications R&D work with that of the corporation; (4) to furnish launching facilities, vehicles, and services in connection with the development and operation of the system; (5) to furnish other services on a reimbursable basis; and (6) to advise the Secretary of State on technical feasibility of the system providing communications service to a particular point in the world.

April 11: USAF launched an Atlas missile from Vandenberg AFB, Calif., in a test flight down the Pacific Missile Range.   A SAC combat missile crew launched the Atlas from a horizontal, semi-hard "coffin" launcher.

- In a recent tightening of DOD policy on information about military satellite launchings, officials were reported to have stated that in future even the Discoverer flights would not be so identified. Any information beyond bare announcement of launch of an unidentified satellite would be considered on a case-by-case basis.
- Atlas-Centaur initial launch from AMR was again postponed by NASA, this time because of fueling problems.
- Robert J. Lacklen, NASA's Director of Personnel, said that the first part of NASA's nationwide recruiting drive to hire 2,000 scientists and engineers had ended on March 15, with the completion of the field work.   During this phase, teams of interviewers contacted 14,000 persons, interviewed 5,000.   The second phase would involve setting up a national job register.
- Dr. Ross A. McFarland of Harvard University was awarded the Walter Boothby Award by the Aerospace Medical Assn. for outstanding research directed at the promotion of health and prevention of disease in professional airline pilots.

April 12: NASA was studying the problem of falling fragments from orbital objects and the possibility of injury or damage to persons or property on earth.   Four fragments from the Atlas booster that put Astronaut John Glenn into orbit were recovered on earth, the first pieces known to have re-entered from an orbiting object without burning up.   Statistically the chances of injury to anyone on earth would be extremely small, especially since the orbits are over water about 80% of the time.   NASA pointed out that some 100 meteorites weighing two pounds or more struck the U.S. every year, yet there had never been a report of anyone being struck by one.

- USAF Blue Scout launched from Cape Canaveral but second stage did not fire.
- NASA would begin an education program in the fall of 1962 to provide financial support for 10 doctoral candidates in science and engineering at each of 10 U.S. universities, NASA Administrator James E. Webb announced to the Institute of Environmental Sciences in Chicago.   Students would receive a stipend of $2,400 per year, expenses up to $1,000 per year, and the university would be reimbursed for tuition, fees, and other expenses.   Grants would be for one year, renewable to a maximum of three years.   Mr. Webb said NASA expected the program "will increase considerably in years to come."
- NASA awarded $1,000 to NASA engineers for inventions contributing to the advancement of aeronautical and space science and technology.   Langley Research Center's Henry J. E. Reid, Jr., and H. Douglas Garner split a $1,000 award for conceiving a simple, lightweight attitude control system to control the axis of a spinning vehicle.   Goddard Space Flight Center's James S. Albus received a $1,000 award for his digital solar aspect sensor that has been flown in several Explorer satellites.

April 12: Army launched the full three-stage Nike-Zeus antimissile missile from Point Mugu, Calif., in its first successful flight test with all stages.

- U.S. Navy claimed a new record for an aircraft speed climb when an F4H-1 Phantom II piloted by LCdr. Del W. Nordberg sped from a standing start to 98,425 ft. (30,000 meters) in 371.43 sec. at Point Mugu, Calif.

- National holiday in the Soviet Union, the anniversary of orbital flight of Yuri Gagarin. In ceremony in the Kremlin's Congress Hall, Cosmonaut Gagarin declared: "We are on the threshold of more new space launchings. When these space ships return to earth, the Soviet people will have more holidays to celebrate." Premier Nikita S. Khrushchev and Cosmonaut Titov also made short speeches, while Matislav Keldysh, President of the Soviet Academy of Sciences, made the major address.

- *Pravda* editorial extolled the future of communism as demonstrated by Soviet space achievements as follows: "The successes of the Soviet conquerors of space reflect the great achievements of the Soviet people in the development of the mighty productive forces of our homeland, the indisputable advantages of socialism, and its superiority over the capitalist system. The most reasonable representatives of the western world cannot fail to admit that socialism, as Comrade Khrushchev put it, is indeed the reliable launching pad from which the Soviet Union launches its spaceships.

  "Only a year has passed since a Soviet man heralded the dawn of the space age by making his first flight into interplanetary space, but this year has been packed with great events. The 22nd CPSU Congress adopted a party program which gave wings to the Soviet people and marked out the clear prospects for building communism. Guided by the historic decisions of the congress and the new party program, our Soviet people will exert every effort to foster the building of a communist society, the bright future of all mankind . . . ."

  In Tass statement, Cosmonaut Gherman Titov described Cosmonautics Day as a "holiday of labor and man's victory over the forces of nature."

April 13: NASA announced the X-15 had been assigned a service function in aeronautical and space research beyond its original role as an experimental test vehicle. The National Research Airplane Committee, composed of NASA, USAF, and USN delegates, had approved a series of experiments including ultraviolet stellar photography, horizon scanning for research on accurate horizon sensing in space vehicles, atmospheric density and micrometeorite measurements, and evaluation of advanced systems and structures.

- In a NASA press conference on the X-15 additional research program, Mr. John Stack, NASA's Director of Aeronautical Research, said: ". . . when I was talking about the total research airplane program, this started really because two men in high positions in a time of great stress also had time to think of the future, and I refer to General Arnold and Dr. George W. Lewis. And when this approach, this sort of thing, was put to them in late '42—and you can readily appreciate General Arnold and Mr. Lewis were pretty busy men at that time—they did under the stress of all that find time to think also of the future. And this kind of a program that has progressed is a kind of a monument to the foresight of those two men."

April 13: NASA Administrator James E. Webb, addressing the National Conference of the American Society for Public Administration in Detroit, said: "No new department or agency in the recent history of the Executive Branch of the Federal Government was created through the transfer of as many units from other departments and agencies as in the case of NASA. Three and one half years ago, NASA did not exist. Today NASA comprises approximately 20,500 employees, ten major field centers, and an annual budget approaching the $2 billion mark."

• Dr. Edward C. Welsh, Executive Secretary of the National Aeronautics and Space Council, was awarded the Arnold Air Society's trophy as the civilian who had made the most outstanding contribution to aerospace science and national security in 1961.

April 13–14: U.S. Army claimed two world flight records for time-to-climb in a helicopter in flights made at Fort Worth, Texas. On April 13, Capt. Boyce B. Buckner (USA) piloted a YHU–1D Iroquois helicopter to 6,000 meters (19,686 ft.) in 5 min. 51 sec. On April 14, Lt. Col. Leland F. Wilhelm (USA) took the turbine-powered helicopter to 3,000 meters (9,843 ft.) in 2 min. 14.6 sec.

April 14: Dr. Homer E. Newell, NASA's Director of Space Sciences, said that in addition to NASA's international program for space science and the European organizations for cooperative space research, "some South American groups are considering the possibility of cooperative sounding rocket endeavors."

• USAF selected proposals submitted by United Technology Corp. for the 120-inch solid motor and STL–ARMA for the guidance efforts to be placed in Titan III which will serve as a large space booster. Contracts were contingent upon final program approval.

April 15: U.S. Weather Bureau began daily international transmissions of cloud maps based on photos taken by TIROS IV. This was the first time that the actual cloud map was sent abroad by radio facsimile, although coded data had been sent previously.

• In an article in *Ekonomicheskya Gazeta,* the economic newspaper of the Soviet Communist Party, Dr. Peter Kapitza, one of Russia's outstanding physicists, said that Soviet science had been hampered by attempts to judge the validity of scientific theories on the basis of Marxist dialectics. If Soviet scientists had listened to Marxist philosophers in the 1950's, he said, the Russian achievements in space would have been impossible.

April 16: Dr. Hugh L. Dryden, NASA's Deputy Administrator, speaking to the Federation of American Societies for Experimental Biology in Atlantic City, said: ". . . the air age brought us great supplies of aluminum and the basis for building lightweight structures, not only for airplanes but also for trains, buses, and ships. The nuclear age brought applications of isotopes in medicine and in inspection of materials. Nuclear developments brought remote manipulators and sealed pumps for hazardous liquids and gases. The space age has brought to maturity the concept of systems analysis and optimization of designs involving many branches of science and engineering. In addition the space age has given us high-temperature ceramics, ablating materials for heat protection, pressure-stabilized lightweight tanks, computers handling large amounts of data, and many other developments which are finding application throughout industry."

April 16: Senate passed the NASA supplemental appropriations bill, with amendments providing $71,000,000 for construction at AMR but contingent on passage of authorization legislation. Bill would now go to Senate-House conference to iron out differences between the versions passed by the two Houses.

- Technical planning and review of the NASA lunar program held with 60 key officials, at MSFC.
- Representatives of Australia, Belgium, France, West Germany, Italy, the Netherlands, and Britain signed an agreement for the establishment of the European Satellite Launcher Development Organization in London.
- In a ceremony at the Smithsonian Institution marking the 95th birthday of Wilbur Wright, the Early Birds—an organization of pilots who soloed before December 16, 1916—presented plaques to its members, about 90 of whom were on hand. Also bronze busts were presented to the Smithsonian of Mario Calderara and Umberto Savoia, Italy's first military pilots and taught by Wilbur Wright.
- AFSC announced that a small, inexpensive radar capable of showing an airport control tower operator whether any of the runways or taxiways are occupied even under conditions of zero visibility or when obstructed by houses, trees, snowbanks, etc., has been under test and would be the subject of a final report in May 1962. The runway radar is known as ASDE (Airport Surface Detection Equipment).
- House of Representatives passed a resolution declaring February 20 to be John Glenn Day, sent it to the Senate.
- AEC-Australian research balloon intended to collect dust samples at 105,000 ft. to determine the density of radioactive fallout was launched from Mildura, Australia, 350 mi. NW of Melbourne, but burst at 100,000 ft.
- Dr. Gerald de Vancouleurs, Professor of Astronomy at the University of Texas, said in NAS Voice of America broadcast that the discovery of clues to the prehistory of the solar system and possible contact with extraterrestrial forms of life "stand out as two of the most challenging and promising goals of space exploration." Dr. de Vancouleurs pointed out that direct sampling and probing of the surface and crusts of the moon and planets may help discover these clues. "Of even greater scientific and philosophic interest," he said, "is the probability that direct exploration of at least one planet—Mars—may place us in contact with extraterrestrial forms of life, whose study may help solve the problems of the origin and evolution of life under different planetary conditions."

April 17: NASA launched a Nike-Cajun sounding rocket from Wallops Station, Va., which detonated 12 grenades from 25 to 57 mi. altitude. Fifteen minutes later a Nike-Asp sounding rocket was launched. It released a cloud of sodium vapor from 26 to 144 mi. altitude in a study of wind velocities and upper-atmosphere densities.

April 17: USAF launched an unidentified satellite from Vandenberg AFB, Calif., using a Thor-Agena B booster.

- Subcommittee on Patents of the House Committee on Science and Astronautics released its report following three years of study of the operation of the patent provisions of the Space Act of 1958. The report of the subcommittee headed by Rep. Emilio Q. Daddario contended the patent provisions that gave the Government title to most inventions resulting from NASA-financed research were damaging to small business, cost the taxpayer money, diluted the national space effort, and made industry reluctant to market the products of such research. Recommended that patent ownership in most cases be turned over to industry, with Government retaining royalty-free use. A minority report by Rep. William F. Ryan pointed out that with almost 95% of Government research money "going to big business, retention of title by the contractors will inevitably bring about concentration of technology and economic power."
- Howard Simons reported that many U.S. space experts feel that the U.S.S.R. has reshaped its space program from a few, spreadout "spectaculars" to a more orderly scientific program. While spectaculars have not been abandoned, it was thought the bulk of the program would be conducted with smaller, cheaper rockets for more useful scientific objectives and with more international cooperation. This was considered to be a victory for Soviet scientists over military and political planners who had favored the spectaculars.
- Reuters reported that the Tower of London had been asked by Airresearch Manufacturing Co., of Los Angeles, for information on the finely articulated joints in a suit of armor once worn by Henry VIII as a possible aid in spacesuit design. The 94-lb. suit of armor was designed for jousting tournaments with the wearer on foot; it covered the whole body and its joints permitted full mobility.

April 18: NASA approved the highest national priority for the Apollo program, covering Saturn C–1, Saturn C–5, Titan II, and Atlas-Agena launch vehicles, as well as all stages and engines connected with the vehicles.

- NASA announced that the first Centaur launch was due no earlier than April 20. Postponement on April 11 was caused by too low calibration of a transducer, which gave a false indication of the pressurization of the fuel systems in the Atlas booster.
- NASA announced it would begin taking applications for additional astronauts now and continue until June 1, 1962. In July persons meeting the requirements would be interviewed, selected applicants given thorough physicals, and final selection of some 5 to 10 astronauts would be made in the fall. Qualifications were also announced; these were slightly less demanding in terms of test-pilot time, but more demanding in that they lowered the age limit to under 35 and required a college degree in science or engineering.
- Jupiter IRBM was fired by a NATO missile crew from AMR.
- USAF acquired its first operational Titan I ICBM site, at Lowry ABF, Colo., as AFSC turned control over to SAC.

April 19: X–15 made its 50th successful flight from Edwards AFB, Calif., with NASA's Joseph A. Walker as pilot. The X–15 No. 1, testing an emergency flight control system, reached a speed of 3,920 mph (mach 5.84) and an altitude of 150,000 ft.

- NASA announced that FRIENDSHIP 7, the Mercury capsule in which Astronaut John Glenn orbited the earth three times, would be lent to USIA for a world tour, with some 20 stops on the itinerary and touching all continents. In mid-August the capsule would be displayed at the Century 21 Exhibition in Seattle, Washington, before being presented to the Smithsonian Institution in Washington, D.C., for permanent exhibition.

- USAF Skybolt missile launched from B–52 bomber over the Atlantic Missile Range in first live flight test; considered successful though second stage failed to ignite.

- House Commerce Committee reported out a commercial communications satellite bill similar to the Senate measure, including a provision "encouraging" communications companies to build and operate their own ground stations.

- State Department received a visa application from Cosmonaut Gherman Titov, who would be a part of the Soviet delegation to the COSPAR International Space Science Symposium to be held in Washington April 30-May 9. U.S. officials indicated that the request would be approved.

- University of Maryland announced that in September it would begin a program leading to undergraduate and graduate degrees in astronomy. NASA has granted the university $192,000, partly for equipment and expenses and partly for 3-year stipends to 10 graduate students recently chosen from among 60 candidates.

- A high-level Presidential study group set up in July 1961 to study the role of nonprofit organizations in relation to Government reported that the U.S. must continue to rely heavily on such organizations and private industry for directing and carrying out Federal research. It further recommended that the Government strengthen its own research organizations and its research management capabilities.

- In opinion poll taken in Ohio Congressional District (23) by Congressman William E. Minshall on whether the U.S. should enter into a cooperative space program with Russia, 47% replied affirmative, 39.6% replied negative, and 13.3% made "no opinion" replies.

- Congressman Olin Teague submitted H. Con. Resolution 461, to express the sense of Congress that the U.S. should not participate in any program for the exploration of space with foreign nations or international bodies which would involve the disclosure of any technical information, unless the Soviet Union by "positive action" participates in an inspection system for armaments and informs the world about hitherto-secret information about its space program. Similar resolutions were introduced by Congressmen Young, Kilgore, Purcell, Roberts of Texas, Riehlman, Roudebush, Casey, and Morris.

- Italian Space Research Commission announced Italy would attempt to orbit its first satellite in 1963. The 200-lb. scientific satellite would be launched from a platform floating in the Indian Ocean and would measure air density variations caused by solar radiation.

April 20: X–15 No. 3 flew to 207,000 ft. and 3,818 mph (mach 5.33) in a test of a special adaptive control system. Flight was made from Edwards AFB, Calif., NASA's Neil Armstrong as pilot.

- Attempt at first flight test of NASA's Centaur was canceled because of troubles in the ground-handling equipment used to pump liquid oxygen into the vehicle.

- NASA published an integrated series of three Quality Publications setting forth NASA's intensified quality assurance program now required in NASA space programs from R&D concept through space operations. These publications required NASA prime and sub-contractors on space system work to establish and maintain a quality program that in many elements was beyond traditional quality control. They also provided guidelines for NASA management to evaluate contractor quality performance both as a factor of current contract performance and as a consideration in award of future contracts.

- White House released report on "Strengthening the Behavioral Science" prepared at the request of the President's Science Advisory Committee by a subpanel headed by Prof. Neal E. Miller of Yale University. The report submitted that "the general issues studied by behavioral scientists are critically important to our national welfare and security," and that "ways must be found to strengthen these disciplines and improve their use." It recommended increasing general and specific education in the behavioral sciences following the lead in the physical sciences. It also suggested that the Social Science Research Council be invited to appoint a standing committee, and that a group of representatives of relevant governmental agencies be created, to review and provide "appropriate advice in the light of current possibilities and needs of behavioral science."

- USAF announced selection of seven USAF officers and one USN officer for the second class in the USAF's space pilot school. The class, to begin on June 18 at Edwards AFB, Calif., would include LCdr. Lloyd N. Hoover (USN), Majs. Donald M. Sorlie and Byron F. Knole (USAF), and Capts. Albert H. Crews, Jr., Charles C. Bock, Jr., William T. Twinting, Robert W. Smith, and Robert H. McIntosh (USAF).

- Jessie G. Vincent, automotive and airplane engine designer and holder of over 400 patents, died in Detroit at age 82. Vice president of the Packard Motor Car Co. from 1915 to 1948, Mr. Vincent was codesigner of the Liberty Marine engine in World War I and as an Army major helped get the Liberty engine into production, at which time it was regarded as the world's finest power plant of its kind.

- U.S. Army claimed a new world speed record for helicopters, when Capt. William F. Gurley (USA) flew a YHU–1D Iroquois helicopter over a closed-circuit course near Fort Worth, Texas, at 133.9 mph, shattering a two-year-old Soviet mark of 88 mph.

April 21: Fifth attempt to launch NASA's Centaur space booster was canceled at the last second at AMR. Engines on the Atlas first stage had been ignited when an automatic detection device spotted a malfunction and shut off the engines.

• Sir Frederick Handley Page, British aviation pioneer, and chairman and managing director of Handley Page Ltd., died in London at age 76. Inventor of the slotted wing, he provided England with its first bombers in 1915 and his Halifax bomber of World War II, of which more than 60,000 were produced, was a key weapon of RAF Bomber Command.

April 22: Jacqueline Cochran became first woman to pilot a jet aircraft across the Atlantic and claimed 49 world records in 5,120-mile flight from New Orleans, La., to Hanover, Germany, with stops at Gander and Shannon, in a Lockheed Jet Star.

April 23: NASA's RANGER IV was launched by Atlas-Agena from AMR, went into parking orbit, and was put into proper trajectory to the moon by restart of the Agena B booster. Failure of a timer in the RANGER IV payload caused loss of both internal and ground control over the vehicle. Analysis of the trajectory indicated that the payload would probably skim the leading edge of the moon on April 25 and be pulled by the moon's gravity to a crash-landing on the far side of the moon, but that none of the experiments would operate and no data would be received.

• U.S. and U.S.S.R. scientists have agreed on a proposal to establish a "world weather watch" for improved collection, analysis, and dissemination of world weather information by means of conventional weather techniques and weather satellites. Dr. Harry Wexler, Director of Research, U.S. Weather Bureau, and Dr. Viktor A. Bugaev, Assistant Director of the Hydrometeorological Service, U.S.S.R., drafted the plan at the World Meteorological Organization headquarters in Geneva. Plan would call for expanded and improved weather observation by both conventional and satellite means, the establishment of world and regional weather centers to collect, analyze, and disseminate weather information, including participation by underdeveloped countries through pooling of their resources to develop regional weather forecasting centers. Soviet participation was indicated by Dr. Konstantin T. Logvinov, Deputy Director of U.S.S.R.'s Hydrometeorological Service. He indicated possibility of Soviet weather satellites, although he said U.S.S.R. was not as advanced as U.S. in this area—"We are just learning from American experience"—and establishment in U.S.S.R. of a ground station to receive cloud-cover photos from satellites. The proposed global weather plan was to be submitted to the World Meteorological Organization May 22 and then to the U.N.

• First International Symposium on Rocket and Satellite Meteorology opened in Washington, D.C. Attended by some 500 delegates from more than 20 nations, the conference was sponsored by the World Meteorological Organization, the International Union of Geodesy and Geophysics, and the Committee on Space Research of the International Council of Scientific Unions. It was hoped that the symposium would lead to an international sounding rocket program during the International Year of the Quiet Sun in 1964–65.

April 23: National Council of the Federation of American Scientists adopted a policy statement that space activities should be classified only when there are compelling military reasons.

- In Voice of America broadcast, Dr. Leo Goldberg of the Harvard College Observatory reported that the accumulation of knowledge in solar physics is proceeding "at a fantastic rate, due to the inherent importance of the subject coupled with a sudden fusing of new instrumental techniques and theoretical ideas." Dr. Goldberg noted that sun spots in the photosphere have been the object of intense study since Galileo but that "observations have not yet been synthesized to provide even an accurate description of their physical nature, let alone the mechanism of their origin." Crediting high-altitude rockets for initial detection and occasional study of hidden solar radiation, detailed study requires continous observation of the sun on a 24-hour basis, which "can only be accomplished with the aid of one or more satellite observatories."

- National Academy of Sciences awarded the J. Lawrence Smith Medal for outstanding achievement in the investigation of meteoric bodies to Dr. Harold C. Urey of the University of California of San Diego. The award was based upon Dr. Urey's many publications on the chemical composition of meteorites, only one of his most recent fields of interest. He received the Nobel Prize for Chemistry in 1934. After World War II he launched into speculations concerning the origin of the solar system, the cosmic abundance of the elements, the composition and structure of the moon and the planets, the latter work depending upon his chemical analysis of meteorites.

April 24: Joint U.S.-Japanese sounding rocket launching at Wallops Station, Va., was postponed because of high winds.

- U.S.S.R. launched COSMOS III, identified by Tass as another of the scientific satellites in a series to study weather, communications, and radiation. Orbital data: apogee, 448 mi. (720 km.); perigee, 142 (229 km.); period, 93.8; inclination, 48°59'.

- DOD removed secrecy classification from the satellite Anna, a joint geodetic satellite designed to enable measurement of intercontinental distances and the shape of the earth.

- First transmission of TV pictures in space, made via orbiting ECHO I. Signals beamed from MIT's Lincoln Laboratory at Camp Parks, Calif., bounced off ECHO I, and received at Millstone Hill near Westford, Mass. Picture telecast was that of the initials "M.I.T." and travelled a distance between 3,000 and 4,000 miles, using equipment built for USAF Project Westford experiment. Since test used misshapen ECHO I, scientists felt that experiment should increase interest in passive satellites for communications once Echo II is orbited.

- President Kennedy gave the order to resume nuclear testing in the atmosphere, to be conducted by Joint Task Force 8 under the command of Maj. Gen. A. D. Starbird.

April 25: Saturn (SA-2) booster rocket was successfully launched from AMR in its second flight test.   Like the first Saturn vehicle launched on October 27, 1961, the Saturn fired only the first-stage engines, generating 1.3 million lbs. of thrust.   Dummy second and third stages filled with water were detonated at 65 miles altitude (Project High Water) and the water ballast formed an artificial cloud.   Maximum velocity was slightly more than 3,700 mph.   Modifications to decrease the slight fuel sloshing encountered near the end of the previous flight test were apparently successful.

* Dr. Edward C. Welsh, Executive Secretary of the National Aeronautics and Space Council, stated that the U.S. program for peaceful development of space was not inconsistent with military space development, in a speech before the National Capitol Section of the ARS.   "The true mission of our military strength is to protect the peace," he said.   "If we do not take adequate care of our national defense, we will not have a chance to do any of the other things in space—at least as free men."

* D. Brainerd Holmes, NASA's Director of Manned Space Flight, speaking before the American Management Association in New York, said: "We believe that the soundness of our present engineering program can be illustrated by the correlation between predicted events and time of occurrence in John Glenn's recent flight.   Indeed, the time from lift-off to impact in this mission was within one second of the pre-flight calculations."

* U.S. resumed atmospheric nuclear testing in Operation Dominic with the detonation of a medium-yield device dropped from an aircraft near Christmas Island in the Pacific.

* American Newspaper Publishers Association meeting in New York, set up a scientific advisory committee to make a year's study of possible application to newspapers of space technology, nuclear energy, computer technology, and other developments.   Members appointed were: Dr. Athelstan F. Spilhaus, Dean of U. of Minnesota's Institute of Technology, chairman; Trevor Gardner, President of Hycon Manufacturing Co.; and Dr. John R. Pierce of Bell Telephone Lab.

April 26: A very eventful day in space exploration: launched into orbit were ARIEL I, COSMOS IV, and two unidentified USAF satellites; RANGER IV impacted the far side of the moon; and a U.S.-Japanese probe was launched from Wallops.

* RANGER IV impacted on the moon at 7:49:53 AM EST, ending a 231,486-mi. journey from AMR that began with its launching on April 23.   Goldstone Tracking Station maintained contact with the 50-milliwatt transmitter in the lunar landing capsule until it passed behind the left edge of the moon.   Impact velocity was 5,963 mph, point of impact 229.3° E, 15.5° S, on a part of the moon never seen by man.   RANGER IV's instrumentation, which ceased useful operation some ten hours after launch, never functioned again.   About the same time as the lunar impact, the Agena B 2nd stage passed to the right of the moon and went into orbit around the sun.

April 26: ARIEL I (S–51), the first international satellite, launched into orbit from Cape Canaveral by Thor-Delta booster. The 132-lb. spacecraft was built by GSFC of NASA and carried six British experiments to make integrated measurements in the ionosphere. Three experiments measured electron density, temperatures, and composition of positive ions in the ionosphere, while two experiments were designed to monitor the intensity of radiation from the sun in the ultraviolet and x-ray bands of the solar corona. The sixth experiment was designed to measure cosmic rays, supported by simultaneous experiments from ground and by aircraft and balloon flights.

- U.S. and Japan launched a joint sounding rocket from Wallops Station, Va. NASA furnished the Nike-Cajun rocket, launch facilities, data acquisition, and a langmuir probe to measure electron temperature. Japan furnished instrumentation, a device designed to measure electron temperature and density simultaneously, functions which up to now have required two separate instruments. Height of the flight was 75.6 mi.

- U.S.S.R. launched COSMOS IV, presumably another in the recent series of scientific satellites. Orbital data: apogee, 206 mi.; perigee, 186 mi.; period, 90.6 min.; and inclination, 65°. No other details were announced.

- In address at Tulane University, Dr. James R. Killian, President of MIT, stated that the Soviet Union had placed "their first rate people" in their space program and thus weakened their high-energy physics program. "We must be very careful in space exploration that we do not tend to draw strength from other technological areas. We need balance," he said.

- NASA deadline for proposals invited from 15 companies for detailed design, manufacturing, and test plan for the Nova concept.

- X–15 flight was canceled because of an engine malfunction seconds before the research aircraft was to be dropped from the B–52 mother aircraft. Pilot was Major Robert White (USAF).

- NASA graduated first group of Project Mercury tracking personnel completing new course at Wallops Station, the seven graduates representing personnel from NASA and DOD contractors. Directed by GSFC, the Mercury Network Training Program consists of specialized courses to support the man in space mission, one requiring that the subsystems at all 18 sites in a global network are continuously monitored and provide precise data concerning the spacecraft's location, altitude, and operational status as well as the astronaut's condition. Real-time data must flow between the sites, the Computer Center at GSFC, and the Mercury Control Center at Cape Canaveral.

- Reported that NASA had drafted for review by other Government agencies a policy statement on measures being studied for the prevention of rocket booster parts surviving re-entry, possibly injuring life and property on earth. Fragments of MA–6 booster had surprisingly survived re-entry. Measures, as reported, included programed booster destruction, perhaps at the cost of payload weight.

- Maj. Gen. O. J. Ritland (USAF), formerly Commander of AFSC's Space Systems Division, was named Deputy to the Commander, AFSC, for Manned Space Flight. Gen. Ritland would be in charge of liaison with and USAF support of NASA's manned space flight program.

April 26: The Senate Committee on Commerce concluded its hearings on S. 2814 to establish a commercial communications satellite system.

- Separation of the disciplines of geology and geophysics in U.S. universities is "nothing short of an intellectual outrage," Dr. Lloyd V. Berkner told the American Geophysical Union in Washington when that organization presented him with its first John Adams Fleming Award. Only three American universities were moving toward broad programs in geophysical study, he said. He urged the NAS and universities to work toward programs that would recognize that geology and geophysics are now "intimately intertwined."

- USAF launched two unidentified satellites, one boosted by a four-stage Blue Scout, the other by an Atlas-Agena B, both from PMR. U.N. Public Registry stated Blue Scout did not attain orbit.

- USAF was awarded the Bendix Trophy for its record-breaking B-58 transcontinental flight in two hours and 56.8 seconds at average speed of 1,214.71 mph on March 5, 1962.

- U.S.S.R. is encouraging "do-it-yourself" science in some 2,000 laboratories located in plants and factories, according to Yevgeny K. Fedorov, chief secretary of the Academy of Sciences, speaking in Moscow. Some 30,000 people with amateur scientific interests were making use of the laboratories, he said, and more would use them as the Soviet work week was reduced in length.

April 27: Press conference held at NASA Hq. on the first international satellite, ARIEL I, with U.K. and U.S. participants. Orbital data was reported: 100.9 min. period; 242.1 st. mi. perigee; 745.4 st. mi. apogee; and 53.87° inclination to the equator. All experiments were reported as functioning. Dr. Robert Seamans, Associate Administrator, outlined NASA's International Program, and pointed to the initiation of the joint U.K.-U.S. S-51 satellite effort in 1959 and the international tracking network (8 non-U.S. stations) participating in the readout of ARIEL I. Sir Harrie Massey, chairman of the British National Committee on Space Research, pointed out the demands of science and space research for international cooperation.

- Langley Gold Medal of the Smithsonian Institution was awarded to Dr. Hugh L. Dryden, NASA's Deputy Administrator, at the annual meeting of the American Philosophical Society in Philadelphia. Cited for his "important applications of experimental science to the problems of flight and for his wise and courageous administration of much of America's research and technical developments that now make possible the conquest of air and space," Dr. Dryden was the tenth recipient of the 54-year-old award.

- Details of Project Anna, flashing-light geodetic satellite, were given to the international Committee on Space Research (COSPAR) meeting in Washington by Mark M. Macomber of ONW. Previously classified, two Anna satellites have been built and when placed into orbit will provide means to calibrate three different satellite tracking systems and provide accurate reference points in space (flashing light photographed at precise times against known star background). NASA will now not have to develop a geodetic satellite to provide open scientific information.

April 27: Dr. Launor F. Carter was appointed Chief Scientist, USAF, to take office in July succeeding Dr. Leonard S. Sheingold. Dr. Carter, Vice President and Director of Research for System Development Corp. of Santa Monica, Calif., has been a member of the USAF Scientific Advisory Board since 1955. A psychologist, Dr. Carter has done much work on leadership, perception, and group behavior.

- George C. Barnhart, pioneer airman and inventor, died in Pasadena. Barnhart held 70 aircraft patents, including one for wing tanks. In 1942, he had turned over his patent for split-edge wing flap, a landing brake used on conventional and jet aircraft, to the Army Air Forces for the duration of the war, thousands of planes being built 1946–51 using the Barnhart flap for which he received no royalties as World War II had not been officially ended. He had intended to sue the Government for $10 million.

April 28: USAF launched unidentified satellite with Thor-Agena booster, the fourth unidentified satellite reported launched from Vandenberg AFB in ten days.

- Second NASA Nike-Cajun probe with Japanese payload was canceled for third time because of high winds at Wallops Island.

- First details of an orbital space station under feasibility study, one shaped like a doughnut and inflated once in orbit, revealed at Langley Research Center. Design of space stations had begun in November 1960 at Langley. Paul Hill, chief of the Applied Materials and Physics Division, stated that structures were now under study which could hold from four to thirty people.

April 29: COSMOS IV, Soviet satellite launched on April 26, was successfully landed in a predetermined area, according to Tass announcement. All equipment for research into the upper atmosphere and space worked well during the 1,250,000-mile flight, Tass said.

- Spokesman at British tracking station at Winkfield Row said that ARIEL I's "signals are good and everything seems satisfactory."

- Melvin N. Gough, former director of NASA's AMR operations and recently chief of the CAB's Bureau of Safety, was named as director of the Federal Aviation Agency's new Aircraft Development Service.

- Cosmonaut Gherman Titov arrived in New York City for 8-day visit to the U.S. in conjunction with COSPAR sessions in Washington.

- U.S.S.R. was considerably behind the U.S. in facilities and equipment for biological and medical research, Dr. A. N. Studitsky complained in an article in *Vestnik Akademii Nauk*, organ of the Soviet Academy of Science.

April 30: X–15 No. 1 flown to an altitude of approximately 246,700 ft. (46.7 mi.) with NASA's Joseph A. Walker at the controls. The record altitude was achieved in a climbing attitude of about 38°; top speed attained was 3,443 mph. Previous altitude record was 217,000 ft., achieved on October 11, 1961, by Major Robert White (USAF). The speed record of 4,093 mph was set on November 9, 1961, by Major Robert White.

- In Voice of American broadcast series on Space Science, Dr. Edward R. Dyer, Jr., of the Space Science Board Secretariat, stated that the history of astronomy is an alternation between major technological and theoretical advances. "The stage is set," he said, "for the next two big steps." The first will be the space-age technology enabling observation of the stars, nebulosities and galaxies using new devices capturing practically all octaves in the scale of radiation ranging from the very high-frequency gamma rays to the slow undulations of the longest radio waves. The second of the anticipated major advances will be a theoretical revolution, the nature of which will depend on results of improved observation techniques. These new findings "will either explain the universe on classical evolutionary grounds or demand a revision of fundamental physical theory at least as far reaching as the Theory of Relativity or the Quantum Theory."

April 30: In address to IAS meeting in St. Louis, George M. Low, NASA Director of Spacecraft and Flight Missions, concluded with a statement of basic organizational concepts: "We believe that we must obtain the very best efforts of the very best people we can find, both in Government and industry, if we are to achieve our National goal. We believe that our organizational concepts and management techniques must be no less excellent than our technical efforts. We believe that with constant attention to these concepts, and with the hard work and dedication of the people involved, we will be able to carry out our responsibility to our Country to be second to none in man's conquest of space."

• Report released by White House of study committee examining nonprofit research groups. Report stated that Government must continue to rely heavily on private institutions, including nonprofit research groups, to carry out Federal research programs. At the same time, the committee recommended that steps be taken to improve the competence of the Government's own scientific staffs and laboratories to direct research programs.

Late April: FAA awarded first seven contracts, totaling $2.66 million, for commercial supersonic transport aircraft. FAA Administrator N. E. Halaby said the action was "the first step leading to the development of a safe, practical, economical airplane to carry the traveling public beyond the speed of sound."

During April: Dr. John C. Houbolt, of NASA's Langley Research Center, writing in the April issue of *Astronautics*, outlined the possible advantages of lunar-orbit rendezvous for a manned lunar landing as opposed to direct flight from earth or earth-orbit rendezvous. Under this concept, an Apollo-type vehicle would fly direct to the moon, go into orbit around the moon, detach a small landing craft which would land on the moon and then return to the mother craft, which would then return to earth. Advantages would be the much smaller craft having to perform the difficult lunar landing and take-off, the possibility of optimizing the smaller craft for this one function, the safe return of the mother craft in event of a landing accident, and even the possibility of using two of the small craft to provide a rescue capability.

• Mach 2.2 windowless supersonic transport A–60 design announced by the College of Aeronautics, Cranfield, England, which featured six buried turbojet engines and 73° sweep slender delta wing. The A–60 design differs from the Sud Super Caravelle being studied as a joint Anglo-French project.

During April: NASA awarded: $447,000 contract to MIT for research on organizational and management concepts suitable for large scale technology-based enterprises with particular application to NASA; and $500,000 contract to Univ. of Calif. (Berkeley) for interdisciplinary space-oriented research in physical, biological, and engineering sciences.

- TIROS IV continued in operation and, to an extremely great extent, provided excellent data. Over 20,000 pictures have been received. 217 nephanalyses had been prepared up to March 26, and 199 transmitted over national and international weather circuits. Nineteen special storm advisories were issued to such countries as the Malagasy Republic, Mauritius, New Zealand, and Australia. In more than seven cases, the Tiros data led to significant readjustments in the analyses of the National Meteological Center. TIROS IV nephanalyses have been used at both Australia and McMurdo Sound in connection with forecasts for Antarctic operations.

- General Electric announced that a laser beam of narrow, coherent light had been used to punch holes in diamonds, the hole being cut in 200 millionths of a second, generating temperatures in the 100,000° F range and without causing structural damage.

- British Broadcasting Corp. announced that 30% of Britain's population—15,000,000 people—listened to its radio coverage of Astronaut Glenn's three-orbit space flight on Feb. 20, comparable to the size of the audiences that listened to Sir Winston Churchill's wartime broadcasts.

- United Technology Corp. fired 35-ton segmented solid booster for a duration of 130 seconds, developing thrust of slightly more than 100,000 lbs for the entire burning time.

- Reported that Electro-Optical Instruments Inc., had developed an ultra-high speed Kerr cell framing camera capable of taking six frames at a rate of 100,000,000 frames per second with exposure times as brief as five nanoseconds. Designated KFC–600, the camera may have wide applications in aerospace studies of physical phenomena such as arc and plasma gas discharge, explosive detonation, hypervelocity impact, shock wave studies, and nuclear instrumentation.

- TIROS II was turned on late in April from unknown spurious source. An engineering investigation was run in early May before turning it off again. An analysis of the data indicated that at least some usable IR data was obtained. It was interesting to note that the IR electronics and tape recorder are still running after 17 months in orbit. Continued operation of the satellite was not feasible because of power problems.

- Tass reported that Soviet astronomers of the Crimean Astrophysical Observatory discovered molecular oxygen in the outer atmosphere of Venus, using 157.6-in. solar telescope and a special spectograph. Also reported were indications of the presence of nitrogen.

MAY 1962

May 1: International Committee on Space Research (COSPAR) meeting in Washington received summary reports of accomplishments of 1961 and plans for 1962 by the U.S. and U.S.S.R. A. A. Blagonravov of the Soviet Academy of Sciences reported that the U.S.S.R. planned to send more men into orbit this year as well as several more unmanned scientific satellites, including one to study "the distribution and production of cloud systems in the atmosphere."

Dr. Richard W. Porter of the U.S. delegation specified that NASA planned to launch 22 instrumented satellites and space probes during the year. During 1961, Porter reported, the U.S. had launched: over 300 high-altitude research balloons; 866 weather rockets; more than 70 larger rockets not meant to carry payloads into orbit; and 31 instrumented satellites and deep-space probes.

- British scientists protested U.S. plans to explode "rainbow" bombs 500 miles above the Pacific test area (called "rainbow" bombs because high-altitude nuclear explosions light up the sky thousands of miles away). Sir Bernard Lovell, director of Jodrell Bank Radio Observatory, Martin Ryle, radio-astronomer, and Sir Mark Oliphant, Australian nuclear physicist, all protested the "probable" distortion of the Van Allen radiation belt by the forthcoming U.S. tests.

- Dr. Richard W. Porter, head of the U.S. delegation to COSPAR meeting in Washington, reported that the moon emits x-rays as a result of solar and cosmic ray bombardment. Rockets shot of October 24, 1961, detected x-radiation from the lunar surface. By future studies of these energies it is hoped to determine the composition of the surface of the moon.

- Secretary of Defense Robert S. McNamara announced in London that the U.S. would contribute over $30 million for the development of the STOL/VTOL fighter aircraft, the P-1127, originated by Hawker Siddeley Aviation Co., as part of the U.S.-British-West German cost-sharing agreement of February. This is the first agreement to develop NATO weapons jointly.

- In television interviews in New York, Cosmonaut Gherman Titov denied reports that any Russian cosmonauts had been killed prior to the orbital flight of Gagarin as reported in the American press. Titov said he also saw the tiny "fireflies" reported by John Glenn.

- In speech to IAS–NASA Manned Space Flight Conference at St. Louis, General Bernard A. Schriever, Commander of AFSC, said that if the Soviets attain a "really significant breakthrough in space technology, they may be able to deny other nations access to space—even for purposes of scientific research. . . . We must have necessary strength to ensure that space is free to be used for peaceful purposes." Schriever said that "both the military and civilian aspects of our space program are vital, and both must be pursued with urgency. They share a common aim—the security and well-being of the United States."

May 1: The Television Academy announced annual Emmy Award nominations. The orbital flight of Astronaut Glenn and Mrs. John F. Kennedy's tour of the White House—regarded as top TV events—were not considered because they had been cooperatively covered by the three major networks.

May 2: Dr. James A. Van Allen termed the planned U.S. hydrogen bomb explosion high over the Pacific Ocean as a "magnificent experiment," one which should bring new knowledge concerning the origin of the inner belt of high-energy protons. He disagreed with scientists abroad who object to the test as a unilateral military experiment.

- NASA scientists reported to COSPAR session that data from EXPLORER IX indicated that the upper atmosphere was heated by sun spot activity. In 1957 and 1958, when solar activity was at the maximum of its 11.1-year cycle, air density at a height of 350 miles was 10 times higher than in 1961 when activity was nearer its minimum. This suggests that "with a decrease in solar activity the upper atmosphere becomes cooler and shrinks to the earth, so that the air density at a given altitude decreases." Report on EXPLORER IX, the 12-ft. inflated satellite sensitive to density changes, was made by William J. O'Sullivan, Claude W. Coffee, Jr., and Gerald M. Keating.

- House of Representatives defeated by voice vote a bill sponsored by Rep. William F. Ryan which would have provided for Government ownership of the U.S. portion of a communications satellite system.

- In news conference at the Soviet Embassy in Washington, Gherman Titov forecast cooperative flights to and around the moon if world leaders could come to an agreement on disarmament. He also stated that Soviet Cosmonaut No. 3 was ready to go.

- Royal Aeronautical Society announced that John C. Wimpenny had made first extended manned flight under his own power in his pedal-powered Puffin. A design engineer, Wimpenny flew 993 yards at an average speed of 19½ mph and an altitude of 5 to 8 feet.

May 2–5: First AFSC Management Conference held at Naval Post Graduate School, Monterey, Calif., to provide exchange of views among leading representatives of industrial, scientific, academic, and financial management areas on the theme of "Systems Acquisition and Management in Today's Environment."

May 3: Two GSFC scientific sounding rockets launched from Wallops Station. An Iris research rocket launched with test instrumentation did not achieve programed altitude and landed 175 statute miles downrange. A four-stage Argo D–4 launch vehicle carried 78-lb payload to an altitude of almost 530 statute miles to record the intensity and distribution of radio noise above the ionosphere at extremely low frequencies.

- Astronaut John H. Glenn, Jr., and Cosmonaut Gherman S. Titov visited the White House and toured the Mall in Washington, D.C. At joint press conference at the National Academy of

Sciences, Glenn and Titov both spoke about possible international cooperation in manned space flight, Titov suggesting that the diplomats would first have to reach agreement on disarmament. Titov said that the Soviet press conference on August 14, 1961, held at Ludnicki, at which the President of the Soviet Academy of Sciences Matislav Keldysh discussed the rocket boosters of VOSTOK II, contained all useful information available on the Soviet rockets. Titov also said that the released photograph of VOSTOK II was of a mockup.

May 3: House of Representatives passed the Communications Satellite Act of 1962 by a vote of 354 to 9, one creating machinery to operate the U.S. portion of a worldwide communications satellite system. Act (H.R. 11040) declared that it is the policy of the U.S. "to establish as soon as practicable a commercial communications satellite system . . . in conjunction and cooperation with other countries . . . to be responsible to public needs and national objectives, serve the communications needs of the United States and other countries, and contribute to world peace and understanding." Act created a private corporation, one half of whose stock would be public owned, the remainder by communications companies. Three directors are to be appointed by the President, six elected by public stockholders and six by industry stockholders.

- Soviet Academician Anatoli A. Blagonravov told the press in Washington that U.S.–U.S.S.R. talks on space cooperation would be resumed at Geneva during the meeting of the subcommittee of the U.N. Committee for the Peaceful Uses of Outer Space on May 28.

- Prof. Vassily V. Parin of the Soviet Academy of Sciences reported that fruit flies carried aloft in Soviet spacecraft had shown a "considerable increase in gene mutations," in a report on space biology to COSPAR meeting in Washington. Evident to some observers was that the Russians have conducted a wide range of biological experiments in their satellite program.

May 4: Astronaut M. Scott Carpenter and Mercury team went through complete simulated MA–7 mission exercise.

- NASA announced appointment of Dr. Raymond L. Bisplinghoff, Professor of Aeronautical Engineering at MIT, as Director of NASA's Office of Advance Research and Technology (OART), effective in July. Dr. Bisplinghoff, who succeeds Ira H. Abbott who retired in January, has been at MIT for 16 years, has served as a consultant to more than a dozen industrial firms, and as a member of numerous DOD and NACA advisory committees.

- Optical and radar observations of Project Highwater experiment, 23,000 gallons of water bursted from upper stages of Saturn test launch at 65-mile altitude on April 25, showed that the burst took 2-3 seconds to expand to about 10 times the size of the moon in a roughly circular pattern, and lasted about 10 seconds. A second fainter trail, about one lunar diameter, followed along the trajectory from the burst and lasted a few minutes, presumably caused by venting out of the incompletely destroyed second stage. Project Highwater was an MSFC responsibility, a bonus experiment to the Saturn test conceived by Dr. Charles Lundquist.

- New York Academy of Sciences conference revealed disagreements on evidence presented on extraterrestrial life. Dr. Bartholomew

Nagy of Fordham University, Dr. George Claus of NYU Medical Center, and Dr. Warren G. Meinchein of Esso Research presented evidence, first reported last year, that they had found "organized elements" in the Orgueil meteorite. Dr. Frank W. Fitch and Dr. Edward Anders of the University of Chicago reported that the Fordham group found "probably nothing more than earthly contaminants such as ragweed pollen and starch grains." Dr. Harold Urey of the University of California (La Jolla) stated that the U.S. is spending $25 billion to go to the moon to answer this and other questions. All agreed that the question of extraterrestrial life was of basic scientific significance.

May 4: NASA's MSFC completed negotiations with Aerojet-General Corp. for the design and development of the M–1 engine, two or more of which will power the second stage of the Nova.

• "Dynamic model" study on one-fifth-scale Saturn rocket at NASA's Langley Research Center offered useful concept for testing structural characteristics of the future Nova rocket. As reported by Homer G. Morgan, application of varying frequencies to the "Baby Saturn" while suspended as if in free flight provided clues on structural vibration capable of transmitting false indications into the vehicle's guidance system. The two successful test firings of the Saturn demonstrated value of this research. After test of a "Baby Nova" for structural efficiency, Morgan said, construction of an operational booster could be expedited. A full-scale Saturn model had been similarly tested at MSFC.

• Launch of USAF Titan I on 5,000-mile flight down PMR marked the 100th missile fired from Vandenberg AFB, Calif., launchings which included 21 Thor IRBM's, 32 Atlas ICBM's, five Titan ICBM's, and 42 satellites.

May 5: 117 members of the International Committee on Space Research (COSPAR) toured Cape Canaveral, and a delegation also toured NASA's Wallops Station.

May 6: USAF Cambridge Research Lab reported record ten-day flight of superpressurized balloon at a constant altitude of 70,000 feet with a 40-lb. payload. Launched from Chico, Calif., on April 26, development ballon landed on command near Cedar City, Utah, and will lead to 20-day balloon flight experiments at a constant altitude later this summer.

• In the Third International Space Science Symposium sponsored by COSPAR, Jet Propulsion Laboratory scientists W. K. Victor and Robertson Stevens reported that radar soundings of Venus suggest that its surface material and roughness are comparable to that of the earth. Radar probes of Venus have demonstrated that UHF signals can be used for space communications over distances of 50–75 million miles.

• Astronaut John Glenn and Cosmonaut Gherman Titov appeared on ABC-TV program, "The Nation's Future." Col. Glenn pressed Major Titov on possible Soviet space mishaps, but no new information was forthcoming.

• AFCRL scientists launched a second artificial meteor by means of a multistage rocket fired from Wallops Island. The meteor traveled at a re-entry speed of 12 kilometers per second, exceeding that attained in the previous experiment of April 21, 1961.

May 6: "American Roulette 500 Miles Up" was the title of the article by Sir Bernard Lovell in the *Sunday Observer* (London), which repeated his arguments against the planned nuclear explosions by the U.S. over Johnston Island in June and July.

- Cosmonaut Titov at the Seattle World's Fair said that he saw "no God or angels" during his 17 orbits of the earth.
- Full flight test of Polaris missile with live warhead, the fifth U.S. atmospheric nuclear test in the current Operation Dominic series, fired from submerged U.S.S. *Ethan Allen* in Pacific test area. Reportedly the first U.S. firing of a missile with a live nuclear warhead, the shot was described by Navy spokesmen as a complete success.

May 7: In speech in New Orleans, NASA Administrator James E. Webb reviewed the National Launch Vehicle Program, which "recognizes all the requirements of our nation in space, and includes the vehicles to fulfill them." He pointed out that of the ten launch vehicles "in the National Launch Vehicle Program, the two smallest rockets [Scout and Delta], and the four most powerful ones [Centaur, Saturn, Advanced Saturn, and Nova], have been, or are being, developed by NASA. The four of intermediate size are adaptations of Air Force missile carriers, Thor, Atlas, and Titan." Mr. Webb pointed out the proximity to New Orleans of NASA's Michoud Plant and the Mississippi Test Facility.

"Now that we have crossed the threshold of space," he concluded, "the benefits that can be gained for all mankind, the assurances to our national security that can be obtained, and the stimulus to our well-being that can be produced, are looming large on the horizon. . . . The important thing now is to make sure we are not behind in any area vital to our national security or our national leadership."

- NASA announced at Cape Canaveral MA–7 launching, scheduled for May 15 or afterward, would be delayed several days because of check-out problems with the Atlas booster.
- In Voice of America broadcast, Dr. Christian J. Lambertsen of the University of Pennsylvania said that there is an increasing recognition of the role man can play in space exploration because "there are aspects of exploration which no machine yet conceived can carry out." He pointed out that "the only computer capable of all functions such as remembering, recognizing, learning, thinking, reasoning, judging, integrating, reacting, communicating, and logically altering a previously programmed sequence of events is man himself." He then reviewed the medicophysiological factors to be overcome for manned space flight, and he warned that "even the practicability of truly extended flight outside the earth's atmosphere is not yet assured."
- Maj. Gherman Titov in San Francisco press conference continued to chide U.S. space program. He said that if he was asked to join a U.S. astronaut in a cooperative space mission: "I would be a bit afraid because there have been quite a lot of failures in your program." And, he remarked, the MR–3 capsule was "not even good enough for flying in orbit."

May 8–10: Second National Conference on the Peaceful Uses of Space sponsored by NASA was held at Century 21 Exposition in Seattle, Wash.

May 8: In message to the Second National Conference on the Peaceful Uses of Space at Seattle, President Kennedy noted that the U.S. was already "working hand in hand with scientific groups of fifty nations.

"Ours is an open society and the benefits of our space program will continue to flow throughout the world. It is my hope that the Soviet Union will cooperate constructively in the proposals which we have made so that all peoples will gain in the improvement of weather observation, communications systems and the manifold output of the peaceful application of space technology."

- X–15 No. 2 flown at 3,511 mph and to about 73,000 feet by Major Robert Rushworth (USAF) with 103 seconds of rocket power in heat resistance test.   Temperature of tail surface was reported to have reached near 1,250° F.
- First launch of Atlas-Centaur was unsuccessful; vehicle exploded 55 seconds after launch over Cape Canaveral.   Flight plan called for starting 15,000-lb.-thrust liquid-hydrogen second stage at 300-mile altitude.
- NASA selected three companies for negotiation of production contracts for major components of the Apollo spacecraft navigation and guidance system in support of MIT's Instrumentation Laboratory initial development of the complex Apollo navigation-guidance system, a contract let last August.   Companies selected from 21 bidders, which will work under Manned Spacecraft Center contract were: A.C. Spark Plug Division of General Motors Corp. to fabricate the inertial guidance with associated electronics, ground support, and checkout system, and to assemble and test all components of the system; Raytheon Company to manufacture on-board digital computer; and Kollsman Instrument Corp. to build the optical subsystems including a space sextant, sunfinders, and navigation display equipment.
- NASA scientists Theodore P. Stecher and James E. Mulligan reported at COSPAR meeting that "a complex series of solar events" in November 1960 caused the inner Van Allen radiation belt to "dump" its captive high energy particles along the earth's magnetic field.   This event disrupted a rocket experiment, but it detected an ultraviolet aurora with almost no visible counterpart.   Indications were that the inner belt was soon refilled and returned to "normal."
- Edmond C. Buckley, NASA's Director of Tracking and Data Acquisition, told the Second National Conference on the Peaceful Uses of Space meeting in Seattle that JPL's successful radar tracking of Venus over a two-month period in 1961, which resulted in a 200-times improvement in the measurement of the Astronomical Unit, was actually a by-product, though an important one, of a rigorous systems test of the Deep Space Net and of several prototype pieces of equipment that had been developed for the Net by JPL.
- Senate Committee on Aeronautical and Space Sciences reported out Communications Satellite Bill S. 2814, which was referred to the Senate Commerce Committee.
- U.S.S.R. turned down proposal to show FRIENDSHIP 7 capsule in the "space medicine" section of the Leningrad Medical Exhibition

next month. It was stated that the spacecraft would have "no purpose in a medicine exhibit."

May 9: In regular press conference, President Kennedy was asked whether "proposed H-bomb explosion 500 miles up" would jeopardize U.S. policy to restrict outer space for peaceful objectives only. He replied that he did not think so and that it was a matter that "we are looking into to see whether there is scientific merit that this would cause some difficulty to the Van Allen Belt in a way which would affect scientific discovery . . .

"I want you to know that whatever our decision is . . . it will be done only after very careful scientific deliberation, which is now taking place . . . Generally what we are attempting to do is to find out the effects of such an explosion on our security, and we do not believe that this will adversely affect the security of any person not living in the United States."

- 18-nation 5th COSPAR meeting in final session voted to establish a consultative group for determining the potential harmful effects of experiments in space. Group of six scientists was charged with recommending to COSPAR's executive council possible action by the international scientific community to sanction or condemn such experiments. Other closing actions: (1) Maurice Roy of the Ecole Polytechnique in Paris was elected President for 1962–63; (2) expressed "satisfaction at the intention of the U.S. to launch a geodetic satellite carrying a flashing beacon for international use"; (3) proposed series of coordinated rocket measurements of winds in the ionosphere from Oct. 1962 to Feb. 1963; and (4) recommended numbering satellites during 1963 by the ordinal number rather than Greek letter.
- President Kennedy in answering a question on a uniform patent policy among governmental agencies said that it was a "difficult problem, because you have to balance off the gains on the one hand and at the same time the incentives to companies to spend their own funds in order to develop patents . . . so that we have some differences in the space agency problems, the Department of Defense and perhaps some other agency of the Government." He said that it was a matter now under review and if changes appear warranted that recommendations would be sent to the Congress.
- USAF announced that it had begun negotiation with United Technology Corp. for development of solid-fuel motors for use in Titan III space booster.
- Army Nike-Zeus test flight from Point Mugu, Calif., a partial success in that "missile accepted and correctly executed control commands transmitted from a ground guidance center," although third stage did not fire.
- A Navy advanced Polaris was test-launched from submerged nuclear submarine U.S.S. *Sam Houston* off the Florida coast.
- Army Pershing missile destroyed in test flight from Cape Canaveral when second stage became erratic, the second failure in last three launchings after 13 straight successes.
- In address to the Operations Research Society, Gen. Bernard Schriever said: "Our continuing national security is dependent on the rapid advance of technology and its adoption in operational

systems. The pacing factor in this advance is not technology—it is management.

". . . two stages occur in the acquisition of new systems—the preselection phase and the implementation phase . . . Not enough attention has been paid to the more crucial phase—the decision to select a specific system from a galloping technology."

May 9: Cdr. George C. Watkins (USN) became the first man to make 1,000 aircraft landings aboard a carrier, when he landed an AD-6 aboard the U.S.S. *Constellation.*

May 10: MIT scientists, headed by Prof. Louis Smullin, fired 13 separate bursts of light (using a ruby crystal maser) through 12-inch telescope at smooth area of the moon, then "caught" the reflection 2½ seconds later in a 48-inch telescope. Ruby crystal rod was about six inches long and the red light would have been visible on the moon. Technique demonstrated was hailed as promising "a means of delivering power over vast distances to orbiting satellites."

- Attempt to launch Anna geodetic satellite by Thor-Able-Star rocket was unsuccessful.

- In address at dedication of the NASA Space Exhibit at the Seattle World's Fair, Vice President Lyndon B. Johnson said that "our entire space program now costs each American about 30 cents a week. During the next few years we plan to spend about 50 cents a week per person on space.

  "However, our space program and its by-products will stimulate a sharp increase in the nation's productive output which in turn will increase our gross national product, our income, and the Federal Government's intake . . .

  ". . . we are hopeful of achieving fruitful cooperation with the Soviet Union in such fields as communications, weather forecasting, mapping the earth's magnetic fields and space medicine.

  "We feel that cooperation in outer space may establish a firm basis for greater mutual understanding—which in turn will help in our efforts to obtain disarmament . . .

  "The responsibility to cooperate also lies heavily on the other great space power—the Soviet Union. I am able to tell you in a spirit of cautious optimism that the Soviet Union appears to realize that—in outer space, at least—there may be something to be gained by cooperating with the rest of humanity."

- In speech to the Oregon Department of Planning and Development, NASA Administrator James E. Webb pointed out that "one of the dominant features of our age is the short time lag between scientific discovery and practical applications." The practical benefits of the U.S. space program would, he said, be early derived in three major ways: (1) satellites will be put to work on a global basis to report the weather, transmit messages and world-wide television programs, and to serve as electronic lighthouses in the sky; (2) in pushing our space program, we are making many technological advances which can be utilized to improve industrial processes and raise our standard of living; and (3) the money we spend on space activities stimulates business in general and our industrial pioneering in particular.

May 10: House Science and Astronautics Committee authorized record $3.7 billion for NASA in FY 1963, cutting $116 million from NASA's request.

- Deputy Secretary of Defense Roswell Gilpatric, at Wilton, Conn., stated that peaceful uses of outer space was still the National goal but that the U.S. would be "ill-advised if we did not hedge our bets." He said that a satellite interception system would get serious attention if another nation was developing such a capability. "We ought to be ready," he said in a press conference, "to anticipate the ability of the Soviets at some time to use space offensively."

- Premier Nikita Khrushchev said in speech to transport workers that despite the claims of American scientists no U.S. rocket has hit the moon: "The Americans have tried several times to hit the moon with their rockets. They have proclaimed for all the world to hear that they had launched rockets to the moon, but they missed every time. The Soviet pennant on the moon has been awaiting an American one for a long time but in vain and is becoming lonesome."

- In answering Premier Khrushchev, Dr. William Pickering, Director of JPL, stated: "On April 26, at 4:47.50 AM, Pacific standard time, RANGER IV was tracked by the Goldstone receiver as it passed the leading edge of the moon. At 4:49.53 it crashed on the moon at a lunar longitude of 229.5 degrees East and lunar latitude of 15.5 degrees South." After 64 hours of tracking, the trajectory of RANGER IV was precisely known and it was only 110 miles from the surface when it vanished behind the moon.

- NASA X–15 pilot, Joseph Walker, said that film taken during recent record altitude flight of 246,700 feet showed five or six mysterious objects. Other space pilots on the panel at the National Conference on the Peaceful Uses of Space at Seattle, all of whom have flown above 100,000 feet, were Cdr. Malcolm Ross (USNR), Capt. Joseph Kittinger (USAF), Col. David Simons (USAF), Neil Armstrong (NASA), Cdr. Forrest Petersen (USN), Maj. Robert White (USAF), and Lt. Col. John H. Glenn (USMC).

- In session on the impact of space exploration on society at the Second Conference on the Peaceful Uses of Space, President Lee DuBridge of the California Institute of Technology said that man is limited as a butterfly: "Man is out of his cocoon [in space] . . . his life span is too short to get very far from home." DuBridge pointed out that a flight to the nearest star would take at least 40,000 years and that a trip to Mars at present speeds would take six months, "but months would run into years even reaching other planets of the earth's system. We are not going to accomplish space travel except to nearby planets for many generations to come."

- In talk at National Conference on Peaceful Uses of Space, William H. Meckling, RAND Corp. economist, stated that space-age economic payoffs are being exaggerated: "The exploration of space is a very exciting affair, indeed. . . . If glamour displaces science in guiding national policy, however, the results may be disappointing. . . .
  "A communications satellite system that charges prices not very much different from present prices and that must be constantly subsidized and protected from competition, is not much of an accomplishment. . . .

"How much is it worth . . . to have improved weather forecasts? In the absence of a market in which weather forecasts or weather information is sold, it is nearly impossible to answer such questions. . . .

"Who is it in foreign nations that is impressed by our space accomplishments? How are their attitudes changed as a consequence? How does that change in behavior affect me as a citizen of the United States? . . ."

May 10: In address to the House of Representatives, Congressman Emilio Q. Daddario pointed to a need for a common patent policy for inventions arising out of research and development financed by Federal funds and cited President Kennedy's statement at his press conference the day before.

- DOD announced that Perkin-Elmer Corp. was completing a new 36-inch airborne telescope with 3–5 times as much resolution as any telescope on the ground. New instrument should finally answer the speculation about the so-called canals on Mars, the mysterious red spot on Jupiter, and perhaps the hidden surface of Venus. "This historic enterprise is only one more step forward, leading to large telescopes orbiting the earth in satellites and eventually constructed on the moon."

- Sparrow III fired from an F4H–1 scored a direct hit in head-on attack on a Regulus II missile while both were at supersonic speed. The interception, made in Pt. Mugu test range, was first successful attack made by an air-launched weapon on a surface-launched guided missile.

- Plans of a non-profit Washington Planetarium and Space Center Corp. were reported at the National Capital Planning Commission by Father Francis J. Heyden, S.J., of Georgetown University Observatory. They called for the construction of a $1.5 million, 85-foot-diameter dome on Daingerfield Island, south of the Washington National Airport. The planetarium would "provide the public with a much needed space education facility" and would be operated by the National Park Service.

May 11: NASA announced establishment of an on-site management unit of some 100 persons at the Space and Information Systems Division Plant of North American Aviation at Downey, Calif. It will be an element of the NASA Western Operations Office of Santa Monica. The Downey office is to expedite NASA contractor decisions and actions essential to development contracts such as the Apollo lunar-landing spacecraft and the second stage for the Advanced Saturn booster.

- Problems with space capsule control system could delay MA–7 launch attempt at least until May 19, it was reported from Cape Canaveral.

- NASA Inventions and Contributions Board made awards of less than $1,000 to the following NASA employees: Wade E. Lanford, Langley Research Center; Billy C. Hughes, John R. Raskin, Robert J. Schwinghamer, Wilhelm Angele, and Hans G. Martineck, Marshall Space Flight Center.

- USAF Atlas ICBM launched from Vandenberg AFB, Calif., launching conducted by 12-man combat crew of SAC's 565th Strategic Missile Squadron from Warren AFB, Wyo.

- USAF Minuteman launched from silo at Cape Canaveral in successful test flight.

May 12: oso 1, the Orbiting Solar Observatory launched March 7, completed 1,000 orbits; approximately 900 telemetry data tapes were acquired and forwarded to all experimenters.

- Lockheed Propulsion Co. test-fired 120-inch diameter solid-propellant, segmented motor for more than 120 seconds and produced a thrust of almost 400,000 lbs. Direction of exhaust jet was repeatedly varied by injection of fluids into the sides of the nozzle.

- Panel of the Federal Council for Science and Technology, headed by Allen V. Astin, submitted report to White House. Report said that Federal laboratories are in imminent danger of losing their best scientists and engineers because present salary scales are not competitive. In a memorandum to Federal department and agency heads, President Kennedy stated that the Administration's proposal for pay increases met one of the recommendations of the panel's report: "With the increasing importance of science and technology in developing our military defense, in achieving our foreign policy objectives, and in sustaining the health and welfare of every citizen, the Federal Government must attract and retain its share of talented scientists and engineers at all levels."

- Report at American Assembly–British Institute for Strategic Studies forum that Western Europe was planning 440 sounding rocket, 22 satellite, and 2 moon probe launchings by 1970.

- Army Nike-Zeus successfully intercepted simulated target over White Sands Missile Range, New Mexico.

- In speech at Lubbock, Texas, Secretary of the Air Force Eugene M. Zuckert said that Armed Forces Day 1962 has a special meaning: "It means John Glenn and our rising capabilities in space. It means our desire to share space for peaceful purposes with the rest of the world. And it underscores our determination to see to it that no nation is disfranchised in space . . .

  "The primary purpose of our arms has always been peace, not war . . .

  "Space is the new measure of our problems as well as our opportunities. The principal responsibilities for the nation's space program are vested in the National Aeronautics and Space Administration and the Air Force. We work together as a team, not rivals. And together we are doing the spadework for the space technology of tomorrow.

  "Today, space is still free and open. It must stay that way. We can not allow it to be pre-empted tomorrow for use as a military staging area by those who hope to dominate the world.

  "I believe with President Kennedy that space is a great opportunity as well as a great task. . . ."

May 13: In address before joint American Assembly–British Institute for Strategic Studies conference at Brighton, England, John J. McCloy urged Western Europe to forego its aspirations to become a third space power and to cooperate with the U.S. in the exploration of space. Mr. McCloy, chairman of the General Advisory Committee to the U.S. Arms Control and Disarmament Agency, appealed to the 70 delegates that once the leaders of the Soviet Union "sensed a united position on the part of the West, they are prepared to moderate their views accordingly." When dealing with the Soviet Union on space issues, McCloy stated: "Questions of international prestige quickly emerge and we are still plagued with the almost frenetic antipathy of the Soviet Union to inspection."

May 13: MSFC Ad Hoc Evaluation Team was established for design review of the Centaur vehicle.

May 14: In NAS broadcast for Voice of America, Dr. Colin S. Pittendrigh, Professor of Biology, Princeton University, said that biological engineering for manned space flight will have an influence on theoretical physiology, but that "space technology holds far less promise of any revolution in biology than in physics." The possibility of extraterrestrial life, he described as "the most exciting scientific question of the day."

Optimum conditions for a manned spacecraft with regard to temperature, moisture, oxygen, and carbon dioxide can be specified. But "we have difficulty in being precise about anything else." He pointed to the possible exclusion of heavy radiation by means of an intense magnetic field, but "can man perform normally in such intense magnetic fields?" Little is known about the reliability of algae, and new studies of photosynthesis and techniques of algoculture are needed. Rhythmic innate oscillations, previously considered superficial to the physiological architecture of organisms, are "a highly significant parameter of physiological organization." And the possible effect of the earth's gravitational field on normal physiological function is not fully known. Dr. Pittendrigh concluded: "Clean questions and clean answers are going to be difficult for some time. When we put organisms into space at present and detect deleterious effects, we are simply unable to disentangle the many variables that exist there and decide which has been responsible."

- U.S. delegate to the 17-nation disarmament conference, Charles C. Stelle, urged early action to ban outer space for nuclear armaments: "This conference could lead to measures designed to insure that outer space can become an impetus to man's peaceful progress and not a battleground in the future."

- Testimony of Feb. 27 by Prof. James A. Van Allen before the House Committee on Appropriations on the relative worth of instrumented unmanned satellites versus manned satellites was released. Van Allen was misquoted on the wire news services to the effect that manned space flight was of limited scientific value in the future.

- FRIENDSHIP 7 capsule went on exhibit at the Science Museum in London in beginning of its tour of European cities.

- Cosmonaut Titov wrote in *Pravda* on his arrival in Moscow: "I returned home, heaved in a breathful of clear spring air, and told the comrades welcoming me: Despite all the comforts of America, there is no land on earth better than our dear, wonderful Soviet homeland." He indicated that American astronauts Alan Shepard and John Glenn have been invited to visit the Soviet Union.

- Drew Pearson referred to Cosmonaut Titov's recent denial of Pearson's report of February 23, 1962, that five Russian cosmonauts had been lost in previous manned space flight attempts.

He pointed out that the unsuccessful attempts were not orbital shot, rather were suborbital "rocket rides" such as that of Cdr. Alan Shepard.

May 15: In testimony before Subcommittee of the House Science and Astronautics Committee, Dr. Homer E. Newell reviewed history of NASA's geodetic satellite project, which had been requested in the FY 1960 budget. In mid-1960, DOD recommended the geodetic satellite program be classified, which was approved by the Unmanned Space Panel of the NASA/DOD AACB in August 1960. Thus was created Project Anna, declassified at the COSPAR meeting in Washington on April 27, 1962.

- NASA witnesses testified before a subcommittee of the House Science and Astronautics Committee that Centaur rocket development administration had in recent months been streamlined. Problems in the program were attributed to too austere a budget at the start, "growing pains" of a new organization, and unexpected technical difficulties in utilizing high-energy liquid hydrogen as a rocket fuel.

- Dr. Knox Millsaps, Executive Director of the USAF's Office of Scientific Research, stated that 547 basic research projects in North America, South America, and Europe have been funded by OSR at more than $23.3 million so far this fiscal year.

- Astronaut Scott Carpenter flew simulated MA–7 mission in "Aurora 7" capsule mated to Atlas booster.

- USAF announced that an unidentified statellite using a Thor-Agena B booster was launched from Vandenberg AFB.

- Prof. Pierre Auger, executive secretary of the Preparatory Committee of the European Space Research Organization (ESRO) and the European Launcher Development Organization (ELDO), stated that plans called for the expenditure of $100 million a year. Program headquartered in Paris would parallel the work of the U.S.'s NASA, sans man-in-space activity. An important part of its program would be cooperation with other international and national organizations, especially NASA, from which satellites and launchers might be obtained. A new firing range at Kiruna, Sweden, is envisaged for firings in the northern auroral zone.

- Naval Ordnance Laboratory announced use of Loki launch tube attached to the barrel of a 5-inch/54 naval rifle to launch the Hasp (high altitude sounding rocket) for weather soundings from ships anywhere at sea.

- USAF announced transfer of AFSC's Ballistic Systems Division this summer from West Los Angeles to Norton AFB, San Bernardino, Calif., to be completed by summer of 1963.

May 16: Second and third joint Japanese-U.S. space probes successfully launched from NASA Wallops Station, the second Nike-Cajun reached 76-mile altitude and the third and last, a night shot, reached 80-mile altitude.

- U.S. delegate to the 17-nation disarmament conference in Geneva, Charles C. Stelle, urged the Soviet Union to divorce the question of preserving space for peaceful uses from its demands for elimination of foreign military bases and rockets designed to deliver nuclear weapons. The Soviets have insisted as a first step of disarmament the total elimination of vehicles capable of carrying nuclear bombs and of foreign bases.

May 16: NASA announced that "unidentified objects" on film of X–15 flight on April 30, mentioned by Joe Walker at Seattle, were actually bits of ice flaking off the fuel tank area.

- Deputy Secretary of Defense Roswell L. Gilpatric, in Armed Forces luncheon address in San Francisco, stated that he was reluctant to identify DOD contributions to the national space program as "The Military Space Program," because "there is only one unified national program, not two or three or four.

   "We are mindful of the stated U.S. national objective of using space for peaceful purposes only, and we [DOD] support this policy completely and wholeheartedly. . . . We are well advised in buying technological insurance even while earnestly hoping that space will be used only for peaceful purposes."

   Since the U.S. must avoid technological surprise in the military space area and there is no guarantee that the U.S.S.R. will cooperate on the peaceful use of space, Gilpatric announced that DOD has decided to develop manned orbital rendezvous space systems for inspecting other satellites to determine intent and neutralizing if hostile, and then land at predetermined locations on earth. These would be separate, he said, from NASA's Gemini and Apollo programs.

   In reviewing the evolution of the U.S. space program, Gilpatric pointed out that the "division of labor" between NASA and DOD in space was based "partly in logic, partly in pragmatic assessment . . ."

- In testimony before a Senate Appropriations subcommittee Air Force Chief of Staff LeMay requested additional funds to speed the development of the RS–70. Asking for nearly three times the Administration's budget request, or $491 million, LeMay said that limiting funds "would lose more than a year" beyond the 1967 target date for the first three prototypes. "This program has been slowed down at least four years. We have delayed and delayed until there is practically no risk in it now."

- House voted approval of plan to create a White House Office of Science and Technology to coordinate Federal research and development programs now approaching expenditures of $12 billion per year.

- Proposals to ban orbiting of nuclear weapons, and to seek international agreement on space activities irrespective of other disarmament measures were included in the final report of the second American Assembly—British Institute for Strategic Studies at Brighton, England.

- Reported that the Soviet Union was seeking through diplomatic channels to establish a space satellite tracking station in Australia, offering assurances that the station would not be used for military purposes or for the introduction of weapons.

May 17: NASA postponed MA–7 flight attempt until at least May 22, 1962, because of modification of altitude-sensing instrumentation in the parachute deployment system. Project Mercury Astronaut M. Scott Carpenter was pronounced in "excellent condition" for his orbital flight after a 5-hour physical examination.

- In an address to the Duke University Symposium on "Regional Implications of Space Research," NASA Administrator James E. Webb stated: "I believe the regional approach is sound. It permits a number of universities to pool resources for research and to plan complementary programs directed to the needs of the region. Industry, too, has much to gain from regional cooperation . . . beginning, more and more, to look to the universities of their region for the most important resources of the age—ideas, scientific brainpower, and advanced technological skill and experience.

"It is not a question today of whether a region can already qualify. . . . The question is whether the region is creatively doing what it can to equip its citizens to serve their area and their Nation in a period when our prosperity and our very existence as a free people depend on scientific and technological leadership. . ."

May 17: First of a series of 80 rocket probes to determine wind patterns over Cape Canaveral initiated by NASA with launch of a single-stage Nike to 80,000 ft. where it laid down a white smoke screen for subsequent photographic study.

- NASA announced that John Stack, Director of Aeronautical Research, would retire June 1 after 34 years of Federal service with NACA and NASA. No successor was announced. Mr. John Stack achieved world-wide eminence for his wind-tunnel research and conception of the rocket airplane as a research tool, beginning with the X–1 and the D–558, and culminating with the X–15. He was awarded the Collier Trophy in 1951 for solution of the transonic wind tunnel problem. In 1947, he shared the Collier Trophy with Maj. Charles Yeager and Lawrence D. Bell for the X–1 which made the first flight through the speed of sound. Stack also received many other awards, including the Sylvanus Albert Reed Award of the IAS and the annual medal of the Swedish Society of Engineers. He was a fellow of the IAS and the Royal Aeronautical Society (England).

- In a speech to the American Ordnance Association, Dr. Glenn Seaborg, AEC Chairman, said that nuclear energy provided the most feasible means of accomplishing long voyages in space: "To perform long-range, manned missions, it is essential that we increase the specific impulse (amount of thrust per pound of propellant) of the propulsion system well beyond the values that can be achieved with chemical combustion. . . ." He stated that a hydrogen-fluorine rocket system, now in research phase, can provide up to approximately 475-lbs. thrust per pound of propellant (kerosene-oxygen—300 lbs.). A nuclear rocket, which utilizes hydrogen (through its heat exchanger system) could provide highest specific impulse of 800 and above. He reviewed progress of Project Rover and Nerva.

- USAF C–130 transport en route with emergency recovery unit for scheduled Mercury MA–7 launching crashed west of Nairobi, Kenya. Thirteen airmen were killed.

- NASA announced that Lockheed Missiles and Space Company had been selected for negotiation of a contract for the design, development, fabrication, and testing of the Rift (reactor-in-flight-test) stage. The Rift vehicle is intended to flight test the Nerva (nuclear engine for rocket vehicle application) being developed jointly by NASA and AEC.

May 17: USAF announced award of a supplemental contract to General Dynamics, San Diego, for work on a global tracking network (GLOTRAC) ground system.

- Reported that White House had endorsed a proposal to abolish the five-man AEC, replacing the Commission with a single administrator.

May 18: NASA launched 76-lb. payload to 83-mile altitude with Nike-Cajun from Wallops Station, a University of Michigan experiment to measure air density and composition.

- In statement before the House Space Subcommittee, Grant L. Hansen, Vice President of the Astronautics Division of General Dynamics and Director of the Centaur program, stated that preliminary data on May 9 explosion indicated that structural failure was caused by the "design of the weather shield between the nose fairing and the Centaur itself" and the design of weather shield "was an engineering mistake." Design of the weather shield was based, he said, upon assumptions that "turned out to be false . . ., very difficult to define by laboratory analysis and without full-scale wind-tunnel tests." In spite of program delays, Hansen said that the Centaur program had been subjected to "unwarranted criticisms."

- In answer to question on Astronaut M. Scott Carpenter's history of motion sickness, John Powers, Mercury information officer, said that Carpenter had no such history. He noted that Cosmonaut Titov had reported that he had only 350 hours of pilot flight time before his space flight, which compares to at least 1,000 hours for each of the seven Mercury astronauts. "Titov never saw his space capsule until 60 days before he flew it," Powers said, "whereas the astronauts have become familiar with their capsules over a period of three years."

- NASA selected General Dynamics/Convair to design and manufacture "Little Joe II" launch vehicle to be used to boost the Apollo spacecraft on unmanned suborbital test flights. Little Joe II will be powered by solid-fuel motors in a cluster.

- Australian National Observatory announced that it would install 40-inch telescope on top of Sidney Smith Mountain at Coonabarabran, New South Wales, one which would have three times the efficiency of its 74-inch telescope at Mount Stromlo near Canberra.

- The Geophysics Corp. of America reported receipt of Weather Bureau contract to study and explain the formation of vast bands of cloud patterns in the upper atmosphere, a phenomenon first revealed in photographs relayed from TIROS I.

- USAF Minuteman ICBM launched from silo at Cape Canaveral in 3,600-mile test flight.

May 19: MA-7 launch attempt postponed to May 24 because of irregularities detected in temperature control device on a heater in the Atlas flight control system.

- Robert R. Gilruth, Director of NASA's Manned Spacecraft Center at Houston, Texas, received an honorary Doctor of Science degree from Indiana Institute of Technology, Fort Wayne, Ind. In address, Dr. Gilruth pointed out that the "American people and the free world would not stand for it, if in 10 years the Russians were flying through space and we were sitting on the gound."

At press conference he disagreed with Dr. James A. Van Allen's recent comment about manned space flight: "I respect Dr. Van Allen's talents but I do not agree with him that manned orbital flights have no value."

May 19: An estimated 7,000 persons visited the opening of the Space Museum at the NASA Marshall Space Flight Center.

May 20: Dr. John F. Clark was appointed Associate Director and Chief Scientist for NASA Office of Space Sciences. He was formerly Director of Geophysics and Astronomy Programs.

- Cosmonaut Yuri A. Gagarin visited Japan, the 30th country Major Gagarin had visited since his space flight.
- Reported that hundreds of Soviet mathematics textbooks are being published in English, and that many of them are being used in American universities.

May 21: In concluding lecture in a Forum series on Space Science broadcast on the Voice of America, Dr. Carl E. Sagen of the Institute for Basic Research in Science at the University of California (Berkeley), explained that "extraterrestrial life and the origin of life are questions intertwined." Recent study on these issues related to four topics: (1) the origin of life on earth; (2) the physical environment of the moon and planets; (3) the present evidence of life beyond the earth; and (4) prospects for future exploration.

He discussed the chemical reactions which must have existed in the primitive terrestrial environment and produced organic matter. The critical event was the origin of a molecular system capable of synthesizing itself, and "this event—which occurred some 4,200 million years ago—can be identified with the origin of life on Earth." From then on the long evolutionary sequence from molecule to Man was under way.

Dr. Sagen asserted that "through the interaction of organic molecules life must arise on countless other worlds, in our solar system, and other solar systems . . . [but] the forms of living organisms on other planets must be intriguingly diverse."

Dr. Sagen reviewed the improbability of life on Venus, the possibility of life on the moon, and the "favorable" environmental features for life on Mars. It is not, he said, a closed question with regard to life on Jupiter, Saturn, Uranus, and Neptune. The presence of extraterrestrial organic matter in meteorites "greatly substantiates the belief that organic matter is an abundant constituent of the solar system." Detailed characterization of extraterrestrial organisms likely to be found in exploration of the moon and the planets "will open up a new era, not only in biology but in all fields of human endeavor."

- William H. Evans was appointed by NASA as Director of the Pacific Launch Operations Office, Lompoc, Calif. In this assignment he represents NASA in its relationships with PMR, coordinates use of range facilities, and provides necessary services and support for NASA technical programs and projects at PMR. Prior to his association with NASA since January 1961, Evans was Head of the Satellite and Space Vehicles Division of PMR, and before 1957 had served with the Naval Air Missile Test Center at Point Mugu since 1946.

May 21: Senate Commerce Committee tentatively approved the space communications satellite corporation bill after adopting 32 amendments, one virtually identical to the bill passed by the House on May 3.

- Supreme Court of the State of New Jersey ruled unanimously that Thiokol Chemical Corporation, producer of the rocket engines for the X–15, was liable for property damage caused in testing the engine. The Court ordered $25,605 damages to be paid to 15 home owners near the Picatinny Arsenal, Rockaway Township, for cracked foundations, floors, walls, chimneys, and fireplaces. Thiokol stated that it would have to pass these costs on to the Federal Government, since it had a cost-plus contract.

May 22: OSO I ceased transmission after 1,138 orbits (launched March 7), having produced for 77 days and provided 200 miles of scientific data tape. It observed and measured over 75 solar flares and subflares.

- X–15 No. 1 flown in air flow study to an altitude of 97,000 ft. and a top speed of 3,477 mph (mach 5.1), piloted by Maj. Robert A. Rushworth (USAF).

- Combined U.S.-U.S.S.R. proposal for $100 million world-wide weather research program, including the use of earth satellites, was presented to a panel of the World Meteorological Organization (WMO) by Harry Wexler, Research Director of the U.S. Weather Bureau, and Viktor Bogayev, Ass't. Director of the Soviet Hydrometrical Service, in Geneva, Switzerland. Proposed plan, if approved by the WMO Executive Committee, would then be considered by UNESCO.

- Navy Astronautics Group established at PMR its first space satellite command, to operate the Transit navigational satellite system. In addition to the headquarters at Point Mugu, the command will eventually encompass a satellite command and ejection station, computer center, operations control center, and satellite tracking facility at Point Mugu, and other tracking facilities at Winter Harbor, Maine; Minneapolis, Minn.; and Wahiawa on Oahu, Hawaii.

- First display of FRIENDSHIP 7 capsule in Belgrade, Yugoslavia. When President Tito viewed it he said: "I was under the impression when I saw the documentary film that the capsule was larger but it's fantastic . . ."

- National Committee for a Sane Nuclear Policy made public the request of eleven scientists who had petitioned President Kennedy to postpone the projected explosion of a hydrogen device in space. They had urged that the project be deferred until it could be reviewed by the Consultative Group on Potentially Harmful Effects of Space Experiments of COSPAR.

- The Air Force Cambridge Research Laboratories launched a 150-foot balloon made of .001-inch-thick plastic. The test, intended to provide data for better research balloons, would send the balloon to an altitude of 117,000 feet for a fifty-hour period.

May 23: NASA and the Swedish Committee on Space Research announced signing of a memorandum of understanding for a cooperative program in space research. Four Nike-Cajun sounding rockets were scheduled for launching during the coming summer from the Vidsel Range in Sweden with scientific payloads to make direct samplings of noctilucent clouds (clouds of unknown origin, faintly luminous, which appear in certain regions of the auroral zone).

May 23: May 24 was declared a day safe from major solar flares for Mercury-Atlas 7 launch attempt by the Sacramento Peak Observatory of the Air Force Cambridge Research Laboratories.

• In regular press conference, President Kennedy answered a question on the possible U.S.-U.S.S.R. cooperation in space because of recent progress in formulating a joint proposal for the use of satellites for weather analysis: "We felt the best place to start was on weather, and any progress we can make on that would be very welcome. I must say that we strongly support any cooperative effort we could make on weather, predictions of storms, and all of the rest, and I hope it will lead to other areas of cooperation in space."

• In testimony on proposed Federal salary reform before the House Post Office and Civil Service Committee, NASA Administrator James E. Webb stressed the requirements of the national space program for adequately paid scientific and technical personnel in government. He pointed out that one third of NASA's personnel were professional scientists and engineers, one third skilled craftsmen and research mechanics and their supervisors, one eighth were engineering and scientific technicians, and one sixth were clerical and secretarial. The remainder were professional-level specialists in management, legal counsel, librarians, editors, etc.

Mr. Webb said that in considering pay raise legislation (H.R. 10480) the Federal Government must be able to compete favorably in the nation-wide labor market and that present salaries must be raised to have "reasonable comparity" outside of Government.

• USAF launched a satellite with a Blue Scout booster, from Point Arguello; according to U.N. Public Registry, did not achieve orbit.

• Congressman John J. Rhodes (Ariz.) polled his district, receiving 5,000 replies to a series of current questions. To the question— "Are you in favor of United States-Russian space cooperation?"— the response was: 19% said "Yes"; and, 70% said "No."

May 24: AURORA 7 with Astronaut M. Scott Carpenter (Cdr., USN) as pilot, launched on MA-7 orbital mission by Atlas booster from Cape Canaveral, the second U.S. manned orbital flight and the fourth U.S. manned space flight. Almost faultless countdown, launch was delayed only briefly by ground fog, and Mercury spacecraft was placed into orbit at 17,532 mph (apogee: 167.4 miles; perigee: 99 miles; period: 88.3 minutes). Astronaut Carpenter performed series of space science and technical development experiments including visual and photographic observations of star fields and "white particles," consumption of solid foods, release of tethered multi-colored balloon to test sighting, and observation of separated Atlas booster. 81,200-mile flight featured attitude stabilization and control pilotage for completion of three orbits, and monitoring of control-systems fuel for proper retrofire attitude. Re-entry caused landing impact point over 200 miles beyond intended area and beyond radio range of recovery forces. After landing, Carpenter egressed through top of capsule and climbed on inflated raft awaiting rescue, to be joined by two USAF paramedics, Airman John T. Heitsch and S/Sgt. Ray E. McClure, who inflated Stulken collar to prevent capsule from sinking. After three hours on water, Astronaut Carpenter was picked up by a helicopter from U.S.S. *Intrepid*, and his capsule was retrieved by the destroyer *Pierce*. Astronaut Carpenter was reported in good physical condition, and flown to Grank Turk Island in the Bahamas for 48-hour debriefings and medical examinations.

May 24: President Kennedy personally congratulated Astronaut M. Scott Carpenter by telephone shortly after the AURORA 7 pilot arrived by helicopter aboard the carrier *Intrepid*, saying: ". . . I wanted to tell you we are relieved and very proud of your trip. I am glad that you got picked up in good shape and we want to tell you that we are all for you and send you the very best of luck to you and your wife."

Carpenter said: "My apologies for not having aimed a little bit better on re-entry."

President Kennedy: "Oh, fine and good. We want to congratulate you and I look forward to seeing you in Washington sometime soon . . . ."

- Saturn SA–3 launch vehicle was static test-fired at MSFC for 119 seconds, producing near 1.3 million pounds of thrust.
- The House of Representatives (342–0) approved the NASA authorization for FY 1963 of $3.67 billion. This was almost three times the NASA budget for FY 1962.
- Research Center for Celestial Mechanics at Yale University announced by the Air Force Office of Scientific Research and the Office of Naval Research, to be headed by Dr. Dirk Brouwer, Director of the Yale Observatory. The facility, to be opened on July 1, will seek new approaches for problems involving earth satellites, lunar and interplanetary probes, and the predetermination of trajectories for interplanetary vehicles.
- X–15 flight postponed so that radar tracking systems would not interfere with the flight of AURORA 7.
- In speech at Cleveland, Miss., General Bernard A. Schriever said: "As long as the Soviets remain committed to their goal of world domination, their accomplishments in space can be a potential source of danger to free men everywhere. In their efforts to bring about the complete victory of their system, they have not hesitated to use any means. There is no reason to believe that space will be an exception. . . .

"National defense missions in space have certain unique requirements, but in many ways they are closely related to other aspects of the national space program. There has been and continues to be close cooperation between the Air Force and NASA.

"The rate of progress in our space program will be determined by two factors. The first of these is research. Development of space technology urgently calls for new knowledge. . . .

"Research alone is not enough. The knowledge we acquire must be applied in timely fashion. The key to this application is management. . . ."
- USAF Titan ICBM destroyed in fueling test at Chico, Calif.

May 24: Astronaut John H. Glenn, Jr., was selected as the National Father of the Year. Earlier this month, his mother had been selected as World Mother of the Year.

- Air Force announced the award of a contract to Pan American World Airways for operation and logistic support of a Rocket Range Test Facility at Fort Churchill, Canada. Operated by the U.S. Army for the past five years, the Fort Churchill Test Range will transfer to USAF operational contract on July 1 of this year and will be managed by the Office of Aerospace Research. The facility was originally established as a joint Canadian-U.S. project to support the International Geophysical Year.

May 25: NASA launched Aerobee research rocket with 213-lb. payload containing four scientific experiments from Wallops Station. Reaching an altitude of almost 125 statute miles, the nose cone was recovered by an S–62 helicopter some 52 miles down range.

- In speech before the Aviation and Space Writers Association meeting in San Francisco, Dr. Harry Goett, Director of GSFC, reviewed the events in the life of scientific satellites after they are placed into orbit. Once in orbit, he said, a scientific satellite "does not have striking news value. . . .

  "Except for manned space projects there is no single personality on whom the spotlight can be concentrated; the job of putting together the cosmic jigsaw puzzle of space from the bits and pieces obtained from our satellites is one that engages the effort of many people throughout the scientific community; and this jigsaw puzzle goes together so gradually that there are no singular events which merit a headline. I suspect that there are Nobel prizes in the making, but it is going to be difficult to determine who should get the medal."

  Dr. Goett, reviewing the scientific accomplishments in space, said that the general public needs to understand this part of the space effort "on the basis of its real accomplishments and potentialities." He said that this was a real challenge to the space writers.

- In speech at the Seattle World's Fair, Secretary of State Dean Rusk urged that activities in space be subjected to international law and supervision before it is too late. Without international regulation, he said, "the frontiers of space might be pierced by huge nuclear-propelled dreadnaughts, armed with thermonuclear weapons. The moon might be turned into a military base. Ways might be found to cascade radioactive waves upon an enemy. Weather control might become a military weapon. . . ."

- One year ago, President Kennedy in his address to Congress declared the national space goal of "landing an American on the moon in this decade."

- Velery Lutsky, Soviet astronomer of the Moscow Planetarium, said in English-language Radio Moscow broadcast to North America, that the flight of AURORA 7 made it "more urgent" that the U.S. and U.S.S.R. cooperate in space exploration.

- Consolidation Steering Committee of the Board of the American Rocket Society (ARS) and the Council of the Institute of the Aerospace Sciences (IAS) completed preliminary "Principles of Consolidation" for a proposed new society to be known as the American Institute of Aeronautics and Astronautics (AIAA).

May 26: NASA's F–1 rocket engine first fired at full power (more than 1.5 million pounds of thrust) for full duration at Edwards, Calif.

- Army Nike-Zeus antimissile missile successfully test-fired all three stages at PMR.

May 27: In special ceremony at Cape Canaveral, Astronaut M. Scott Carpenter and Walter Williams, Director of Operations for Project Mercury, were both awarded NASA Distinguished Service Medals by NASA Administrator Webb at the direction of President Kennedy.

- At NASA news conference at Cape Canaveral, Astronaut M. Scott Carpenter gave a detailed chronological account of his three-orbit flight in AURORA 7. He quoted Mercury training psychologist Dr. Robert B. Voas, who had told Carpenter that data-collection is the "heart and soul of science." Carpenter said: "And that's true. Isolated data points mean nothing. We now have two data points on nearly everything. His [John Glenn's] flight confirms [mine [and [mine confirms his . . . ."

  Carpenter said of weightlessness that "there was the sensation of floating . . . Every function is easier when you are weightless. It's the only time I've ever spent in that [space] suit that was comfortable."

  He reported briefly on his launch into orbit, his scientific observation, re-entry, and recovery, full scientific reports of which would be forthcoming in future weeks.

- Cosmonaut Yuri Gagarin in Tokyo said: "The U.S.S.R. is at present constructing a giant spaceship which cannot be compared to those I and Titov rode in size."

- DOD announced establishment of the Defense Industry Council to provide a focal point for the review and discussion of the findings of industry study groups as well as a forum for mutual exchange of logistics management objectives and views.

May 28: In speech on the floor of the House of Representatives, Congressman George P. Miller, Chairman of the Committee on Science and Astronautics, paid tribute to Astronaut Scott Carpenter, stating in part:

  "Science is the beneficiary of his operation. . . . Commander Carpenter's success is a compliment to the program; it is a tribute to American ingenuity and American ability to solve the mysteries of outer space. . .

  "To Scott Carpenter, to Alan Shepard, to Gus Grissom, and to John Glenn, this House and the American people owe a great debt of gratitude. From a scientific standpoint, I am happy to announce to you today the great success of this orbital flight."

- Senator Howard W. Cannon of the Committee on Aeronautical and Space Sciences spoke on the floor on the short supply of scientists available to NASA and to its private industry contractors. He proposed a closer relationship between NASA and the universities: "We need to develop and develop from the beginning, space-oriented physicists, space engineers, space chemists, and space biologists." He cited NASA's cooperative program with universities established without the benefit of legislation, and said that he proposed to introduce legislation to amend the NASA Act for the purpose of: "(1) to increase the number of undergraduates who will make space science and space engineering-their primary field of work; (2) to provide NASA with clear directive and the necessary funds to establish an integrated and intimate program of research, fellowships, grants and curriculum specialization with universities . . .; and (3) to establish one or more university-operated space laboratories which will stimulate basic research and train professionals. . . ."

May 28: Metal fragments resembling top of gasoline drum, thought to have fallen from the MA–7 Atlas booster, were discovered near Barkly East, South Africa.

- U.S.S.R. launched COSMOS V (apogee: 1,000 miles; perigee: 126 miles; inclination: 49°; period: 103 minutes), and said that it was an instrumented payload. Weights of the payloads of the Cosmos series initiated on March 16 have not been announced, presumably to veil the reported new launch rockets being used.

- In an impromptu speech at opening of an Italian industrial exhibition in Moscow, Soviet Premier Khrushchev acknowledged that the U.S. had achieved "notable successes" in space and saluted the courage of Mercury astronauts Glenn and Carpenter. Commenting on M. Scott Carpenter's MA–7 flight in AURORA 7, Khrushchev declared:

  "I congratulate him especially because he gave proof of great courage in a situation in which he could have been burned in his ship or drowned in the sea. All the scientists could do from the earth was to advise him to drink water and to land. But he had the presence of mind and the courage to continue his flight."

May 29: M. Scott Carpenter Day in Boulder, Colo. Presented with an engineering degree from the University of Colorado, Astronaut Carpenter said that he hoped the degree, which he missed in 1949 because he failed a course in heat transfer, could be justified by the "unique" education in the subject received during his re-entry in AURORA 7. At a news conference, Carpenter said: "All mankind stands to gain more as a result of these flights— and mark my words, man will be going to the moon and neighboring planets in the near future—than in any previous exploration in the history of the world."

- Before the Scientific and Technical Subcommittee of the U.N. Committee on the Peaceful Uses of Outer Space, NASA's Dr. Homer E. Newell of the U.S. delegation proposed that an international launching site for space research be established near the Equator. Such a launching site would allow study of the equatorial electrojet (electric current in upper atmosphere which follows the geomagnetic Equator) and of the ionosphere without interference from the Van Allen radiation belts.

- Joint DOD-NASA Agena D Agreement signed which would authorize the DOD to develop a standardized Agena D stage for joint use with the Atlas and Thor first stages. Agena D is designed to use present flight-proven equipment, to emphasize simplification in vehicle design, to employ production techniques that will allow adaptability to advanced components without basic design change, and to permit lower cost and firm scheduling of space shots. Agreement supplemented basic DOD-NASA agreement effective Feb. 23, 1962, on the National Launch Vehicle Program.

May 29: NASA announced that ARIEL I, the U.S.-U.K. ionosphere satellite launched on April 26, was functioning well except for one experiment, the solar ultraviolet detector. Telemetry signals were being received by fifteen ground stations around the world, and the tape recorder, which recorded on each orbit, was being commanded successfully about twelve times a day.

- At opening session of the Legal Subcommittee of the United Nations Committee on the Peaceful Uses of Outer Space, U.S. delegate, Leonard Meeker, proposed that all members of the U.N. undertake to aid space vehicles or astronauts in distress, and to return to the launching countries any that are forced to land elsewhere.
- USAF announced launching of an unidentified satellite from Vandenberg AFB, Calif., using a Thor-Agena B booster.
- FRIENDSHIP 7 space capsule arrived in Ghana in its global exhibition tour.
- Sir Bernard Lovell of England described the high-altitude nuclear tests (Project Fishbowl) as a "black moment for humanity and an affront to the civilized world . . ."

May 30: M. Scott Carpenter Day in Denver as an estimated 300,000 persons witnessed the Memorial Day parade honoring the Mercury Astronaut from Colorado.

- An Australian rocket with five cameras was launched from Woomera as TIROS IV passed overhead. The rocket's camera took photographs of clouds from 80-mile altitude while TIROS IV made simultaneous photography of cloud cover from 400-mile altitude.
- The Science Minister of France, Gaston Palewski, outlined plans for orbiting a satellite at a Cabinet meeting. The date had been set "almost to the hour" for the launching of the first French space satellite, although the actual date was kept secret, according to Information Minister Alain Peyrefitte.
- Pioneer naval aviator, Vice Admiral Patrick N. L. Bellinger, died. In 1915, Bellinger became the second American flier to be launched by catapult, established flying boat distance record, established altitude record of 10,000 feet, and participated in aerial spotting for artillery fire. In 1916, he made naval tests of live bombing and practical tests of air-to-ground radio communication. In 1917, he made machine-gun firing tests from seaplane and first night seaplane flights using floodlights. In May 1919, he commanded the NC-1 in transatlantic flight to the Azores, damage to aircraft ended the flight there. During World War II, he was commander of the Atlantic Fleet Air Force.

May 31: Astronaut M. Scott Carpenter returned to work at Langley AFB facilities of the Manned Spacecraft Center after a four-day trip to Colorado. All seven Mercury astronauts were scheduled to receive Peninsula Distinguished Service Awards, a tribute from several civic groups on Virginia's lower peninsula, on June 1.

- AEC announced that U.S. would begin "rocket borne" nuclear tests at an altitude of "tens of kilometers" shortly after nightfall on June 1 at the Johnston Island test area.
- Air Force Cambridge Research Laboratories announced series of three high-altitude balloon astronomy experiments—Star Gazer, Sky Top, and Balast—to be conducted later this year. The primary objective would be to obtain more and complete information about the space environment with emphasis on the moon, Mars, and Venus.

May 31: A Blue Scout booster combination space probe was launched from Point Arguello, Calif., by the Air Force. No details were released.

- USAF superpressure balloon made a record-breaking nineteen-day flight at a constant altitude of 68,000 feet in a test conducted by Cambridge Research Laboratories. The balloon was launched from Kindley AFB, Bermuda, and landed by radio command in the Pacific, near Iwo Jima, 3,600 miles west of Hawaii, covering a total of 9,300 miles.

During May: Checkout was completed and the University of Alaska assumed responsibility for the Alaska Data Acquisition Facility near Fairbanks. Part of the GSFC system, the Alaskan facility is an 85-ft. dish and its associated electronics system will be used on tracking and data acquisition of polar-orbiting Nimbus, Ego (Eccentric Geophysical Observatory), and Pogo (Polar-Orbiting Geophysical Observatory) satellites of GSFC.

- Announced that the Nippon Electric Co. of Japan (NEC) was constructing several NEC-1 active repeater communications satellites to be launched by NASA Scout or Thor-Delta boosters during the Olympic Games in Japan in 1964.

- NASA Langley Research Center announced contract with Space-General Corp. for the development of an inflatable paraglider to measure micrometeroid flux in the lower regions of space up to 700,000 feet. Inflatable Micrometeroid Paraglider (IMP) will test new sensor system consisting of alternate layers of mylar and aluminum and will determine suitability of inflatable re-entry paraglider.

- Prime Minister Jawaharlal Nehru of India visited the tracking station built by GSFC and located at Ahmedabad, 300 miles north of Bombay.

- High interest in the Rogallo-wing concept among amateur aircraft-builders reported to have concerned NASA developers since apparent simplicity of the flexible wing offers serious trouble.

- Rep. Edgar Hiestand asked citizens of the 21st Congressional District of California—"Do you consider scientific information resulting from a U.S. moon flight worth its $40 billion cost?" Results of the poll were: 44% replied "no"; 35% replied "yes"; and 21% replied "undecided."

- Project 60 established by Secretary of Defense to improve field operations involved in management of DOD and NASA contracts.

- Awards for 1961 announced by the American Helicopter Society: first Igor I. Sikorsky International Trophy to Mikhail L. Mil of the Soviet Union; Dr. Alexander Klemin Award to Brig. Gen. Robert R. Williams, U.S. Army Aviation Center Commander; Capt. Wm. J. Kossler Award to Air Rescue Service, MATS; Frederick L. Feinberg Award to Lt. Col. Francis M. Carney (USAF); and the Grover E. Bell Award to the engineering organization of Sikorsky Aircraft.

- The FAI certified that a Soviet E-166 jet fighter piloted on October 7, 1961, by A. Fedotov had set a new world speed record of 1,491.9 mph over a 100-km. closed course. This record was previously held by the McDonnell F4H-1 Phantom II which flew the course at 1,390.24 in September 1960. The FAI also certified several speed, altitude, and altitude-with-payloads records the Russians had claimed for the Beriev M–10 jet seaplane.

During May: Swedish industrial group formed in Stockholm with aim of studying peaceful applications of space research, a branch of Sveriges Mekanforbund, a general industrial organization. Among those participating are Saab Aircraft Co., Telefon AB LM Ericcson, Svenska Flygmotor AB, and Svenska AB Gasac-cummulator (Aga).

- National Society of Professional Engineers Award of 1962 announced for Dr. Charles Stark Draper of MIT for his "outstanding contributions to the military affairs of the country . . ."
- The Fédération Aeronautique Internationale (FAI) received the Edward Warner Award granted every second year by the International Civil Aviation Organization (ICAO) Council, for "outstanding contributions to the development of international civil aviation."
- IAS awarded its Flight Test Engineering Fellowship to Raymond P. Boyden, Director of Engineering, U.S. Army Hq., St. Louis. Gene J. Matranga of NASA's Flight Research Center and Maj. James C. Wayne (USAF) were named runnerups. All will attend Princeton's Aeronautical Engineering Department.

JUNE 1962

June 1: X–15 No. 2 flown in evaluation test of an alternate stability augmentation system and returned to earth's atmosphere at a 23° angle of attack, Maj. Robert White (USAF) as pilot. Flight reached maximum speed of 3,750 mph (mach 5.53) and a peak altitude of 129,000 feet.

- USAF launched unidentified satellite with Thor-Agena B booster, from Vandenberg AFB, Calif.
- Deadline for applications for NASA's new astronaut positions; NASA began screening the more-than-250 applications received. The armed services supplied 53 of the applications while more than 200 civilians submitted applications. Once selected, the new astronauts would undergo intensive training at Manned Spacecraft Center for Projects Gemini and Apollo.
- First African showing of FRIENDSHIP 7, Astronaut John Glenn's space capsule, drew record crowds. USIA reported that over 40,000 persons had viewed the capsule in Accra, Ghana, in three days, more than at previous showings in London, Paris, Madrid, and Belgrade.
- The seven Mercury astronauts were presented with Distinguished Service Awards from the Lower Virginia Peninsula in ceremonies at the Manned Spacecraft Center Hq. at Langley AFB, Va.

June 2: USAF announced that OSCAR II satellite had been launched piggyback on an unidentified satellite on June 1, on behalf of the American Radio Relay League. With a 92-minute period, the 10-lb. satellite broadcasts "Hi" in Morse code on 144.993 megacycles for use by amateur radio operators. Unlike OSCAR I launched on December 10, 1961, advance notice was not given the Project Oscar Association on the launching of OSCAR II.

- Navy successfully fired a Polaris missile from submerged nuclear submarine U.S.S. *Thomas Edison*, off Cape Canaveral. The Polaris flew some 1,500 miles.
- Scientists of the National Center for Atmospheric Research, at Boulder, Colo., after intensive interviews with Astronaut Scott Carpenter, concluded that the layer of "haze" reported on the horizon by both Carpenter and Glenn was in reality the phenomenon known as airglow—the emission of light of various colors and wavelengths, caused by chemical reactions in gas molecules in the upper atmosphere.
- NASA reported it had received more than 700 requests for personal appearances by the Mercury astronauts for the July 4th holiday, but there was little likelihood that any would be accepted. Astronaut John Glenn had received more than 8,000 requests for appearances since his orbital flight on Feb. 20, 1962.

June 3: Soviet Government issued a statement which warned that high-altitude nuclear tests by the U.S. would "carry the nuclear arms race into outer space." Addressed to "all states and peoples," the statement said that the test could disrupt radio communication, endanger future flights by cosmonauts, and "change the weather." The statement further warned that the U.S. would not be permitted to gain any military advantages from the tests.

- Worst single-aircraft disaster in commercial aviation history, when Air France Boeing 707 crashed and burned on takeoff at Orly Field, Paris, killing 130 passengers and crew.

June 4: High-altitude nuclear test over Johnston Island in the Pacific failed when Thor rocket and the thermonuclear device were destroyed by the safety officer "because the tracking system was malfunctioning." Debris fell into the ocean well within the 520-mile-radius safety area around Johnston Island.

- Polaris missile launched from AMR was destroyed 90 seconds after launch when it veered off course.

- U.S.S.R. charged in newspapers and radio broadcasts that purpose of U.S. high-altitude nuclear tests over the Pacific was to prepare for a surprise nuclear attack against the Communist bloc, contended the tests were designed to perfect ways of blacking out radio and radar communication, thereby preventing a retaliatory strike at the U.S.

- Boeing-Vertol was selected by the Air Force to build long-range helicopter capable of carrying a payload of at least 5,000 lbs. for 200 nautical miles or 2,400 lbs. for 700 miles. Modification of Vertol 107 helicopter, now in production for civil use, would satisfy USAF's Specific Operational Requirement 190.

- Army fired Pershing missile from AMR. The 200-mile flight was a severe test of the missile's guidance, with the launch made in the teeth of 50-mph winds; the missile rose with a "definite tilt" before the guidance corrected its attitude.

- House of Representatives passed a bill (H.R. 10618) granting the consent of Congress to the Southern Interstate Nuclear Compact, which would establish a regional cooperative effort to acquire nuclear resources and facilities. NASA is one of the Federal agencies authorized under the bill to cooperate with the Compact.

June 4–5: Thirty-fifth anniversary of the second nonstop transatlantic flight, by Clarence C. Chamberlin and Charles A. Levine. Taking off from Roosevelt Field, L.I., on June 4, 1927, they followed much the same route taken in May of the same year by Charles A. Lindbergh, but flew on to Eisleben, Germany, for a nonstop distance record of 3,911 miles. Mr. Chamberlin, now 68, recalled in an interview that they ran into bad weather over England, climbed their single-engine Wright-Bellanca to over 20,000 ft., although "now they tell you you must have oxygen above 10,000 feet."

June 5: Astronaut M. Scott Carpenter and his family were received by President Kennedy at the White House, then flew to New York for the city's reception and to receive New York City's highest award, the Gold Medal. The reception committee included, in addition to Mayor Wagner, former Presidents Herbert C. Hoover and Harry S. Truman.

June 5: NASA explained the reasons for the 250-mile overshoot of the landing by AURORA 7, attributing most of the overshoot to a 25° error in the capsule's yaw attitude at the time the retrorockets were fired. This caused the thrust of the retrorockets to be exerted on the capsule at an angle rather than in line and resulted in less slowing of the capsule and consequently a shallower descent curve than planned. This accounted for some 175 miles of the 250-mile overshoot as well as a deviation of 15 miles to the north of the orbital track. Other factors that contributed to the overshoot were the retrorockets delivered about three per cent less than the specified thrust, and the retrorockets fired about three seconds late.

- A cooperative program for testing Relay and Telstar communications satellites by NASA and communications organizations in the U.S., Europe, and South America was announced by NASA. Providing ground stations and conducting the communications experiments would be AT&T, IT&T, British General Post Office, French National Center for Telecommunications Studies, West German Post Office, Brazilian Department of Posts and Telegraphs, and Telespazio of Italy. The foreign organizations would participate on a voluntary basis with no exchange of funds by the governments. NASA would provide ground stations with orbital data on the satellites.

- U.N. Committee on Peaceful Uses of Outer Space, meeting in Geneva, set up a ten-nation working group under the Scientific and Technical Subcommittee to examine international exchanges of scientific information on outer space. Takeo Hatanaka of Japan was named chairman; other nations represented were the U.S., U.K., U.S.S.R., Belgium, Chad, France, Hungary, Romania, and the United Arab Republic. Another group, chaired by W. A. Sarabhai of India, was set up to examine the U.S. proposal of an international research rocket range to be established near the Equator.

- Space Vehicle Panel of the President's Science Advisory Committee, headed by Dr. Jerome V. Wiesner, visited MSFC.

- Greenhut Construction Co. was awarded $903,158 contract for modification of a Saturn static test stand at MSFC, Huntsville, Ala. Built in 1956 for Redstone and Jupiter missile test-firings, the stand was later altered to accommodate the huge Saturn S–I stage. Under this contract, the test stand would be modified to provide two test positions capable of holding the Saturn stage, with work to be completed by February 1963.

June 6: Three sounding rockets were launched from NASA's Wallops Station, Va.: The first, a Nike-Apache, was launched at 7:40 PM (EDT), with a 70-lb. payload containing a pitot-static probe, reaching 78 miles altitude; the second, a Nike-Cajun, was launched at 8:05 PM, consisted of 11 explosive charges and a balloon, released from 25 miles to 64 miles altitude; the third, a Nike-Asp, was launched at 8:56 PM, consisted of sodium vapor clouds to measure atmospheric winds and diffusion, ejected at about 20 miles and extending to a peak altitude of about 100 miles.

- NASA announced that a contract would be negotiated with the Linde Co., a division of United Carbide Corp., to supply liquid hydrogen for use on the West Coast, particularly for development of RL–10 engine and for the 1.2-million-pound-thrust M–1 engine under development by Aerojet-General. Linde would build a plant at Sacramento, Calif.

June 6: U.S. patent (No. 3,008,154) was granted to Dr. Vladimir K. Zworykin for a rocket to be used in detecting strength and polarity of electric fields in clouds. Such data could then be used in control of cloud movements and in effecting weather changes. Patent was first applied for in 1948 but until now was kept secret at request of DOD. Dr. Zworykin is a consultant to the Princeton Laboratories of RCA and honorary vice president of RCA.

June 7: X–15 No. 1 flown to 3,716 mph by NASA pilot Joseph Walker and put through a series of high-speed sharp-angle maneuvers involving banks and testing at three angles of attack. Skin temperatures may have reached 1,000°; the rocket engine developed some roughness in the course of the flight.

• Nike-Cajun vehicle with an experiment to measure winds and temperatures in the upper atmosphere was launched from NASA's Wallops Station. In the night flight, 12 special explosive charges were ejected and detonated at intervals from about 25 st. mi. altitude up to about 58 st. mi.

• Pakistan made its first space experiment, the launching of a Rehbar I sounding rocket from a site near Karachi. Part of a cooperative program between NASA and the Pakistan Upper Atmosphere and Space Research Committee, the Rehbar I was a Nike-Cajun rocket supplied by the U.S., carrying a sodium vapor payload to an altitude of about 80 miles to measure upper-altitude winds. Several more such launches are programed for the next few months. Prior to the launches, Pakistani scientists and technicians were given training in NASA research and launch centers.

• NASA announced selection of Bendix Corp.'s Radio Division, Towson, Md., for contract to operate five of NASA's worldwide Project Mercury tracking and communications stations. Two-year, $10 million contract also called for engineering and operations services for all Mercury stations, beginning January 1, 1963. Under incentive contract, Bendix would have opportunity to earn bonus payments above its fixed fee by demonstrating superior performance; this contract probably first awarded by any agency to provide monetary incentive for outstanding performance in service-type work.

• USAF Titan II ICBM launched from AMR on a planned 5,000-mi. test flight, the missile falling short of its target but achieving most of its test objectives.

• USAF B–52H claimed a world record for nonstop, non-refueled distance flight over a closed circuit, having flown 11,400 miles in 22 hours, 38 minutes, 41.8 seconds, with an average speed of 510 mph. The closed-circuit course ran from Seymour Johnston AFB, N.C., to Bermuda; Sondrestrom, Greenland; Anchorage, Alaska; Los Angeles, Calif.; Key West, Fla.; and back to Seymour Johnson. Pilot was Capt. William M. Stevenson (USAF); copilot was Capt. Floyd J. Schoendiest (USAF). Present record is held by Lt. Col. J. R. Grissom (USAF), who on December 14, 1960, flew a B–52G a distance of 10,078.84 miles.

June 7: NASA announced the appointment of Dr. Richard B. Morrison as Director of Launch Vehicles and Propulsion Programs, Office of Space Sciences, in NASA Hq., succeeding Col. Donald Heaton (USAF). Dr. Morrison was previously professor of aeronautical engineering and supervisor of the Aircraft Propulsion Laboratory at the University of Michigan.

- ONR's 85-ft., steerable radiotelescope dedicated at Hat Creek Radio Astronomy Facility, Hat Creek, Calif. Designed for extensive mapping of the galaxy—and particularly the hydrogen-gas clouds of the galactic nucleus—the Philco-built telescope would be operated by University of California.

- U.S. Embassy in Warsaw announced withdrawal of U.S. offer to display the FRIENDSHIP 7 capsule in Poland during its international goodwill tour. Reason given was that the "Polish authorities are unable to give a positive reply to our offer."

June 8: Dr. Hugh L. Dryden, Deputy Administrator of NASA, announced jointly with Soviet representative Anatoly Blaganravov in Geneva that a U.S.-U.S.S.R. agreement had been reached to coordinate launchings of weather satellites to provide information for the "world weather watch" of the World Meteorological Organization (WMO). Also announced was agreement on a joint effort to map the earth's magnetic field; talks were continuing on joint cooperation in the field of communications satellites. Announcement was made after two weeks of talks on U.S.-U.S.S.R. space cooperation in Geneva, which grew out of the exchange of letters between President Kennedy and Premier Khrushchev in March.

- White House Reorganization Plan No. 2 became effective, establishing the Office of Science and Technology in the Executive Office of the President. Dr. Jerome B. Wiesner was appointed Director, retaining duties of the special assistant to the President for science and technology and receiving certain functions transferred from the National Science Foundation.

- Chimpanzee "Zena" ejected in capsule from USAF B-58 flying at 1,060 mph. at 45,000 feet in escape capsule test, over Edwards AFB, Calif. W/O E. J. Murray (USAF) had been first man ejected in the capsule on February 28, at speed of 565 mph.

- DOD announced that it was consolidating all military basic research overseas. The first step would be the opening of a Defense Research Office in Rio de Janiero, Brazil, on July 1, 1962, to coordinate the military basic research program throughout Latin America. NSF and NIH would open offices in the same building for their independent programs, with all efforts coordinated by the State Department.

- USAF fired a Minuteman ICBM from a silo at AMR and sent it 3,000 miles downrange. In an associated test, a single-stage Nike was fired to trail the Minuteman up to 12 miles to chart wind patterns.

- In address at Ohio State University, General Curtis E. LeMay (USAF) said: "It is in the fields of science and technology that the greatest explosion of problems is occurring. We are in the midst of an age of technological change that will make the industrial revolution seem dull indeed. The problems associated with technical change are producing opportunities in greater abundance than ever before. Opportunity is not only knocking at the door, it is beating gongs all around the house . . . ."

"In science and technology, there is no ceiling. We are going into space. This will be man's greatest adventure—not only for the men who crew the ships and make the journey, but for the people on the ground who support the effort. . . .

"The free world faces a grave threat. In rapidfire fashion, new discoveries and developments are being made almost daily that prevent us from relaxing our vigilance. Our future is directly dependent on having an educated and informed people who are in tune with the world tempo and alert to the myriad applications and possibilities in all fields of human endeavor. . . ."

June 8: The Rev. Theodore M. Hesburgh, President of the University of Notre Dame, speaking at the commencement exercises at MIT, urged scientists to develop greater moral awareness and a greater use of science and technology "for the true service of mankind." He told the science graduates: "You should be concerned, you should lead, but the beginning of significant human leadership involves a deep respect for the totality of man's intellectual and moral heritage, an active cultivation of the wide areas of wisdom above and beyond your science and technology."

June 9: Titan I liquid-fueled engine strapped to the side of a 175-ton solid-fuel rocket engine and the configuration test-fired by Aerojet-General Corp. to evaluate the compatibility of liquid and solid engines for the USAF's Titan III space booster. The configuration generated 700,000 lbs. of thrust.

• Joseph A. Walker, Chief Research Pilot for NASA's Flight Research Center, was awarded an honorary Doctor of Aeronautical Sciences degree from Washington and Jefferson College, Washington, Pa. Walker had been graduated from Washington and Jefferson in 1942 with a major in physics.

• Astronaut Scott Carpenter received a B.S. degree in aeronautical engineering from the University of Colorado, a degree he missed in 1949 by not taking the final examination in a course in heat transfer. He also received the school's Norlin Award, given annually to an alumnus for outstanding achievement.

June 10: Astronaut Alan B. Shepard, Jr., was awarded an honorary Master of Arts degree by Dartmouth College.

June 11: Rehbar II sounding rocket launched from site near Karachi, a continuation of U.S.-Pakistan joint program. First launching of Nike-Cajun sounding rockets carrying sodium vapor payloads was June 7.

• DOD announced reassignment of responsibilities for the DOD Communications Satellite Program. The Air Force was assigned responsibility for the development, production, and launch of all space devices, consistent with the general policy for the development of military space systems announced in March 1961. The Army was assigned responsibility for the development and operation of the ground environment except for that portion of the ground environment associated with launch and control of satellites themselves. Defense Communications Agency (DCA) was assigned task of integrating ground and space components. First effect was the transfer to the Air Force of overall systems management of Project Advent, a responsibility of the Army since September 1960 when it was transferred from ARPA.

June 11: USAF Office of Scientific Research Summer Seminar in Geophysics, in which some 200 scientists participated, convened at Cloudcroft, N.M, the seventh annual such review of progress in the theoretical and technical aspects of geophysics.

June 12: X–15 No. 3 research airplane flown to an altitude of 180,000 ft. and a speed of 3,445 mph. (mach 5) by Maj. Robert White (USAF). Primary objective of flight was checkout of X–15 with adaptive control system.

- Piloting a NASA F–106 supersonic jet airplane from Pope AFB, N.C., Astronaut Virgil I. Grissom notified Jacksonville, Fla., air route traffic control center that his Tacan radio navigation aid had gone out. Flying above clouds and around thunderstorms, Grissom began operating an emergency radio beacon signal and was directed to a safe landing under a thunderstorm at Patrick AFB by Miami air traffic control center.

- FAA disclosed that a tiny cotter pin and a bolt less than an inch long may have caused the March 1 disaster—95 people killed in Boeing 707 jet airliner seconds after takeoff from Idlewild Airport.

- USAF C–123 transport fitted for aerial spray operations returned to Langley AFB, Va., after spraying 17,000 acres of Iran and Afghanistan to bring locust plague under control.

- Egypt President Gamal Abdel Nasser was among those visitors "intrigued" by Mercury capsule FRIENDSHIP 7, on display in Cairo. "How do you get out of the thing?" President Nasser reportedly asked.

June 13: United Nations' scientific and technical subcommittee of the Committee on the Peaceful Uses of Outer Space agreed on a world-wide technical program in Geneva. Program prepared by 28-nation meeting provided for the exchange of scientific data, encouraged a series of international space research projects (world-wide survey of earth's magnetic field, polar cap experiments, cooperation in fields of space communications and weather satellites, and the IQSY program), and established international launching ranges for sounding rockets near the equator.

- In testimony before the Senate Committee on Aeronautical and Space Sciences, NASA Administrator James E. Webb indicated that decision on the number of orbits, three or six, planned for the next Mercury flight had not been decided and that the selected pilot-astronaut would also be announced soon. On the relationship between civilian and military space activities: "President's [space] policy has been, in accordance with the law, to develop the space program as a civilian peaceful effort to the fullest extent possible, but always pressing with the kind of technology that would permit us to move rapidly in the military field if we were required to do so. It is a little bit like 'keep your powder dry' with respect to the military side . . ."

  Deputy Secretary of Defense Roswell L. Gilpatric stated to the Committee that DOD "was interested in developing the technology" but we [DOD] have as yet no approved program for the development of a manned orbital system." Gilpatric said that DOD was "very conscious of the need to take out technological insurance" so that the U.S. could be "prepared and not surprised" in the event that the U.S.S.R. attempted to make "hostile use of space."

June 13: NASA Manned Spacecraft Center awarded $1 million contract to North American Aviation, Inc., for a paraglider development program.

- Reported that Neil Armstrong, NASA X–15 test pilot, was studying approaches and abort maneuvers for Dyna Soar project by flights in an F5D aircraft.
- Full-scale model of Mercury spacecraft dropped into Trinity Bay in parachute system design test.
- Capt. Richard H. Coan (USAF) established a new world's record for a helicopter distance flight over a closed course, flying an H–43B Huskie 656.258 miles on a 25-mile course near Mono Lake, Calif. Previous record of 625.464 miles had been held by a Russian military helicopter Mi–1 (June 1960).
- Army Transportation Command awarded a contract to Ryan Aeronautical Company of San Diego for the design of a flexible wing cargo glider capable of carrying heavy payloads when being towed by a helicopter.
- James V. Bernardo, Director of NASA's Educational Programs Division (AFE), was awarded the Frank G. Brewer Trophy for 1961 by the National Aeronautic Association, for the development of a plan for national spacemobile demonstration units to supplement high school science programs and for organizing 32 aviation education workshops in 21 U.S. colleges prior to his joining NASA.

June 14: In regular press conference, President Kennedy was asked for his assessment of the U.S. space effort and "whether you plan any major realignments, such as a military role?" In answer, he said: ". . . the military have an important and significant role, though the prime responsibility is held by NASA and is primarily peace, and I think that that proportion of that mix should continue. I think the American people have supported the effort in space, realizing its significance, and also that it involves a great many possibilities in the future which are still almost unknown to us and just coming over the horizon.

"As far as where we are, I don't think that the United States is first yet in space, but I think a major effort is being made, which will produce important results in the coming months and years."

- In Paris, delegates from 10 nations—France, Britain, West Germany, Italy, Spain, the Netherlands, Belgium, Switzerland, Austria, and Sweden—signed a convention creating the European Space Research Organization (ESRO). Denmark and Norway have indicated their intention of joining later. Prof. Pierre V. Auger of France was named Director-General of the new organization. With an eight-year budget of $300 million, ESRO plans to carry out some 500 space experiments beginning with 10 sounding rockets in 1963, and the first European satellites in 1967. Launching vehicles will be supplied by another organization, the European Council for the Construction of Spatial Vehicles, which was created last March.
- TIROS IV weather satellite was no longer transmitting pictures usable for global weather forecasting, although TIROS IV was still taking "direct" pictures on command which were suitable for limited U.S. weather analysis. Infrared instrumentation was still providing information on the earth's heat balance. Launched on

February 8, 1962, TIROS IV had exceeded its designed performance and clarity of pictures from the new (Tegea) lens was outstanding. It had transmitted some 30,000 pictures to date.

June 14: Astronaut Alan B. Shepard, Jr., was awarded the Aero Club's Godfrey L. Cabot Award in Boston, as America's first man in space.

- Astronaut Gus Grissom was awarded an honorary Doctorate of Space Science degree by the Brevard Engineering College, Melbourne, Florida.
- House Committee on Science and Astronautics reportedly ordered an investigation of alleged government discrimination against women, naming in executive session Congressman Victor L. Anfuso as chairman of a special nine-man, two-woman investigation subcommittee.
- Army Pershing test missile destroyed itself shortly after second-stage ignition following launch at AMR.
- Army announced the activation of the first Pershing missile battalion at Fort Sill, Oklahoma.

June 14–September 15: GSFC Test and Evaluation Division conducted "simulation" summer workshop to promote exchange of technical information between scientists and the GSFC staff. Selected subjects on space environment simulation will be presented by GSFC scientists before professor–graduate-student teams.

June 15: Two-stage Nike-Apache sounding rocket launched from Wallops Island with 95-lb. payload to 89-mile altitude with GSFC experiment to measure electron density and electron collision frequency in the ionosphere under undisturbed conditions.

- NASA launched first two of six tests of the performance of the Canadian Black Brant sounding rocket. The first carried a payload to an altitude of 58 miles above Wallops Station, the second reached 62 miles altitude.
- U.S. Weather Bureau announced that the formation of the season's first hurricane would be detected by one or all of its battery of ships, planes, radar, and Tiros weather satellites. Last year TIROS III discovered Hurricane Esther as it was forming in the Atlantic Ocean. Altogether, the Weather Bureau said, TIROS III had seen 5 hurricanes and one tropical storm in the Atlantic, two hurricanes and a tropical storm in the eastern Pacific, and 9 typhoons in the central and western Pacific. In preparation for 1962 season, the Weather Bureau arranged to transmit satellite cloud-photographs by photo-facsimile to warning centers in San Juan, New Orleans, and Miami, where they will be used in forecasting and tracking tropical storms.
- At Senate Committee on Aeronautical and Space Sciences hearing on the Administration's FY 1963 budget request for the space program, Dr. Harold Brown, DOD Director of Research and Engineering, testified that DOD is "developing the technologies which contribute to military and unmanned orbital systems able to rendezvous with satellites and then land at pre-selected locations on earth." Dr. Brown assured the committee that DOD has "no intention to pre-empt those areas which are within the proper purview" of NASA.
- President Kennedy congratulated labor and management at U.S. missile sites for achieving a new low in work stoppages. Ac-

cording to the President's Missile Sites Labor Commission, only 1 man-day per 1,100 man-days worked was lost during the year ending this date, while 1 man-day per 96 man-days worked was lost during the 1956–1961 period. The President expressed confidence that the commission would continue to receive co-operation of labor and management "in making sure that our missile and space programs go forward . . . as economically as possible."

June 15: Senate Committee on Aeronautical and Space Sciences unanimously approved authorization for NASA in FY 1963 of $3,749,515,250. Also authorized was $71 million supplemental authorization for FY 1962, including $55 million for expansion of Cape Canaveral and $16 million for real estate at the Mississippi test facility.

• NASA announced appointment of Bob P. Helgeson as Chief of the Nevada Extension of NASA–AEC Space Nuclear Propulsion Office (SNPO–N), effective August 1. As Chief of the Nevada Extension, Helgeson will direct construction and operation of Nuclear Rocket Development Station (NRDS), national site for testing nuclear rockets.

• Astronaut Donald K. Slayton underwent physical examination by heart specialist Dr. Paul Dudley White in Boston.

• Static testing of U.K.'s Blue Streak was completed after more than 400 tests since August 1959.

June 16: Virgil I. (Gus) Grissom Day in Mitchell, Indiana, the home-town of the second Mercury astronaut into space.

• Senate debate on legislation to permit private ownership of communications satellite systems began, with Sen. John O. Pastore as floor manager of the White House-backed bill.

• Sir Robert Watson-Watt, British radar pioneer, said: "Even in physics, it is undesirable that technological fools should rush in where scientific angels fear to fly," referring to the high-altitude nuclear test planned by the U.S. Sir Robert was speaking to the 3-day "Scientists for Survival" meeting in New York.

June 17: Testimony by NASA Administrator Webb before House Appropriations subcommittee was released in which he declared that the U.S. has given far more than it has received in exchanges of space flight findings with the Soviet Union. By publicly releasing information gained from space flights, the U.S. is " . . . cooperating with them, but not getting very much in the way of return cooperation," he said. Webb defended U.S. policy of making space research results generally available:

"We are, in this open way, exposing the problems just ahead of us to the largest number of able minds all around the world. This is really the way mankind has made its forward advance. No one can tell from which mind comes the solution to a problem.

"I think the progress which can come from this can never be matched by Russia and the nations which do these things in secret."

• USAF announced launching of an unidentified satellite with an Atlas-Agena booster from Point Arguello, Calif.

• In an address at Northeastern University in Boston, NASA Administrator James E. Webb told the graduating class that "change and the accelerating rate of change will be the dominant features of your existence." In discussing NASA's program, he pointed out that in 1961 the nine largest contractors, who received 62 per cent of NASA contracts, actually purchased or subcontracted well over half this work to 10,989 first-tier suppliers or subcontractors located in 46 states and the District of Columbia. While 56 per cent of NASA direct purchases and contract awards ($25,000 or more) went to states west of the Mississippi, the award of subcontracts so shifted the distribution that 53 per cent of the dollars were actually spent east of the Mississippi. All regions of the nation can contribute to "the full development of the possibilities inherent in the application of scientific and technical advances," and share in "the potential for a period of economic growth that will bring a flowering of human progress, education, and culture."

June 17: Senator Hubert Humphrey spoke on the floor on the coordination of information between DOD, NASA, and AEC, and inserted in the record a memo of June 13, 1962, in which he criticized NASA's information exchanges. He stated: "NASA has developed highly advanced blueprints, so to speak, for information exchange. . . . The blueprints are a long way from realization. . . . [NASA] possesses, moreover, in-house personnel and a principal information contractor with a very high degree of professional competence."

- Through the official Soviet newspaper *Pravda*, the U.S.S.R. asserted that the U.S. was endangering an international program for cooperation in the use and exploration of outer space by failing to accept the Soviet drafts of codes to regulate activities in space, and by planning to undertake a program for the military use of space.

    Pravda reported: The "United States thus had left the door open for carrying the armament race into outer space. It seems that American representatives have come to Geneva not so much to negotiate on the peaceful exploration of space as to justify by legal pettifogging the Pentagon's militaristic plans in space and these plans already are too well known."

- Dr. Wernher von Braun, MSFC Director, presented cash invention awards totaling $1,750 to 5 employees on behalf of the NASA Inventions and Contributions Board. Award of $750, highest at MSFC to date, went to engineer Billy C. Hughes for feeder restrictor for gas-bearing gyros to be used for Saturn guidance platform. Other inventors honored were John R. Rasquin and Robert J. Schwinghamer ($500 jointly) for device that electronically measures roll during rocket assembly; Wilhelm Angele and Hans G. Martineck ($400 jointly) for an electrical cabling system; and Martineck ($100 singly) for plug and connector for miniaturized circuits.

- In address to Loyola University in Los Angeles, Gen. Bernard A. Schriever (USAF) said: "For the first time in history, scientific and engineering knowledge gives man a realistic means of world wide attack on the age-old problems of hunger, poverty, and disease . . . Now science has opened the way to space.

    "There is every reason to be grateful for this new knowledge and power. But some have gone too far. Instead of merely accepting and using the fruits of scientific discovery, they have made a religion of science. This, in essence, is what has happened under Communism—and it has turned out to be the most false, deceptive, and cruel religion the world has ever known. It means that men and women are treated the way a biologist treats plants and animals. It is power without principle . . ."

June 17–August 10: Space Science Summer Study, sponsored by National Academy of Sciences Office of Space Sciences (funded by NASA grant), held at State University of Iowa, Dr. James Van Allen as chairman. Scientists represented NASA, DOD, AEC, NSF, private industry, and research organizations. Review of NASA's space science program was contained in a detailed NAS report released January 1963.

June 18: NASA selected Hughes Aircraft Co. for negotiating a $2.5 million, 6-month study contract on advanced Syncom satellite. The contract covered satellite subsystems which would require long leadtime developmental and feasibility work. Second-generation Syncom would be 500-lb., spin-stabilized satellite capable of relaying hundreds of telephone calls or several TV channels. (First-generation Syncom, for which Hughes is prime contractor, is limited to single telephone channel relay.) Syncom project is under technical direction of GSFC.

- USAF announced the launching of an unidentified satellite with a Thor-Agena D booster from Vandenberg AFB, Calif.

- NASA's OSO I, launched March 7, 1962, had provided 60,000 minutes of data on its solar-pointed experiments prior to failure of its real-time telemetry on May 22, 1962. On May 15 the tape-recorded playback system had malfunctioned. Data received from OSO I included information on more than 75 solar flares and subflares.

- In extended action on the floor of the Senate, Senators Russell B. Long, Albert Gore, Estes Kefauver, and Wayne Morse opposed the passage of the Administration's amended communications satellite system bill (S. 2814) previously brought out of the Committee on Aeronautical and Space Sciences. Senator Long charged that the Senate had been subjected to lobbying "the like of which the Congress had never seen before" on the private ownership of the proposed communications satellite system, and he was questioned by Senator John O. Pastore, the bill's floor manager.

- In an address before the American Management Association Forum in New York, Maj. Gen. Robert J. Friedman, AFSC Comptroller, spoke on "Management of the Decision-Making Process." He pointed out that the Air Force Systems Command was an organization of 60,000 employees with an annual budget of seven billion dollars to administer. AFSC has, he said, over 33,000 active contracts involving companies of all sizes and in all states.

  Gen. Friedman discussed the "boundaries" within which AFSC must manage its decision-making processes: (1) actions of the military competitors of the U.S.; (2) the constraints of limited funds; (3) the number of alternative courses available in the solution of technical, strategic, and tactical military problems; (4) limitations of time; and (5) the availability of information and competence at various organizational levels within one jurisdiction.

- Aerobee sounding rocket launched to 140-mi. altitude at White Sands Missile Range (WSMR), N.M., by USAF Cambridge Research Laboratories, with objective of studying x-radiation from the moon.   X-rays expected from the moon were blanketed by much stronger source in general vicinity of constellation Scorpius. Analysis of data from sensors indicated the radiation is in form of soft x-rays that do not penetrate earth's atmosphere below 50-mi. altitude.   Data also seem to indicate the sensors may have detected a second x-ray source on the galactic rim.   AFCRL planned a second rocket flight for October to verify results of this experiment.

June 18: Aerospace Research Pilot School began a seven-month course to train seven Air Force and one Navy officers for future space missions and projects.   This was the second space-training course given at Edwards AFB, Calif., but the first for potential "operational" personnel.

- At Seventh Military-Industry Missile and Space Reliability Symposium at San Diego, Gen. Bernard A. Schriever (USAF) stated that it had become clear that "performance has outstripped reliability in a number of areas" and that this "imbalance" needed correction since "overall systems effectiveness implies considerably more than performance."   He pointed out that AFSC had attacked this problem by setting up reliability offices in AFSC headquarters, in each of the four development divisions, and in the three contract management regions.   Representatives from each of the offices serve on the Reliability Task Force to achieve AFSC-wide focus and action.   The publication of Military Specification (MIL–R–27542) on "Reliability Requirements for Aerospace Systems, Sub-systems, and Equipments" replaced previous documents and as a standard section in all new systems contracts ensures the "greatest possible understanding between the Air Force and industry with regard to specific reliability requirements."   Gen. Schriever pointed out that quantitative reliability figures were being included in system inception documents (SOR's and OSR's).   Over the long term, research and training of personnel include reliability considerations.

June 19: TIROS V launched into orbit by Thor-Delta booster from Cape Canaveral, faulty guidance system placing it into elliptical orbit (apogee: 604 mi.; perigee: 367 mi.; period: 100.5 min.) instead of 400-mi. circular orbit.   Cloud-cover pictures transmitted to tracking station at Wallops Station on early orbits were of excellent quality.   TIROS V was expected to chart the origin, formation, and movement of hurricanes, typhoons, and other storms during the August-September peak tropical storm period.

- In speech to the National Rocket Club in Washington, George Low, NASA's Director of Spacecraft and Flight Missions, outlined the three methods of manned lunar missions (direct, earth-orbit rendezvous, and lunar-orbit rendezvous) and indicated that a NASA decision would shortly be forthcoming as to which would be used in the Apollo program.

- Mercury Astronaut Virgil I. (Gus) Grissom was presented the General Thomas D. White Space Trophy by Secretary of the Air Force Eugene M. Zuckert, at the National Geographical Society in Washington.

    In his presentation remarks, Secretary Zuckert said: ". . . The National Aeronautics and Space Administration carries responsibility for the development of space technology and, in company with like-minded peoples of the world, the utilization of space for the benefit of men. Captain Grissom and the other astronauts are playing a vital role in the national space effort. NASA's program is of vital importance to the Air Force. We must build the nation's defense capability in space on the foundation which NASA lays . . .

"The feat for which he [Capt. Grissom] is recognized," said Secretary Zuckert, "is one step, a very, very early step, toward mastery of space for constructive purposes. The American people have launched a tremendous program to acquire the capability to master the aerospace just as we have mastered the seas and the air, and kept them free . . ."

June 19: USAF brought down a research balloon after a record 19-day flight 9,300 miles from Bermuda to near Iwo Jima. Launched at Kindley AFB, Bermuda, on May 31, the balloon carried instruments measuring wind direction and velocity, humidity, pressure, and temperature. One of a series of high-altitude flights, the 34-ft.-diameter balloon flew at a constant altitude of 68,000 ft.

* OSCAR II (Orbital Satellite Carrying Amateur Radio) re-entered earth's atmosphere and disintegrated, after 295 orbits since launch June 1.

* AFSC announced development of "self-maneuvering unit" (SMU), a 160-lb. pack which converts a pressure-suited spaceman into a one-man space vehicle. Designed to position and stabilize an astronaut or crewman outside a spacecraft and to give him mobility in an orbiting assembly area, SMU contains its own stabilization, control, propulsion, and power systems.

* Position of Special Assistant for Astronaut Affairs created by NASA under the Director, Manned Spacecraft Center. Temporarily filled by Ford Eastman, the position provides central coordination of commitments to extra-program activities of the astronauts.

June 20: Aerobee 150A launched from Wallops Station with 271-lb. payload boosted to 97-mile altitude; carried camera to study behavior of liquid hydrogen under conditions of symmetrical heating and zero gravity. Lewis Research Center payload was recovered 25 minutes after liftoff.

* Legal Subcommittee of the 28-nation United Nations Committee on the Peaceful Uses of Outer Space concluded its first session of three weeks' duration with a factual report. The report indicated that "no agreement has been reached on any proposals" but that a "most useful exchange of views" had been held.

* Thor missile with nuclear warhead launched for high-altitude nuclear test (Operation Dominic), destroyed over Johnston Island two minutes after launch.

* Army Corps of Engineers issued a call for bids to modify Launch Complex 19 at Cape Canaveral to accept the Gemini manned spacecraft. The Corps estimated that work would cost about $1.5 million and bids would be opened August 16.

* Radarastronomy observations of the moon taken by Millstone radar, MIT Lincoln Laboratory, revealed unusually strong echo from region of crater Tycho. G. H. Pettengill and J. C. Henry, who reported on experiments in *Journal of Geophysical Research*, concluded the enhanced radar reflectivity may be explained by un-

usually rough crater floor and by crater walls consisting of material characteristic of lunar subsurface.

June 21: X–15 No. 3 was announced to have been flown by Maj. Robert M. White (USAF) to 250,000 ft. (47.3 mi.) altitude, a new unofficial world record for manned aircraft (later revised to 247,000 ft. or 46.8 miles). Goal of the flight, which attained speed of 3,682 mph (mach 4.99), was to check the adaptive control system at design specifications, and Maj. White said the plane performed as expected.

- NASA announced that Dr. George L. Simpson, Jr., University of North Carolina sociologist, had been named Assistant Administrator for Public Affairs, effective September 1. Dr. Simpson had been Executive Director of the Research Triangle Committee of North Carolina, a regional development organization, since 1956. As a representative of the social sciences, he would bring them into working relationship to NASA's physical science effort, and underline NASA's awareness of the social and economic impact of the space program. Dr. Simpson succeeds Dr. Hiden T. Cox who resumed his post as Executive Director of the American Institute of Biological Sciences on July 1.

- Neil A. Armstrong, NASA research pilot, awarded the 1962 Octave Chanute Award by the Institute of the Aerospace Sciences (IAS), as the pilot who had contributed most to the aerospace sciences during the year. He was cited for "outstanding contributions in both engineering and piloting capacity in the development of an experimental adaptive control system and in the flight testing of that system in the X–15, No. 3 aircraft."

June 22: NASA's Manned Space Flight Management Council held meeting at NASA Headquarters.

- Plans for first live transatlantic TV programs in mid-July announced. To be produced by 3 U.S. networks and the 16-nation European Broadcasting Union (EBU), the programs would be transmitted by AT&T's Telstar satellite.

- Last USAF B–52 strategic bomber, produced by the Boeing Company, left final assembly area at Wichita, Kansas. The B–52H model was the 744th in the series since the first B–52 came from Boeing plant on November 29, 1951. With numerous improvements throughout its production, B–52 had eight versions—from B–52A to B–52H (speed of 650 mph and unrefueled range of 12,500 miles). Models G and H were capable of carrying Hound Dog 500-mile-range missile and Skybolt 1,000-mile-range missile.

- Jacqueline Auriol set a new women's international air speed record for 100 kilometers (62.10 miles) at Istres, France, flying a French-built Mirage III jet at 1,149.23 mph. Previous record, held by American aviatrix Jacqueline Cochran, was 783.02 mph.

- NASA Group Achievement Award presented to staff of Wallops Station by Thomas F. Dixon, NASA Deputy Associate Administrator, in ceremony at Wallops. Award scroll signed by Administrator James E. Webb cited the 400 employees for:

"Service of the highest order in providing rescue services, communications, transportation, food, lodging and medical care to the people of Chincoteague, Virginia, during the storms of March 6 to 9, 1962;

"For alertness and attention to their obligations in carrying out the mission of the Wallops Station despite emergency conditions;

"For courage and fidelity in protecting and restoring facilities and equipment of Wallops Station to operable condition with such speed and efficiency as to permit scheduled launchings to be met."

June 23: Langley AFB celebrated its 45th anniversary with the TAC Thunderbird team, a band concert, displays of aircraft and missiles, and displays of NASA's Langley Research Center highlighting the "open house" day.

- USAF launched from Vandenberg AFB, Calif., a satellite employing a Thor-Agena B booster combination. NASA *Satellite Situation Report* indicated this satellite attained orbit and decayed on July 7, 1962.

- *Pravda* announced opening of exhibition "in honor of the designer of the U–2." No relation to U.S. high-altitude Lockheed airplane, Soviet U–2 was a trainer plane designed in 1927 by Nikolai N. Polikarpov.

- Virginia State Exchange Clubs awarded Distinguished Virginian Award to William J. O'Sullivan, head of the Space Vehicle Group at Langley Research Center. Representative Thomas N. Downing, member of House Committee on Science and Astronautics, made the presentation to O'Sullivan in recognition of his concept, design, and development work with the Echo satellite.

June 24: OSO I began transmitting real-time data on solar observation after five weeks of intermittent transmittal. NASA scientists speculated that the cause of the reduction in the satellite's rate of spin from 50 rpm to 42 rpm, which caused the transmitter to restart, was either a result of bearing friction within the turning wheel or the slowing down of the satellite's rotation because of its position with respect to the Earth's magnetic field.

June 25: NASA announced appointment of Walter L. Lingle, Jr., as Special Assistant to the Administrator and, temporarily, as Deputy Assistant Administrator for Public Affairs. Formerly executive vice president of Proctor and Gamble, Mr. Lingle served three months as Deputy Administrator of AID just prior to NASA appointment.

- NASA Hq. requested MSFC to redirect the Centaur program toward primary mission of Surveyor lunar landing.

- USAF Office of Aerospace Research announced establishment of a basic research laboratory at the Air Force Academy at Colorado Springs, to provide in-house OAR research capability and enable teaching faculty and honor students to conduct research.

June 26: USAF 564th Strategic Missile Squadron of SAC successfully launched an Atlas ICBM from Vandenberg AFB.

- USAF designated Dyna Soar manned space glider as the X–20.

- In speech to the ARS Capital Section, Rear Adm. J. P. Monroe (USN) stated that the Navy had an interest in space exploration and operations as essential as its need for aircraft was in the past. He pointed out that DOD's responsibility for developing a military capability in space was not inconsistent with the nation's peaceful objectives.

June 26: Dr. Frank E. Bothwell, Chief Scientist of the Center of Naval Analysis, Franklin Institute, received the Navy's highest civilian award, the Distinguished Civilian Service Award, for his work on design and development of the Polaris ballistic missile.

June 27: X-15 No. 1 flown to unprogramed but record speed of 4,159 mph (mach 6.09) by NASA's Joseph A. Walker at Edwards, Calif. Record speed was achieved at 96,000 ft. on climb to an altitude of 120,000 ft. In test of X-15 re-entry stability, Walker glided in at 23-degree angle of attack, highest angle flown to date, and placed X-15 into 80-degree bank.

- NASA announced that flight of X-15 No. 3 on June 21 by Major Robert White (USAF) reached an altitude of 247,000 feet (46.8 mi.) instead of the 250,000 feet as first announced.

- Thor-Agena D booster launched USAF satellite from Vandenberg AFB, Calif.

- NASA Director of Manned Space Flight, D. Brainerd Holmes, announced that Mercury-Atlas 8 manned flight would be programed for as many as six orbits late this summer. Astronaut Walter M. Schirra (Cdr. USN) was announced as prime pilot, with Astronaut L. Gordon Cooper (Major USAF) as backup pilot. Capsule No. 16 was scheduled as the mission spacecraft.

    Holmes said: "We believe that another three-orbit mission will increase considerably our growing knowledge of space flight. Anything more than three orbits should be considered a bonus." A four-orbit mission would bring the spacecraft down about 200 miles east of Midway Island in the Pacific, while a five or six orbit flight would cause it to land 300 miles northeast of Midway.

- "DOD-NASA Guide: PERT Cost System Design," a joint set of management principles for common use by NASA and DOD contractors, was issued. The basic PERT (program evaluation and review technique) was first introduced for government-industry use in 1958 and has been applied largely to evaluation of schedules and time-related problems with wide success. Since NASA and DOD use virtually the same industrial base, adoption of a single approach as contained in the NASA-DOD Guide will minimize differences in application of PERT techniques to cost applications.

- NASA named Charles H. Zimmerman as Director of Aeronautical Research in the Office of Advanced Research and Technology. Zimmerman, who had joined NACA in 1929 and served as associate head of the Aerospace Mechanics Division of Langley Research Center since August 1959, succeeded John Stack who retired last month. Zimmerman would be responsible for the expanding NASA aeronautical program, which supports and conducts research for national goals in military and commercial aircraft, including V/STOL aircraft, supersonic transports, long-range flight at hypersonic speeds, technical support of the USAF X-20 (Dyna Soar) and full responsibility for the X-15 research airplane program.

- Dr. Eugene B. Konecci appointed NASA's Director of Biotechnology and Human Research in the Office of Advanced Research and Technology. Dr. Konecci will be responsible for directing research and development of future life support systems, advanced systems to protect man in the space environment, and research to assure man's performance capability in space.

In announcing the appointment, Associate Administrator Seamans said: "Our success and progress in manned space flight in the next 10–20 years depend on the human research we do today. The human, man-machine and man-system requirements must be determined through research prior to the design of any manned system. The human capabilities and limitations will directly influence various subsystems of the space vehicle. It is therefore important that work in Biotechnology and Human Research be conducted at an accelerated rate, in order to have the necessary answers for the design of future aero-space vehicles."

The Life Sciences Research Group at Ames Research Center will have a major role in carrying out the programs of Dr. Konecci's office, along with the other NASA centers and full utilization of the Nation's military, research, and industrial resources and personnel.

Dr. Konecci was Chief of the Life Sciences Section of the Missiles and Space Systems Division of Douglas Aircraft Co. and had previously served as a research scientist at the USAF School of Aviation Medicine and as Chief of Physiology and Technology in the USAF Directorate of Flight Safety before 1957.

June 28: DOD announced successful static firing of the Thiokol-developed acceleration rocket for USAF's X–20 manned space glider. X–20 previously known as Dyna Soar.

June 29: X–15 No. 2 was flown by NASA pilot John B. McKay in a relatively low-level flight to obtain information on aerodynamic heating. Maximum speed was 3,204 mph (mach 4.85) at 73,000 ft.; maximum altitude was 82,000 ft.

- First USAF Minuteman ICBM launched by military crew from Cape Canaveral, successfully reaching intended target area downrange.

- Navy fired a Polaris missile from AMR and sent it 1,400 miles downrange. The missile featured a new nose cone in shape and substances of construction as well as parts that would be used in more advanced models.

- Two Aerobee 150A sounding rockets with Johns Hopkins spectrophotometric instrumentation for measurement of day and night airglow successfully launched from NASA Wallops Station, reaching 129- and 131-mile altitudes.

- USAF awarded a contract for two high-speed X–19 VTOL aircraft which would be used to explore VTOL characteristics on behalf of the triservice V/STOL Transport Development Program. The X–19 is a twin-engine, tandem, high-wing aircraft with four tilting propellers mounted in nacelles at the wingtips, and is designed to cover a range of speed from 0 (hovering) to 400 knots in conventional flight.

June 30: University of California's sixty-inch cyclotron, basic research tool that led to construction of the first atomic and hydrogen bombs in the U.S., was closed down and dismantled to make room for more modern equipment. Invented and first operated by Ernest O. Lawrence in 1939, this cyclotron was instrumental in the discovery of neptunium, plutonium, astatine, curium, berklium, californium, and mendeleevium. In addition to these seven transuranium elements, the machine also contributed to the discovery of carbon-14 and produced radioisotopes for a variety of medical experiments.

June 30: U.S.S.R. announced launching of COSMOS VI into earth orbit (apogee: 250 mi.; perigee: 126 mi.; period: 90.6 min.; inclination: 49°). As with other unmanned satellites in the Cosmos series which began on March 16, 1962, COSMOS VI was stated to be a scientific satellite instrumented to explore radiation and other hazards to manned space flight.

- AEC and Columbia University announced the discovery of the existence in nature of two different type of neutrinos (the smallest atomic particle), one connected with mu-mesons, the other with electrons. Neutrinos were first proposed to explain a loss of energy not otherwise accounted for.

- By this date, 57 nations had joined the U.S. to support the development of peaceful uses of outer space, uniting with NASA in joint-flight, flight-support, or training programs.

- An estimated 23,000 visitors toured Marshall Space Flight Center during the "Family Day" observance of MSFC's second birthday.

- Article in Soviet periodical *Ekcnomicheskaya Gazeta* stated U.S.S.R. had 84 All-Union and centralized professional organs, 94 central offices of technical information at the Councils of the National Economy, more than 4,000 offices of technical information, 3,000 "homes" and "rooms" of technology at industrial enterprises, and more than 16,000 scientific and technical libraries. More than 60,000 people were employed in this network.

- During Fiscal Year 1962, ending this date, North American Aviation, Inc., received highest dollar-value contracts from NASA—$199.1 million worth of contract awards. Other top contractors were McDonnell Aircraft Corp., with $68.4 million; Douglas Aircraft Co., $68.3 million; and Aerojet-General Corp., $66.3 million.

During June: FAA proposed a regulation to restrict firings of the 5,000 amateur rocket clubs in the U.S. Beyond firing safety precautions, FAA is concerned with airlane safety. Reportedly some amateur rockets weigh up to 75 lbs. and can reach an altitude of over five miles.

- Series of full-scale wind-tunnel tests of VTOL fan-in-wing model vehicle at NASA Ames Research Center for Army lift-fan flight research program. Part of General Electric (Cincinnati) research contract, model and test hardware were fabricated by NASA, while GE fabricated and tested propulsion systems including two lift fans, two diverter valves, and two YJ85 engines. Wind-tunnel tests proved acceptability of inlet temperature, control characteristics, transition capability, and aircraft stability.

- Analysis of six photographic emulsion blocks carried aloft by DISCOVERER XVIII on December 7, and recovered on December 10, 1960, revealed that inner Van Allen radiation belt dips as low as 180 miles. According to Herman Yagoda of the USAF Cambridge Research Lab., the production of star tracks and "enders" by energetic protons of the inner belt increased exponentially with altitude from 180 up to 340 miles. Thus manned space operations would probably be restricted to lower latitudes rather than high inclination orbits because it is not feasible to shield against the energetic protons of the inner belt with present payload limitations.

During June: NASA reportedly called a "tempest in a teapot" the U.S.S.R. criticism of U.S. scientists for not crediting Soviets with discovery of the maximum electron flux in the earth's outer radiation belt. U.S. scientists freely credit U.S.S.R. for determining experimentally that maximum flux was 10^{-8} particles per square centimeter per second, but Soviet conclusion was based on unconfirmed assumption which was not authenticated until U.S. scientists analyzed definitive measurements from EXPLORER XII.

- United Technology Corp. experimented with hypergolic ignition system, spraying highly reactive fluid into nozzle end of solid-propellant rocket motor. Fluid immediately ignited the motor, which developed more than 100,000-lbs. thrust in the test.

- Bell Aerosystems engineers, John N. Cord and Leonard M. Seale, recommended in paper to the Institute of the Aerospace Sciences meeting in Los Angeles that a U.S. astronaut be landed upon the moon with the means of survival, perhaps for as long as three years, until an Apollo vehicle could be sent to return him to earth. The astronaut could do valuable scientific work and would be on "a very hazardous mission but it would be cheaper, faster, and perhaps the only way to beat Russia."

- Air Force Office of Scientific Research (AFOSR) began negotiating a new type of three-year contracts for basic research with Westinghouse Electric, IBM, and RCA in the solid-state field. This was the first time that basic research had been contracted on any other basis than one year at a time, is designed to provide more continuity and certainty of completion of research projects. As under the previous system, OSR and industry each bear roughly half of the costs of the research.

- An astronautics ship together with a floating drydock has been proposed as a launching platform for ICBM's and large space boosters, including Saturn. LCdr. Burton Edelson (USN), writing for the Bureau of Ships *Journal*, contended that a 10,000-ton ship, such as the seaplane tender *Curtiss*, could be specially outfitted as an integrated astronautics ship and could handle all solid-fuel launchings, including Minuteman. An accompanying floating drydock could handle large liquid-fuel rocket launchings up to and including Saturn.

- New camera developed for space use by the National Bureau of Standards can be operated at a rate of more than one million frames per second. Camera was developed to record luminous phenomena of shock waves in the upper atmosphere as they pass through the so-called "plasma" (highly ionized gases). Such plasma, said the report, has been produced at the Bureau of Standards by sending high-velocity shock waves through helium gas.

- NASA Wallops Station awarded contracts over $1 million during the month for essential services and construction work.

- Cost Reduction Task Force of the National Security Industrial Association concluded that DOD was paying more for its procurement because of "cost-oriented" practices. Undertaken at the request of the Secretary of Defense, the released report recommended that fixed-price contracts be used more often and earlier in the procurement cycle, and that price analysis rather than cost analysis be emphasized in buying.

During June: Twentieth Anniversary of Army aviation was highlighted by a two-week celebration in St. Louis, Mo.

- GE's J-93 turbojet engine, being developed for the B-70 aircraft, completed its preliminary flight rating test (PFRT) on schedule. The engine will propel B-70 to speeds greater than 2,000 mph and altitudes higher than 65,000 ft.

JULY

July 1: oso I was transmitting continuous signals and 20% of real-time data was being acquired from each 95-min. orbit. During 11 weeks of near-perfect operation from launch on March 7 to May 22, oso I transmitted 1,000 hours of scientific information. Before oso I, less than an hour of solar phenomena data had been collected above the earth's atmosphere by all previous rocket-flight observations. oso I had begun transmitting again on June 24, 1962.

- State Department announced plan to incorporate its office of the Science Advisor into the main policy-making elements within the department. Based on recommendations by scientist Lloyd V. Berkner in his report "Science and Foreign Relations," the reorganization would make science a more important factor in the formulation and execution of U.S. foreign policy.

- DOD opened Defense Research Office in Rio de Janeiro, Brazil, to coordinate basic scientific research throughout Latin America. Col. Leonard M. Orman (USA) headed the office and the Army Element, and Lt. Col. Charles J. Lyness (USAF) headed the Air Force Element. Also established, near the Defense Research Office, were NSF and NIH offices. U.S. Department of State would coordinate all the offices' activities.

- About 350,000 undergraduates and graduate students attended college through the benefits of the National Defense Education Act during its 4-year program, Secretary of Health, Education, and Welfare Abraham A. Ribicoff announced.

July 2: NASA signed three letter contracts with NAA's Rocketdyne Division for further development and production of the F-1 and J-2 rocket engines. The contracts provided: (1) $1 million for long lead-time items in F-1 engine R&D; (2) $3.4 million for early production effort on 55 F-1 engines; and (3) $1.7 million for early production work on 59 J-2 engines. Ultimate value of the final contracts, extending through 1965, would be about $289 million.

- House Committee on Science and Astronautics issued report based on NASA's authorization hearings and two days of specific testimony on the Centaur program.

- NAA announced that the 1961 Robert J. Collier Trophy, U.S. aviation's highest honor, was awarded to four test pilots of the X-15 rocket research airplane: Scott Crossfield (NAA), Joseph Walker (NASA), Major Robert White (USAF), and Cdr. Forrest Petersen (USN). Co-sponsors of the annual award, National Aeronautic Association and *Look* magazine, said the Collier Trophy was awarded to the pilots "for invaluable technological contributions to the advancement of flight and for great skill and courage as test pilots of the X-15."

- House of Representatives passed a bill (H.R. 9485) to amend the Space Act of 1958 to designate March 16 as National Goddard Day, commemorating the date in 1926 when Dr. Robert H.

114

Goddard launched the first successful liquid-fuel rocket. The bill was then referred to the Senate Committee on Aeronautical and Space Sciences.

July 2: Lt. Col. John A. Powers, addressing the Texas Associated Press Broadcasters Association, said that the next manned space flight (MA–8) might be as many as seven orbits. "We'll go for the full seven if all systems work perfectly," the Project Mercury spokesman stated.

• Reported that the Air Force Logistics Command (AFLC) civilian retraining program begun in October 1957 has proved to be highly successful. According to AFLC program director Eric W. Jordan, 80,000 civilian employees, whose skills have become obsolete in today's rapidly changing technology, will have been retrained and placed in new jobs by the end of 1962.

July 2–5: Spacecraft FRIENDSHIP 7 was publicly displayed in Rangoon, Burma; it left Rangoon on July 5 for Thailand.

July 3: Stratoscope II balloon failed to reach planned altitude and ejected dummy payload prematurely. Sponsored by NASA, National Science Foundation (NSF), and the Office of Naval Research (ONR), the 77-story-high balloon was on its final test launching for use as a carrier of a 36-in.-lens telescope to an 80,000-ft. altitude to photograph stars and the planets.

• ARIEL I's discovery of a new ion belt, at an altitude of 450 to 500 miles, was announced at the International Conference on the Ionosphere, London, by Prof. James Sayers of Birmingham University. Previous measurements had led physicists to believe that the ionization levels declined gradually above 200 miles.

• NASA Administrator James E. Webb announced establishment of NASA's Northeastern Operations Office, to be located in the Boston, Massachusetts, area. The office will coordinate NASA's liaison with university and business contractors in the northeast United States.

• Reported that the U.S. would send a Mariner B spacecraft to Mars in late 1964, to land on the red planet in early 1965. The launching of this heavily instrumented probe would mark the beginning of the scientific search for extraterrestrial life.

July 4: American Meteorological Society issued statement on the implications of the control of weather and climate, which endorsed the resolution adopted in the United Nations on December 20, 1961. U.N. resolution called for a cooperative international effort to improve weather prediction and to explore the possibilities of weather control and climate modification.

July 5: H–43B Huskie helicopter piloted by Capt. Chester R. Radcliffe, Jr. (USAF), set world's distance record in a 900-mi. flight from Hill AFB, near Salt Lake City, to Springfield, Minnesota. Previous world's record of 761.027 miles had been set by U.S.S.R. helicopter in September 1960.

• Poet Robert Frost, in speech to the students of Middlebury College's Bread Loaf School of English in Vermont, said: "I've been thinking in a scientific way lately. Someone in Washington wants me to do something about glorifying America—something about the space age.

"The glory of it is that we're making Promethean defiance against the unknowable—space. That's glorious—and that's as far as I go.

"I don't know what will come of it.   Maybe better communi-cations between the stock markets of the world."

July 6: Civil Service Commission ruled that NASA was authorized to hire an "aerospace engineer and pilot," one whom NASA would designate an astronaut, at maximum salaries for Government service grades GS 13, 14, and 15.

- Last of 4 solid-fuel Arcas rockets fired at White Sands in NASA Langley Research Center photographic study of deployment characteristics of 15-ft.-diameter parachute for use in measuring atmospheric winds at altitudes from 100,000 to 240,000 ft. Other 40-mi.-high Arcas shots were on June 30, July 2, and July 5, 1962.

July 7: U.S. detonated a low-yield nuclear device in the first aerial test within continental U.S. since 1958.   AEC and DOD said the purpose of the explosion, which occurred a few feet above ground at the Nevada test site, was to test the effects of nuclear wea-pons.

- Bell Telephone Laboratories (BTL) announced that Telstar, first privately owned communications satellite, would be launched July 10 from Cape Canaveral.   BTL built the satellite, purchased the Thor-Delta launch vehicle from NASA, and would reimburse NASA for launching and tracking costs.

- Ryan project engineer Willis F. Everest was slightly injured in crash of Ryan Flexwing aircraft undergoing flight-test research at Langley Research Center (LARC).   40-ft.-wingspan aircraft based on the Rogallo concept had completed aerodynamic tests in Langley Research Center's full-scale wind tunnel and was to be flight tested by LARC and the Army Transportation Research and Engineering Command.

- DOD released a report indicating that underground nuclear explosions were more easily detected than had been known previously.   This new knowledge was based on seismic studies of current U.S. underground tests in Nevada.

- Soviet E-166 aircraft was flown an average speed of near 1,660 mph in a two-way flight at 47,000 ft. over a 15-to-25-km. course near Moscow, by Lt. Col. Georgy Mosolov according to Tass. If confirmed by FAI, the flight would break record set in Nov-ember 1961 by McDonnell F4H of 1,606.342 mph flown by Lt. Col. Robert G. Robinson (USMC).

July 8: Megaton-plus hydrogen device exploded at more than 200-mi. altitude, lofted by carrier Thor rocket from Johnston Island in the Pacific as a part of Operation Dominic.   It was the highest thermonuclear blast ever achieved and lighted up the Pacific sky from Wake Island to New Zealand, causing some communi-cation disruption but less than predicted.   Success followed two previous attempts on June 4 and June 19.   It was expected that TRAAC satellite launched in November 1961 would be in a position to measure the influence of the blast upon the Van Allen belt.

July 9: TIROS V stopped transmitting pictures from the Tegea-lens, medium-angle camera.   The Tegea camera system transmitted 4,701 pictures of which 70% were considered excellent quality. The wide-angle Elgeet-lens camera, which is still functioning, had transmitted 5,100 pictures to date, some of which aided in the analysis of Typhoon Joan over the Western Pacific.

July 9: NASA scientists concluded that the layer of haze reported by Astronauts Glenn and Carpenter was the little-understood phenomenon called "airglow." Using a photometer and other instruments, Carpenter was able to measure the layer as being 2-degrees wide. Airglow accounts for much of the light of the night sky.

- MIT's Lincoln Laboratory announced development of the gallium arsenide diode, capable of generating light at wave lengths in the near-infrared region. Use of this diode was expected to be useful in closed-circuit television and in communication with re-entering spacecraft, since the infrared beam may be able to penetrate the ionized plasma sheath built up around a spacecraft as it re-enters the earth's atmosphere. Development was reported by R. J. Keyes and T. M. Quist of the Lincoln Laboratory's applied physics group.

- General Thomas D. White, former USAF Chief of Staff, wrote in *Newsweek* article: "There are military requirements in space which this nation can fail to fulfill at its grave peril. . . . I wish we would move faster on the satellite inspector and interception. We soon may need to verify what the Soviets have put into space; we may someday want to shoot it down. . . . Another intriguing concept . . . ought to have more steam and money behind it. This is the true 'space plane.' "

July 9–14: Communist-led World Peace Congress opened in Moscow with more than 2,000 delegates from 101 countries attending. Purpose of the Congress was to line up as much world opinion as possible behind U.S.S.R. foreign policy, particularly in the area of disarmament.

July 10: TELSTAR, the first privately financed satellite, launched into orbit from AMR by NASA Delta booster. TELSTAR (apogee: 3,503 mi.; perigee: 593 mi.; inclination: 44.79 degrees; period: 157.8 min.) was funded by the American Telephone and Telegraph Co. (AT&T) under a NASA-AT&T agreement of July 27, 1961: Bell Telephone Laboratories design and build satellites at own expense; AT&T reimburse NASA for Delta launch vehicles, launch, and tracking services (approximately $3 million per launch); Bell System conduct the communications experiments and NASA provide telemetry; and both NASA and AT&T analyze data and results, to be made available by NASA to the world scientific community.

- First commercial transmission of live TV via satellite and first transatlantic TV transmission, when TELSTAR experimental communications satellite of AT&T demonstrated vast new capabilities. Pictures were telecast from AT&T center near Andover, Me., to TELSTAR, then received and placed on all three major TV networks in the U.S. TV signals also were relayed from Andover, Me., to TELSTAR, and then relayed to French antenna at Pleumeur-Bodou on the Brittany peninsula and the British station at Goonhilly, Cornwall.

In American relay experiment via TELSTAR, AT&T Board Chairman Fred Kappel in Maine called Vice President Johnson in Washington.

In the first successful transatlantic TV transmission, picture of waving American flag was transmitted via TELSTAR to both

France and Britain, while transmission was also picked up at both Andover and Holmedel, N.J., and relayed to American TV networks.

First voice transmission from space was the Christmas message of President Eisenhower which was broadcast from USAF Project Score satellite on December 18, 1960. ECHO I, passive communications satellite, provided a reflector for a host of communication experiments after its launch on August 12, 1960. The moon also had been used as a reflector in communications experiments, the first being a radar reflection on January 11, 1946, by the Army Signal Corps.

July 10: President Kennedy nominated Dr. Jerome B. Wiesner to be director of the Office of Science and Technology.

- Launching of TELSTAR marked tenth straight successful flight of the 3-stage Delta rocket. The history of Delta goes back to the Thor-Able and the earlier Vanguard, from which it acquired its upper stages. Originally designed as an interim booster when NASA ordered twelve Deltas from Douglas Aircraft in April 1959, it achieved what NASA Administrator Webb called "the greatest level of reliability of any of our launch vehicles . . . ." The following satellites were orbited by Delta boosters: ECHO I, TIROS II, III, IV, and V; EXPLORERS X and XII; OSO I; ARIEL I; and TELSTAR.

- NASA Langley Research Center (LARC) announced plans to build a cyclotron laboratory on 1,300 acres of USAF-owned land, a former Bomarc missile site, halfway between LARC and Williamsburg, Va. The laboratory will include two accelerators to generate electrons and protons of the energies spacecraft will encounter on deep space flights.

- A U.S.S.R. Tu–114, world's largest commercial airplane, left Moscow on a survey flight for direct air route to Havana, Cuba. When begun, the Moscow-Havana commercial service would mark the first regularly scheduled flights of Soviet aircraft into the Western hemisphere.

July 11: NASA press conference explained basic decision on Apollo manned lunar exploration program: to base the next phase of its planning, research and development, procurement, and space flight program on the Saturn (C–5) to accomplish the initial manned lunar landing and recovery, using the lunar orbit rendezvous as the prime mission mode. Based on more than a year of intensive study, the basic decision on the lunar orbit rendezvous (LOR) mode enables immediate industrial consideration of the lunar excursion vehicle and firm planning on early employment of two-stage Saturn (C–1B) to test-flight the Apollo configuration in earth orbits.

It was also announced: (1) an in-depth study of an unmanned lunar logistic vehicle would be undertaken; (2) continued feasibility studies would be made of the earth-orbit rendezvous mode using Saturn (C–5) with a two-man spacecraft, with the possibilities of a direct flight with this spacecraft; (3) continued study would be made of the Nova vehicle (two to three times capability of C–5).

NASA Administrator Webb said: "We are putting major emphasis on lunar orbit rendezvous because a year of intensive study indicates that it is most desirable, from the standpoint of time, cost and mission accomplishment. In reaching this decision, however, we have acted to retain the degree of flexibility vital to a research and development program of this magnitude . . .''

LOR was unanimously recommended by NASA's Manned Space Flight Management Council because it: (1) provides highest probability of mission success with essentially equal mission safety; (2) provides mission success months earlier than other modes; (3) will cost 10 to 15 per cent less than other mode; and (4) requires least amount of technical development beyond existing commitments.

July 11: Second successful test of USAF Titan II rocket, 5,000-mi. flight from AMR. Second generation ballistic missile with storable liquid fuels will be used as a booster in NASA's Project Gemini.

- First east-to-west transatlantic TV transmission, French station at Pleumeur-Bodou reflecting eight-minute telecast off 170-lb. TELSTAR satellite in 15th orbit to AT&T facility at Andover, Me. Program featured the French Communications Minister Jacques M. Marette and several entertainers.

  British Broadcasting Company announced that French transmission violated agreement with Britain and other European Broadcasting Union nations.

- White House Press Secretary, Pierre Salinger, said that successful performance of TELSTAR indicated that a full study of international communications involving all concerned governmental agencies would be initiated by the White House. He said that a preliminary working paper by Tedson J. Meyers, administrative assistant to FCC Chairman Newton N. Minow, was already under study by the State Department and other agencies.

- The Senate passed NASA authorization for FY 1963 (H.R. 11737) of $3,820,515,520. Amendments submitted by Senator William Proxmire to require competitive bidding on space projects and to establish a presidential commission to study the impact of the space program upon the nation's manpower were both defeated by lopsided votes. Senate authorization restored $116 million cut by the House.

- NASA announced that Mercury Astronaut Donald K. Slayton (Major USAF) would assume new operational and planning responsibilities. Extensive medical observation and examination of Slayton's irregular heart action (atrial fibrillation) by MSC medical staff, by groups of Air Force specialists, and Dr. Paul Dudley White agreed that he should not be considered for solo space flights.

- NASA awards of less than $1,000 for inventions were made to the following NASA employees: D. H. Buckley, R. L. Johnson, Earl W. Conrad, R. J. Weber, F. D. Kochendorfer, John C. Nettles, Glen E. McDonald, Gerald Morrell, L. A. Baldwin, V. A. Sandborn, C. J. Blaze, C. G. Richter, and Marcus F. Heidmann of Lewis Research Center; and David H. Schaefer of Goddard Space Flight Center.

- NASA's Director of Manned Space Flight, D. Brainerd Holmes, briefed the House Committee on Science and Astronautics on the lunar-orbit rendezvous (LOR) procedure which NASA announced as the prime mission mode for the initial manned lunar landing.

July 11: Record altitude flight attempt by X–15 canceled for a second time because of technical difficulties.

- USAF superpressure balloon brought down 1,200 miles northwest of Honolulu after remaining aloft for a record-breaking 30 days since launch by AFCRL from Kindley AFB, Bermuda. 34-ft.-diameter mylar research balloon had been under constant surveillance by FCC tracking and data network, and had maintained more than 66,000-ft. altitude day and night since June 10 with 50-lb. payload.

July 12: USAF Titan II flown 5,000 miles down AMR on third test flight.

- Leading editorial of *New York Times* said, "The tendency of man's technological progress to outdistance his institutions for coping with that progress is strikingly indicated by current developments in space. . . ." Pointing to the "brilliant initial successes" of TELSTAR, it stated: "The clear lesson of these technological accomplishments is that space is now a full-fledged area of human activity for a wide variety of purposes, and will increasingly be employed for men's ends in the years immediately ahead. Yet the cosmos today is a lawless dimension and there is no universal agreement even on so elementary a question as where space begins—no boundary line between the region in which existing national and international law holds sway and the region in which it does not. . . ."

    Editorial called for fulfillment of the responsibility resting on the space law committee of the United Nations' Committee on the Peaceful Uses of Outer Space since "the cosmos bears some resemblance to a jungle."

- Walter C. Williams, associate director of NASA Manned Spacecraft Center, defended orbital flights by astronauts that apparently are duplicates of previous flights, saying that the only way the U.S. can have a space program in depth is to repeat some of the operations. He stated that new data were gathered from each flight, replying to critics of the second manned 3-orbit spaceflight mission (MA–7).

- Statement issued by the European Broadcasting Union declared that France had "contravened" an international agreement by transmitting the first television broadcast from Europe to the United States via TELSTAR. Under an agreement including the French, "no television material of entertainment or informative character" was to be broadcast to the U.S. before the joint "Eurovision" program scheduled for July 23. The appearance of Yves Montand and two other singers on the French transmission on July 10 was considered the case in point. In response, French officials said they had not transmitted a program but merely some experimental sequences.

- Reported that at NASA meeting of lunar scientists the view presented by astronomers, a geologist, and geophysicists that scientists can learn to become astronauts as well as an astronaut can make scientific observations. Which scientific discipline should be represented as a member of the 3-man Appollo crew did not receive a unanimous endorsement.

- Atlas ICBM launched by USAF squadron from Vandenberg AFB, two Polaris A–2 IRBM's successfully launched from submerged U.S.S. *John Marshall* off Cape Canaveral, and a USAF Minuteman exploded 50 seconds after launch from silo at Cape Canaveral.

July 13: Transatlantic telephone conversation via TELSTAR satellite, from AT&T President Eugene McNeely to Jacques Marette, French Communications Minister. On next orbit, McNeely spoke with Sir Ronald German, director-general of the British Post Office. Technicians had unofficially talked on transatlantic circuit via TELSTAR the day before, July 12.

- United Aerospace Workers and the International Association of Machinists served a notice of intention to strike on July 23 "unless an honorable settlement is reached by that time" against five major space companies—Lockheed, North American, Douglas, General Dynamics, and Aerojet. Fifty-one plants, test sites, and missile bases would be affected by walkout of 125,000 union members.

- Indian Government concluded agreement with the U.S.S.R. for the manufacture in India of jet aircraft engines (Soviet RD–9). Engines will be used to power Indian-designed HF–24 supersonic jet fighter now being developed at the state-owned Hindustan aircraft factory at Bangalore. Indian Government had previously planned to equip the HF–24 fighter with the British Orpheus engine.

- USAF Atlas E successfully test-fired from PMR to impact near Wake Island, the second successful firing of an Atlas E from Vandenberg AFB.

July 14: NASA launched large balloon from site near Goose Bay, Labrador, carrying bioscience payload including two rhesus monkeys and four hamsters for primary cosmic radiation experiments. Balloon was expected to reach maximum altitude of 128,000 feet and to float some 2,000 miles to a landing near Edmonton, Alberta, Canada.

- At Hyannis Port, Mass., President Kennedy issued a statement on disarmament calling for the Soviet Union to join in a "creative search for ways to end the arms race and to devote our common skills and resources to the enlargement of the peaceful opportunities to mankind. . . . In a nuclear age, all nations have a common interest in preserving their mutual security against the growing peril of the arms race. . ."

- Navy Hydrographic Office accepted delivery of modified Lockheed Super Constellation WV–2 to be used in Project Magnet, a worldwide survey for the improvement of sea and air navigation charts. Rear fuselage section was demagnetized and electrical circuits had been required to eliminate magnetic fields.

- Cleveland educators pointed out that the new transatlantic communications prompted by TELSTAR and other satellites would be a tremendous incentive to foreign language study. Supervisor of Foreign Languages Eugene K. Dawson said: "I can think of nothing more exciting for our students than to hear direct broadcasts in the language they are studying."

- Cosmonaut Yuri Gagarin was promoted from major to lt. colonel in the Soviet Air Force.

July 15: First NASA balloon carrying bioscience payload of two rhesus monkeys and four hamsters traveled faster and higher than planned after launch from Goose Bay, Laborador, requiring alternate plan for early jettisoning to effect daylight recovery. Payload was recovered 45 miles north of Prince Albert in Saskatchewan, all animals reported dead upon recovery. Conducted by Ames Research Center Life Science Laboratory, experiment was designed to check primary cosmic radiation.

July 15: Reported from Stockholm, Sweden, that NASA team had arrived to launch four Nike-Cajun rockets for cooperative investigation with the Swedish Committee for Space Research of high-altitude, bright night clouds. Rockets would be launched from Jokkmokk in northern Sweden.

• NASA announced that two technical notes propose new concept on the mechanics of solar heating of the upper atmosphere. Dr. Isadore Harris (TN–D–1443) and Dr. Wolfgang Priester (TN–D–1444) of Goddard Space Flight Center proposed that corpuscular radiation (i.e., solar wind) in association with hydromagnetic waves may be the energy source required to explain the results obtained from orbiting satellites. Harris' note provided for the first time a diurnal picture of the thermosphere (120 to 2,050 kilometers above the earth's surface). Priester's note postulated the physical properties scientists are likely to encounter during a complete 11-year solar cycle when upper atmospheric conditions change according to solar flare activity.

• Resumption of 17-nation Disarmament Committee session meetings in Geneva. Appeal of President Kennedy for Russians to "join in a creative search for ways of ending the arms race" was rejected in Tass statement published in Moscow—the U.S. was attempting to "conceal its rejection of general disarmament."

• NASA announced that nine selected college graduates had begun a year of intensive training in NASA's first Management Intern Program. After four training periods of three months (3 in Headquarters offices and 1 at field center), successful interns will be offered permanent NASA employment in administrative positions.

• Radio Corporation of America announced that its Radar Division had developed a new technique applying the doppler effect to monopulse radar, to allow for velocity measurements of a space vehicle up to 100 times more precise than by conventional methods. Developed for the Army Signal R&D Lab at Fort Monmouth, the new technique could be applied to radars already in service.

• Astronaut Virgil I. Grissom (Captain, USAF) promoted to Major, USAF.

• Pioneer flight surgeon, Dr. Bernard L. Jarman, who was appointed the first medical examiner of the Civil Aeronautics Agency in 1927, died in Washington.

July 16: X–15 No. 1 flown to 107,000 feet at speeds up to 3,733 mph by Joseph A. Walker, in series of seven "roller-coaster dips" in test of the alternate stability augmentation system.

• British transmitted first transatlantic color television during two orbits of TELSTAR, twice transmitting still photographs to Andover facility of AT&T.

• *Aviation Week* reported that NASA had delayed announcement of its lunar-orbit rendezvous (LOR) decision for a week to allow for its study by the President's Science Advisory Committee.

• In Cleveland ARS meeting, Associate NASA Administrator Robert Seamans stated that decision on the lunar-orbit rendezvous (LOR) would allow time to develop a larger Nova booster than previously contemplated.

July 16: Douglas Aircraft Co. reached separate agreement with two labor unions on a three-year contract, thus avoiding threatened strike of aerospace workers.

- In an interview with fourteen American newspaper editors, the text of which was released, Premier Nikita Khrushchev asserted that showing of documentary film on the Soviet antimissile missile to the Communist-sponsored Peace Conference last week had been overruled by those who felt its showing "might have been misunderstood" as a warlike gesture. "Had the people been shown this film," Khrushchev said, "they would have seen what kind of a machine it is. You can say our rocket can hit a fly in outer space."

July 17: X–15 No. 3 flown to record altitude of 58.7 miles (314,750 feet) by Major Robert White (USAF). Flight was programed for only 282,000 feet but maximum speed of 3,784 mph was 284 mph faster than planned. White's flight made him eligible for astronaut wings, reserved for those who have flown over 50 miles high, previously held only by four Mercury astronauts. Flight was also the first to achieve the original design altitude of the X–15 of 50 miles.

- NASA Ames Research Center reported that a number of beetles had survived the 25-mile-high balloon flight in which the monkey and hamster passengers in the bioscience payload perished because of a failure in the life-support system supplying oxygen and heat.

- In reviewing technical considerations leading to NASA's selection of the lunar orbital rendezvous mode (LOR) for Project Apollo, D. Brainerd Holmes, Director of Manned Space Flight, told ARS lunar conference in Cleveland:

  "The mission I have described has been widely reported in the newspapers and technical journals, with a liberal use of superlatives in assessing its magnitude and complexity. In this case, resort to superlatives is well-advised—this is truly a staggering undertaking. Entirely new concepts of component and system reliability must be developed and proven. Extensive tests must be carefully planned and conducted, and results must be exhaustively studied. Crew capabilities must be developed and meshed with proven automatic systems so that the two work together with Swiss-watch precision.

  "It is a challenging task, studded throughout with difficult decisions which must be soundly based and properly made. . . .

  "With the decision as to the method by which we will go to the moon we think we have taken a giant step forward. Essentially, we have now 'lifted off' and are on our way. . . ."

- In Cleveland, Dr. Joseph F. Shea, NASA's Deputy Director of Manned Space Flight for Systems, told ARS conclave that the manned lunar landing was greatly dependent upon information acquired by the unmanned Ranger and Surveyor payloads. The landing zone for Apollo flights (strip 20° wide along lunar equator) is about the size of Alaska. Maps of two-fifths-of-a-mile to an inch are required, which must be based on photographs which can identify five-or six-ft. objects for navigational fixes and initial exploration. Smallest objects now discernable in the photographs of the moon's surface by the test telescopes on earth are about the size of a football field.

July 17: In address to ARS lunar meeting in Cleveland, Dr. James A. Van Allen said that protons of the inner radiation belt could be a serious hazard for extended manned space flight, and that nuclear detonations might be capable of cleaning out these inner belt protons, perhaps for a prolonged period, for making manned orbits about 300-mi.-high above the earth. Van Allen said that the recent shot over Johnston Island was far too small and too low to have such an effect.

- Subcommittee of the House Committee on Science and Astronautics, headed by Rep. Victor Anfuso, opened hearings on the role of women in the space program. Jane B. Hart and Jerrie Cobb, representing the 13 women who completed unofficial space qualification tests at the Lovelace Foundation, testified that women had a real contribution to make and were qualified as astronauts. Jacqueline Cochran, well-known flier, said that "there is no doubt in the world that women will go into space."

- British astrophysicist, Zdenek Kopal, suggested in ARS lunar conference at Cleveland that the moon may have an abundant water supply, including geysers and under surface glaciers or permafrost.

- In remarks referring to the editorial in the NASA issue of *Aviation Week and Space Technology*, Representative George P. Miller, Chairman of the Committee on Science and Astronautics, said that "the national space program continues to fire the imagination of the American people. The spectacular space achievements are but the top of the iceberg. Underneath is a rapidly growing base of solid scientific knowledge which will benefit the Nation in thousands of different ways. The agency spearheading the massive scientific effort is the National Aeronautics and Space Administration . . ."

- In speech to the National Rocket Club, Senator Barry Goldwater said: "I am convinced that the American people fully endorse space preeminence for the United States. This is our long-range goal as a nation. . . . Space superiority in all its scientific, technological, and military aspects is fundamental to the future well-being, security, and prosperity of the United States . . .

  "As a new agency, NASA has picked up a momentum, a direction and a purposefulness directed toward long-range research and development which is one of impressive merit . . .

  "From all I can observe, the relationship between NASA and the Air Force in the space program has proceeded fairly well to date. But I would point out that the Department of Defense is responsible for military affairs, not NASA. NASA is not a military service, and its broad-based program will undoubtedly contribute greatly to our own national military requirements. . . . The point I am trying to make is that the requirements of the United States for military programs in space should not be neglected in any vital aspect . . ."

- France announced plans to establish rocket launching facility along the Atlantic coast near Mont-de-Marsan in southwestern France. Station will be ready by 1967, when France has agreed to evacuate the rocket research center at Colomb-Bechar in Algeria.

July 17: Wash. *Evening Star* reported that NASA's Director of Manned Space Flight, D. Brainerd Holmes, had announed in Cleveland that NASA X–15 pilot Neil A. Armstrong would become the first civilian astronaut as a part of the new selection of astronauts for Projects Gemini and Apollo.

July 18: NASA launched rigidized Echo-type balloon on Thor booster to 922 mi. in inflation test. Nicknamed "Big Shot," the 13-story balloon was inflated successfully and was visible for 10 minutes from Cape Canaveral. Movie film capsule parachuted into sea northeast of San Salvador was recovered by three pararescue men of the Air Rescue Service. This was the largest manmade object sent into space, the previous record being held by the 100-ft ECHO I.

- President John F. Kennedy presented the Robert J. Collier Trophy, aviation's most distinguished award, to four X–15 pilots: Major Robert M. White (USAF), A. Scott Crossfield (NAA), Joseph A. Walker (NASA), and Cdr. Forrest S. Petersen (USN). Following the White House ceremony, the pilots received personal miniatures of the Collier Trophy at a luncheon co-sponsored by the Aero Club of Washington and the National Aeronautic Association.

- In X–15 awards ceremony before distinguished guests and Headquarters personnel, at NASA Headquarters, tribute was paid to the team of governmental and industrial persons responsible for the contributions of the X–15 program to aeronautics and space flight. Associate Administrator Seamans reviewed the objectives and achievements of the X–15 program. NASA Outstanding Leadership Awards were presented to Paul F. Bikle (Director of the NASA Flight Research Center) and Hartley A. Soulé (Langley Research Center X–15 project manager) by Administrator Webb. After a short address on the pride of the nation concerning NASA's achievements, Vice President Lyndon B. Johnson (Chairman of the National Aeronautics and Space Council and a key figure in the Congressional enactment of the National Aeronautics and Space Act of 1958) awarded NASA Distinguished Service Medals to the X–15 pilots Forrest Petersen (Cdr., USN), Robert M. White (Major, USAF), and Joseph A. Walker (NASA).

- Astronauts John Glenn and Scott Carpenter, and George M. Low, NASA Director of Spacecraft and Flight Missions, testified before House Subcommittee on qualifications for astronauts. Low testified that qualifications would be raised rather than lowered for Project Apollo, that the sex of pilots had never been a requirement, but that if any resources were diverted for a woman in space effort "we would have to slow down on our national goal of landing a man on the moon in this decade." Astronaut Glenn said on women astronauts: "I couldn't care less who's over there [in the next seat] as long as it's the most qualified person. . . . I wouldn't oppose a women's astronaut training program; I just see no requirement for it."

  NASA had not found one woman to date who met all astronaut requirements: American citizenship; excellent physical condition; degree in physical or biological sciences or engineering; and experimental jet flight-test experience.

July 18: USAF launched unidentified satellite with an Atlas-Agena B booster from Point Arguello, Calif.

- DOD canceled construction of Navy's 600-ft.-diameter radiotelescope at Sugar Grove, West Virginia, because major and unforeseen scientific advances provided better methods to acquire information. Initiated in 1954 and costing $47 million to date, the Sugar Grove dish had as its primary purpose military research in ionospheric physics, space communications, navigation, and radio astronomy. Since then, parametric amplifiers and masers revolutionized space communications, while more inexpensive acquisition methods had been developed for ionospheric data. DOD decision did not affect National Science Foundation's 140-ft. steerable radiotelescope at nearby Green Bank, W. Va., which would be used to scientifically map radio and thermal sources with greater efficiency.
- Senate confirmed Dr. Jerome B. Wiesner as Director of the Office of Science and Technology.

July 19: NASA test pilot John B. McKay flew X–15 No. 2 in sustained heat test, reaching peak altitude of 84,500 feet and speed of 3,375 mph (mach 5.11) in flight which created an estimated temperature on the leading edge of the wings of 1,100° F for more than a minute, and 1,290° F for a few seconds when leaving and re-entering the atmosphere.

- First known antimissile missile interception of ICBM nose cone announced by DOD, the interception of an Atlas nose cone launched from Vandenberg AFB high over the Pacific by a 3-stage Nike-Zeus fired from Kwajalein. Interception of 16,000-mph nose cone by experimental Nike-Zeus was regarded as comparable to a bullet hitting a bullet, although DOD did not state whether physical contact had been made.
- Newsmen in London and New York exchanged news items and conversations in the first two-way transatlantic telephone connection via TELSTAR communications satellite. Reuters transmitted its first news report to the world press via satellite.
- In *Izvestia*, Soviet correspondent V. Matveyev said that the background of U.S. satellite TELSTAR showed that progress in the development of productive forces in the U.S. was being usurped by the modern capitalist states in the interests of the monopolies.
- In Pentagon ceremony, USAF Chief of Staff LeMay pinned astronaut wings on Major Robert M. White, the first man to fly a winged aircraft, the X–15, into space.

July 20: NASA Administrator Webb announced that the Mission Control Center for future manned space flights would be located at Manned Spacecraft Center (MSC), Houston, Tex. The Center, including its computer complex, communications center, flight simulation facility, and flight operations displays, would be operational by 1964 for Gemini rendezvous flights and later Apollo lunar missions.

- USAF launched an unidentified satellite employing a Thor-Agena B booster from Vandenberg AFB.
- U.S. Weather Bureau transmitted TIROS V photographs to Australia from Suitland, Md., the first time that Tiros photographs had been transmitted abroad for current weather analysis by a foreign country. Photographs were of cloud formations west of Australia.

July 20: USAF-NASA 30-member ad hoc committee established to plan a "joint hypersonic research program for the next couple of years." The committee would consider two primary missions for the Aerospace Plane (ASP), or hypersonic aircraft: (1) vehicle system to provide earth-to-orbit-and-return capability; and (2) earth aircraft capable of 5,000-mi. unrefueled flight. ASP would be follow-on program to X–15 research aircraft program.

July 21: NASA selected design for the Advanced Saturn launch complex northwest of Cape Canaveral, featuring a 2,500-ton crawler-mode vehicle recommended by the Launch Operation Center (LOC). Saturn C–5 launch vehicles (350 ft. high) would be erected and checked out vertically and then transported to the Complex 39, consisting of 4 launch pads about 9,000 ft. apart. Managed by LOC, Complex 39 construction supervision had been assigned to the Army Corps of Engineers.

- President Kennedy sent telegram to aerospace unions and industries stating that a strike would "substantially delay our vital missile and space programs and would be contrary to the national interest." The International Association of Machinists, one union involved in threatened strike of 150,000 workers in major aerospace companies, immediately accepted the President's request to postpone strike for 60 days.

- From Hyannis Port, President Kennedy announced that Dr. Robert R. Gilruth was one of the five Federal career officials who would receive the highest Government award, the 1962 President's Award for Distinguished Federal Civilian Service. Director of NASA Manned Spacecraft Center and of Project Mercury since its inception, Gilruth, the announcement said, "successfully carried out one of the most complex tasks ever presented to man in this country—the achievement of manned flight in orbit around the earth." The awards would be presented later at White House ceremony.

- Launch of NASA Venus probe (Mariner) postponed because of malfunction in command destruct system of the Atlas-Agena B.

- Reported that Premier Khrushchev had witnessed test firing of Soviet Polaris-type missile from submerged submarine in the Arctic Ocean, believed by Western observers to be the first such Soviet test.

- Soviet News Agency Tass announced that the U.S.S.R. was again resuming nuclear tests in the atmosphere, quoting an official Soviet Government statement, in part: "The explosions of the American nuclear bombs above Christmas and Johnston Islands have produced their echo—they have made Soviet nuclear tests in reply inevitable."

- Egyptian armed forces test-fired four single-stage rockets, which witness President Nasser said could hit any target in Israel. Two of the rockets, named "El Zahir" (Victorious), had a reputed range of 222 miles; the other two, named "El Kaher" (Conqueror), were claimed to have a range of 360 miles. Rockets were launched from a secret base 48 miles north of Cairo, and were built entirely by Egyptian technicians, according to President Nasser. After the firings, Cairo Radio announced: "The U.A.R. has entered the space age."

- NASA said by press to have "put alphabet in orbit" with abbreviated names such as WOO (Western Operations Office), PLOO (Pacific

Launch Operations Office), and recently NEOO (Northeastern Operations Office).

July 22: NASA Mariner Venus probe launched by Atlas-Agena B from AMR, the booster deviating from course at 212 seconds and commanded to be destroyed by the range safety officer of 290 seconds and nearly 100 miles high. Work immediately began to launch another Mariner spacecraft before the end of the 50-day Venus window on September 10, 1962, hopefully within several weeks if difficulty with the Atlas could be ascertained. Mariner had been planned to reach the vicinity of Venus about December 8 and to pierce the dense cloud layers hiding the surface of Venus from observation.

- Chairman of FCC, Newton N. Minow, said on "Meet the Press" TV program: "There are some optimists who think we can see the Japanese Olympics in 1964. I would personally think that is stretching it a bit, although the technology certainly will permit us to have a working communications satellite system within three or four years. This is the reason why we are are pushing to have legislation passed—it has passed the House and is now pending in the Senate—to determine national policy as to how the communications system will work. As always in this country technology is ahead of public policy. This time we want to keep our national legislative policy abreast of science."

- United Auto Workers (UAW) of North American and Ryan Aviation voted to delay walkout for 60 days as requested by the President, thus joining International Association of Machinists (IAM) which had previously accepted the 60-day delay in planned strike against General Dynamics (Convair) and Lockheed.

July 23: TELSTAR relayed two 20-minute live TV shows, the first formal exchange of programs across the Atlantic. The first U.S. program to the Eurovision network of stations in 18 nations included sequences on the Statue of Liberty, a major league baseball game in Chicago, President Kennedy's news conference, Astronaut Walter M. Schirra from Cape Canaveral, and the Mormon Tabernacle Choir from Mt. Rushmore. Three hours later on another orbit, the Eurovision program was beamed to the U.S. where it was carried by all three networks. It included scenes of Big Ben in London, the Colosseum in Rome, the Champs Elysées in Paris, reindeer in the Arctic Circle region of Sweden, Sicilian fishermen tending their nets, and a scene in the Sistine Chapel of Vatican.

- In regular press conference, President Kennedy said: "I understand that part of today's press conference is being relayed by the Telstar communications satellite to viewers across the Atlantic, and this is another indication of the extraordinary world in which we live. This satellite must be high enough to carry messages from both sides of the world, which is of course a very essential requirement for peace, and I think this understanding which will inevitably come from the speedier communications is bound to increase the well-being and security of all people here and those across the ocean. So we're glad to participate in this operation developed by private industry, launched by Government, in admirable cooperation."

July 23: NASA supplemental appropriation for FY 1962 was approved by the House and the Senate. It carried $82,500,000 for research and development and $71,000,000 for construction and facilities at Cape Canaveral.

- USIA reported that the U.S.S.R. had been invited to participate in the TELSTAR broadcasts but had never answered the invitation.

- Atomic Support Agency's Project Banshee balloon carried payload of instruments and 500 lbs. of high explosives to 15-mile altitude over White Sands, explosion four hours after launch providing test of high-altitude effects of large explosions; the 200-foot-long string of instruments which parachuted to earth was recovered.

- In interview with Science Service, Dr. John O'Keefe of NASA said that pattern of re-entry of pieces of MA-6 booster rocket reinforced the theory that tektites originate from the moon. Tektites are black, glossy rocks scattered on the earth's surface, and the wide area covered by MA-6 rocket pieces was a similar pattern.

- Report by Dr. James R. Killian, Jr., released by the National Civil Service League, stated that the U.S. Government is "unusually desperate" for more and better scientists and engineers. Education is not keeping pace "with the growing size of our national commitments in science and technology" and Government is being outbid by industries and even some universities for scarce talent. Greatest need, said the former scientific adviser to President Eisenhower and now President of MIT, is for scientists and engineers "with managerial ability." "Already there is a dangerous weakening of supervising technical leadership."

- U.A.R. held three-hour parade in Cairo celebrating the tenth anniversary of the Egyptian revolution. Twenty of the new Egyptian rockets were displayed.

July 24: USAF Blue Scout, Jr. space probe launched from Point Arguello, Calif.

- Senate and House space committee conferences approved compromise legislative authorization for NASA in FY 1963 of $3,744,115,250. Bill provided $43,160,750 less than NASA requested. Nearly $1.3 billion was assigned to manned space flight, including the lunar program.

- Three major U.S. TV networks telecast separate five-minute newscasts via TELSTAR, each featuring their respective Paris news correspondents.

- Edward R. Murrow, Director of USIA, quoted on floor of the House by Congressman Ryan as stating that the Voice of America cannot financially afford to use TELSTAR satellite for its overseas transmissions if the rates charged were the same as rates charged for use of undersea cables.

- AFSC announced that Colonel Charles E. "Chuck" Yeager (USAF), first man to fly faster than the speed of sound (X-1, October 14, 1947), was named Commandant of USAF Aerospace Research Pilot School at AFFTC, Edwards AFB, Calif.

July 25: NASA Wallops Station launched Aerobee sounding rocket with GSFC-University of Colorado 208-lb. payload to an altitude of 68 miles, experiment orienting an ultraviolet spectrophotometer in the direction of the sun to study wave-length profile as a function of attitude and to calibrate instrumentation for future satellite flights.

July 25: NAS Committee on Atmospheric Sciences submitted three-volume report on "The Atmospheric Sciences, 1961–70" to Dr. Jerome Wiesner, Special Assistant for Science and Technology to the President. Prepared under Dr. Sverre Petterssen of the University of Chicago and Dr. C. Gordon Little of the National Bureau of Standards, the report recommended: (1) a tripled funding of scientific research in the atmospheric sciences over the next 10 years; (2) university output of doctorates in the atmospheric sciences be increased by a factor of at least four or five; (3) universities must broaden and strengthen their programs to become national centers of academic and scientific excellence; and (4) Government research agencies must develop more rigorous research and educational programs.

• Two-stage Caleb rocket successfully air-launched over PMR from Navy F4H Phantom jet piloted by Lt. A. Newman (USN), Caleb reaching an altitude of 1,000 miles with 120-lb. NRL payload designed to measure ion composition of the earth's upper atmosphere. Project Hi-Hoe, of which this shot was a part, was a series of inexpensive air-launched high-altitude probes under development by the Naval Ordnance Test Station, China Lake.

• NASA Manned Spacecraft Center in Houston invited eleven firms to submit research and development proposals for the lunar excursion module (LEM) intended for use in lunar orbital rendezvous (LOR) flights of the multimanned Apollo spacecraft.

• Thor booster carrying nuclear warhead for atmospheric test (Operation Dominic) was destroyed on pad at Johnston Island; reportedly there were "no injuries to personnel" and "no hazard from radioactivity."

• Lockheed Propulsion Co. announced successful static firing of 12-inch-diameter, solid-propellant (polycarbutene) motor at −75°. Spokesman said: "In more than 100 firings of polycarbutene motors ranging from 4 inches to 120 inches in diameter, and under environmental conditions from plus 200° to minus 100° we have a record of 100 percent success."

• In speech to the 1962 Boys Nation, NASA Administrator Webb spoke of the future of the space age: "There is an intimate connection between the space effort and the future of this nation." Mr. Webb was made an honorary member of Boys Nation and was presented a plaque for his "interest in the youth of the nation."

• Senator John L. McClellan addressed the Senate on S. 2631, designed to prevent strikes which would "obstruct our vital missile, space, and other programs, the success of which are indispensable to national prestige, security, and well-being."

• Pope John XXIII at Vatican summer palace stated to several thousand pilgrims that U.S. TELSTAR satellite had "helped strengthen brotherhood among peoples," its use "marked a new stage of peaceful progress."

• President Gamel Abdel Nasser formally opened a jet aircraft factory, according to official Egyptian sources.

July 25–26: NASA witnesses testified before the House Committee on Science and Astronautics on the subject of economies in the space program.

July 26: Acting mayors of New York and West Berlin, Paul R. Screvane and Franz Amrehn, exchanged greetings during six-minute transatlantic telephone call via TELSTAR on its 152nd orbit. Call had been arranged by USIA in cooperation with AT&T.

- AFSC established permanent Research and Technology Division, located at Bolling AFB, to provide centralized planning and direction of applied research and advanced technology programs. Provisionally activated on April 4, 1962, the new division would be headed by Maj. Gen. Marvin C. Demler and, in addition to supporting development of advanced aerospace systems, it would be the AFSC central point for scientific and technical liaison with universities, the USAF Scientific Advisory Board, NSF, AFOAR, and the scientific community; and in maintaining liaison with Army, Navy, and NASA installations.

July 27: X–15 No. 1 flown to near 100,000 feet to begin descent maneuvers with yaw damper off, NASA's Neil Armstrong as pilot, in test of re-entry control with electronic equipment turned off. X–15 No. 1 was grounded after this flight for installation of telescopic cameras for future research flights.

- NASA-JPL-USAF Mariner R–1 Post Flight Review Board determined that the omission of a hyphen in coded computer instructions transmitted incorrect guidance signals to Mariner spacecraft boosted by two-stage Atlas-Agena from Cape Canaveral on July 21. Omission of hyphen in data editing caused computer to swing automatically into a series of unnecessary course correction signals which threw spacecraft off course so that it had to be destroyed.

- GSFC awarded contract to IBM Corporation's Federal Systems Division for computer support services for Project Mercury flights, nonrendezvous Gemini, and unmanned lunar flights scheduled for Project Apollo.

- USAF launched unidentified satellite (Alpha Theta) from Vandenberg AFB with Thor-Agena booster.

- Reported that Smithsonian Astrophysical Observatory was establishing a midwestern network of 16 observing stations with four automatic cameras to photograph the night sky to locate meteorites quickly after they fall. Located in 7 midwestern states, the stations will scan a total possible recovery area of two and one half billion acres.

- NASA Administrator Webb named Franklyn W. Phillips, Assistant to the Administrator (October 1, 1958-present), to establish and direct NASA's Northeastern Operations Office in Boston, Mass. Phillips served as Acting Secretary of the National Aeronautics and Space Council from January 1959 to February 1960.

July 28: U.S.S.R. launched COSMOS VII into orbit (apogee: 299 miles; perigee: 130 miles; period: 90.1 minutes; inclination: 65° to the equator), announcing that satellite would gather data on the "radiation hazards for long space flights."

- In address at Wheeling, West Virginia, Thomas F. Dixon, NASA Deputy Associate Administrator, pointed out that NASA and DOD "cooperate closely. For example, we have established a national launch vehicle program to provide the rocket power that both civilian and defense activities require. United States policy demands that every possible effort be made to perserve space as a peaceful resource for all mankind. We must not lose sight, however, of the fact that manned space flight involves facilities, vehicles, and techniques that may well be significant for national defense. . . .

"Our landing on the moon will be a 240,000-mile step forward in the great adventure of space conquest. No one can say how far we will ultimately go, but I assure you of this—we are on our way to making the United States pre-eminent in space as well as on earth."

July 28: National Science Foundation issued "Providing U.S. Scientists with Soviet Science Information," a publication which listed seven firms in Chicago, New York, and Washington which offer for sale the bulk of openly published Soviet scientific and technical literature, including 120 journals available in translation.

• Day-long procedural quorum calls in the U.S. Senate prompted by filibuster tactics of liberal opponents of the Administration's communications satellite bill.

• James H. "Dutch" Kindelberger, President of North American Aviation (1928–48) and chairman of the Board (1948–date), died in California. After Armistice in 1919, he joined Glenn L. Martin plant in Cleveland with Donald W. Douglas, later following Douglas and supervising the engineering of the DC–1 and DC–2, the first Douglas Aircraft passenger planes. During World War II, NAA built 14% of the U.S. military aircraft including the B–25 and P–51, later the famed F–86 Sabre, and more recently the X–15. Under Kindelberger, NAA's Rocketdyne Division pioneered the rocket propulsion development for Navaho, Atlas, Jupiter, Thor, and Redstone engines. NAA is presently prime contractor on NASA's Apollo program and the S–II.

• Maj. Robert M. White (USAF) was awarded the Flying Tiger Pilot's Trophy in recognition of his record 314,000-ft.-altitude flight in the X–15 rocket research aircraft.

July 29: UCLA scientists reported that they had developed training devices to allow men to contract individual muscles, which during the stresses of liftoff or re-entry in space flight would permit trained astronauts to operate controls by twitching single muscles.

• Reported that NASA's two-man Gemini spacecraft would probably begin flying and making landings in West or South Texas in 1963, using Rogallo wing and landing skids.

• Disclosed that the U.S. and U.S.S.R. differed on the total width of microwave bands to be assigned to communications satellites. U.S. proposed two bands with total width of about 3,000 megacycles, but U.S.S.R. favored a much narrower band of frequencies which would include those used by military and would be subject to considerable interference. To date, only informal proposals were made, and the negotiations to work out frequencies for satellite communications were still in the future.

• Soviet Navy Day, Admiral of the Fleet Sergei G. Gorshkov declaring in article in *Pravda* that rocket-firing atomic submarines are the backbone of the Soviet Navy; "The Soviet fleet is now more modern than the navy of any capitalist country."

July 30: House of Representatives passed compromise NASA authorization bill for $3,744,155,250, accepting a nearly $22 million increase voted by the Senate for lunar and planetary exploration. Bill was sent to Senate for final action.

- Sky Shield III would ground more than 1,800 airplanes for 5½ hours on Sunday, September 2, it was announced. For the military exercise to test U.S. air defense, DOD selected middle of a three-day holiday period as the date having the least effect on civil air traffic.

July 30–31: Filibuster against Administration satellite bill continued on the floor of the Senate.

July 31: Dr. Jerome B. Wiesner, Scientific Advisor to the President, testified before the House Committee on Government Operations subcommittee that getting better scientists in government appeared to be the "most important single problem as I've tried to understand why so many of our programs go badly." Declining to comment on specific failures, Dr. Wiesner pointed out that the Administration sought "to redress unbalances between Federal and general industrial salary scales. . . . Such reform is essential if we are ever to reverse the present deterioration in the quality of the technical force in the government service."

- In testimony before House Science and Astronautics Committee, Richard B. Morrison, NASA's Launch Vehicles Director, testified that an error in computer equations for Venus probe launch of Mariner R–1 spacecraft on July 21 led to its destruction when it veered off course.

- Former President Dwight D. Eisenhower spoke on the people-to-people benefits to be gained by live international communications, broadcast televised to U.S. via TELSTAR from Stockholm, Sweden.

- British Postmaster General Reginald Bevins told Parliament that Britain would spend $1,960,000 in the coming year on development of satellite communications.

- Replying to a charge that U.K. was playing a declining role in development of satellite communications, Postmaster General Reginald Bevins told Parliament: "I assure you that the British Government is fully alive to the possibilities of satellite communications and when it comes to an operational system, we shall not be left out in the cold." He predicted a global system of satellite communications within this decade.

- Donald F. Conaway, national executive secretary of American Federation of Television and Radio Artists, announced he had requested a conference with European counterparts to discuss minimum pay scales for performers appearing on future Telstar satellite relays. He added: "Telstar is a great challenge. We cannot realistically set an international rate for performers until many problems are settled with the countries."

During July: Outfitted in pressure suits in a simulated space environment, five NASA scientists of MSFC and MSC completed study determining how an astronaut could work on J-2 rocket engine during actual space flight, at Rocketdyne. Maintenance, repair, replacement, and adjustment of components of the hydrogen-fueled J-2 were performed along with research on the design of special tools and pressure suits.

- NASA Marshall Space Flight Center announced first static test of P&W liquid-hydrogen engine. In first firing of series, which was run in vacuum chamber to simulate space environment, RL-10 ran nine seconds and generated a rated 15,000 pounds of thrust.

- First successful reactor tests using liquid hydrogen as coolant were performed by Los Alamos Scientific Laboratory at the Nuclear Rocket Development Station (NRDS), Jackass Flats, Nev. "Cold flow" series of reactor tests involved operation of Kiwi-B reactor, previously conducted only with gaseous hydrogen. Liquid hydrogen as a propellant is planned to power the Nerva nuclear engine being developed for Projects Rover and Rift.

- NASA announced that it was establishing computer center at Slidell, La., to service the Michoud Operations by November. Facility will be one of the nation's largest computer centers and will perform engineering calculations arising in the development, fabrication, and static testing of the Saturn S-I and S-IC first stages.

- Research Institute for Advanced Study of the Martin Company was awarded contracts by NASA, AFOSR, and ONR for basic research in nonlinear mathematics, an area relating to problems in space flight, long-range communications, and automation.

- Samuel K. Hoffman, President of Rocketdyne Division, NAA, received the Spirit of St. Louis Medal from the American Society of Mechanical Engineers for his "contributions to aviation, aerospace and rocket industries through the development of the first major long-range rocket engine."

- Karel J. Bossart of General Dynamics/Astronautics received the Astronautics Medal of the British Interplanetary Society for his "individual contribution to the science of astronautics, particularly in the engineering sciences."

- The Messerschmitt factory, Augsburg, Germany, celebrated the 20th anniversary of the construction of its Me-262, the world's first operational jet aircraft.

- U.S.S.R.'s Aeroflot inaugurated two-hour helicopter sightseeing flight over Moscow on Sundays, the popularity of which required advanced bookings of passengers for this "fashionable form of transport."

- SRN.2 Hovercraft was publicly demonstrated at the Isle of Wight, a 27-ton watercraft powered by four-paired Blackburn Nimbus gas-turbine engines. SRN.2 research vehicle developed by Hovercraft Development Ltd. has already traveled more than 8,000 miles and up to 62 knots speed. Four-ton SRN.1 Hovercraft has operated for more than 300 hours and the speed range has been increased from 25 to 65 knots.

During July: USAF announced that largest chimpanzee science research center was under construction at Aeromedical Research Laboratory, Holloman AFB, a doubling of facilities and professional staff possible through funds ($500,000) made available by NASA.   50 anthropoid apes are currently in various stages of training, ranging from space veterans Ham and Enos to preschool adolescents.

• FAA awarded seven additional contracts for research on supersonic transport aircraft, bringing the total of supersonic transport study awards to $6.1 million.

• North American Aviation expected to pass its peak World War II employment figure of 93,000 persons within the next 90 days, its current corporate employment now being at 90,000.   Rocketdyne Division has grown from 11,000 to 13,000 employees within the past year.

AUGUST 1962

August 1: Senate voted NASA FY 1963 authorization of $3,744,115,250, representing $5,400,000 less than the amount originally agreed to by the Senate and about $2 billion above the amount authorized for NASA in FY 1962. The bill (H.R. 11737) was sent to the White House for the President's signature.

- House passed Independent Offices Appropriation Bill for 1963 (H.R. 12711) containing an amendment providing for 25% indirect costs on research grants made by the 26 executive agencies of the bill, including NASA. Appropriation for NASA was $3,644,115,000, representing $143,161,000 less than budget estimates.

- USAF launched unidentified satellite with Thor-Agena booster from Vandenberg AFB.

- Five-day filibuster in Senate against Administration-supported communications satellite bill was shelved under compromise agreement between bill's supporters and opponents. Bill was referred to Foreign Relations Committee, which was ordered to report it back to the Senate not later than August 10.

- Rocketdyne Division of NAA announced plans to expand its Canoga Park, Calif., facilities to manufacture F-1 and J-2 rocket engines for NASA's Advanced Saturn launch vehicle.

- Four American scientists led by Dr. William A. Cassidy, research scientist at Lamont Geological Observatory of Columbia University, left New York on three-month expedition to Argentina, where they hoped to find and unearth a 13½-ton meteorite. Supported by National Science Foundation grant, the expedition would seek the huge meteorite which was reported in late 1700's by Indians of north central Argentina.

- Gen. Bernard A. Schriever, AFSC Commander, told House Military Operations subcommittee that the Midas early warning satellite would "take longer to develop than initially forecast." He said that USAF and DOD had been overly optimistic about the reliability that could be expected from components and about technical features of the system.

- USAF Atlas F ICBM was launched from a silo at Vandenberg AFB in successful 5,000-mi. flight to vicinity of the Marshall Islands in the Pacific, the first Atlas F launch from an underground silo.

- AEC–DOD jointly announced that repair work on launch facilities at Johnston Island would delay further high-altitude nuclear tests for some weeks. Damage occurred July 25 when Thor rocket and its nuclear warhead were deliberately destroyed on the pad.

- Recent Soviet article discussed various methods the U.S.S.R. has been studying for sending a man to the moon in the current decade. The earth-orbital rendezvous method was reported as considered the most reliable, but consideration also has been given to the direct ascent method, using the "Mastodon" rocket. No decision has been made public.

August 1: Soviet instrument ships observing U.S. nuclear tests for weeks had left the Pacific test area on July 10 or 11, following U.S. high-altitude test of July 9.

• Newly-created Army Materiel Command took over the Army Missile Command (formerly Army Ordnance Missile Command), responsible for management of 19 missile systems.

August 2: Joseph A. Walker (NASA) flew X–15 No. 3 in test of the research aircraft's ability to automatically correct undesired yaw. In nine-minute flight, the X–15 reached maximum altitude of 147,000 ft. and speed of 3,445 mph (mach 4.99).

• First full-scale research model of an inflatable space station was displayed at NASA Lewis Research Center. Developed by Goodyear Aircraft Corp. for NASA, the doughnut-shaped, three-story-high structure was made of rubberized fabric and equipped to accommodate three to ten astronauts.

• Senate Foreign Relations Committee (Sen. John J. Sparkman of Alabama as acting chairman) announced plans for hearings on international aspects of the communications satellite bill. Plans called for testimony by Secretary of State Dean Rusk, other State Department officials, and representatives of Defense Department, Commerce Department, and Justice Department, as well as NASA.

• British Parliamentary Secretary for Science, Denzil Freeth, said that ARIEL I had produced interesting and valuable information about the ionosphere and the higher atmosphere. He added: "ARIEL at first behaved almost perfectly but recently transmission of data has been interrupted irregularly. The reason is not known but is being investigated."

• U.K. Defense Minister Peter Thorneycroft announced that Thor missile bases in Britain would be abolished by October 1963. British long-range defense would be assumed by U.S. fleet of Polaris-firing submarines and, later, by Skybolt air-to-surface missiles carried on British bombers.

• When questioned at press conference regarding the British decision to discontinue Thor missile bases, President Kennedy stated: "Our ability to meet our commitments to the defense of Western Europe in the conventional and in the nuclear field remain unchanged . . ., and the United States' commitment remains unchanged."

• Aerojet-General Corp. proposed a booster two to three times larger than Nova, to be launched from the sea. Two-stage Sea Dragon with takeoff weight of 20 to 100 million pounds would be assembled in existing drydocks, towed to point of launch, and launched from the water without gantry cranes or launch pads. Both stages would be recovered in the water by hydrodynamic deceleration.

• FAA announced its airport aid program for FY 1963, a record $74.2 million to be spent on improvement or construction of 419 airfields.

• French physicist Duke Armand de Gramont, founder of the Institute of Theoretical and Applied Optics and president of Optique et Precision de Levallois, died in his chateau near Paris. He had helped develop the electronic microscope.

August 2–3: In meeting at NASA Langley Research Center, Langley and MSC personnel presented reports on space station studies to officials from NASA Hq. and other interested agencies.

August 3: NASA radiation-research balloon released biological payload near Prince Albert, Saskatchewan, after 130,500-ft.-high, 51-hour flight from Goose Bay, Labrador. Two monkeys survived the flight in good condition, but four hamsters could not survive the day-long recovery operation in the Canadian wilderness; fate of the flour beetles was not yet known. The animals and insects were flown to the University of California at Berkeley for scientific study. This was the second in a series of four balloon experiments conducted by NASA Ames Research Center.

- Attorney General Robert F. Kennedy and FCC Chairman Newton N. Minow testified before Senate Foreign Relations Committee on the Administration-backed communications satellite bill. Urging prompt passage of the bill, Kennedy said it was "one of the most important pieces of legislation offered by this Administration" and that he was "perfectly satisfied it will protect the public interest." This was the eleventh time Minow had testified for the bill, which the House had passed on May 3 and which previously had been approved by two other Senate committees.

- Advanced Syncom satellite, being developed for NASA by Hughes Aircraft Co., probably would carry four radio signal repeaters and could provide up to 300 two-way telephone channels or one TV channel. In 24-hour stationary orbit, one Advanced Syncom would be sufficient to link four continents (North America, South America, Europe, and Africa); three such satellites would provide worldwide coverage.

- USAF Cambridge Research Laboratories announced that it would conduct a series of high-altitude balloon astronomy experiments during 1962 and early 1963. Primary objective would be to get more complete and accurate information about the moon, Venus, and Mars. Scientific director of the project would be Dr. John Strong of Johns Hopkins University.

August 4: Senators Joseph S. Clark of Pennsylvania and Wayne Morse of Oregon, leaders of the Senate opposition to pending communications satellite bill, stated on radio and television interview that they still insisted the communications satellite system must be owned and controlled by the Government. However, they added that they were not committed to Government operation, and that they were "perfectly willing to consider working out a lease or a license" with AT&T, RCA, or any other communications company.

- NASA General Counsel John A. Johnson, speaking before the American Bar Association's Section on International and Comparative Law in San Francisco, stated: "It appears that the existing state of the law is that we have an area of space extending upward from the surface of the earth for an indefinite distance which is exclusively controlled by the underlying State, and above that, beginning at some undefined point, lies the 'free' realm of outer space. Whether there is or should be an intermediate third realm to which the exclusive power of the underlying State does not extend, but in which the full freedom of outer space may not be enjoyed, is an interesting item for speculation. . . ."

The problem is of practical significance, "because all spacecraft, before injection into orbit, must be launched through the air

space.   Likewise, all space missions involving re-entry and landing require that the spacecraft move back through the air space
on their return to earth. . . .

"It now appears that the manned vehicles which will be
developed over the next five to ten years will enter the atmosphere
rather steeply, level out, and glide at altitudes ranging from
about 25 to 60 miles for distances perhaps as great as 7,000 to
10,000 miles before landing.   Inevitably, it will become necessary
to know in advance whether any portion of the re-entry phase
of a manned space flight violates the territorial air space of
another State because of the altitude at which its land or territorial waters may be overflown. . . .

"It would be most unwise to attempt to reach a solution to the
so-called 'boundary' problem on the basis of the difference
between the regimes of aerial flight and space flight.   But even
if it were possible to define a boundary between those regimes on
scientific and technical grounds, it would bear no necessary
relationship to the national interests which the principle of
territorial sovereignty is designed to serve.

"I think it is evident that if this problem is to be solved it will
be done on the basis of an accommodation of the political interests
of the States concerned, and not on the basis of scientific or
technological criteria."

August 4: James A. Martin, NASA X–15 Program Manager, in letter to
   the Washington *Evening Star* clarified the method of computing
   mach number in high-speed flight.   "Mach number is a direct
   function of atmospheric temperature at the altitude at which the
   vehicle is flying.   This temperature varies from day to day for
   the same altitude.   Because of this, it is possible that the same
   number of feet per second at the same altitude can be more or
   less than a given Mach number."

   He pointed out that on November 9, 1961, Major Robert
   White flew the X–15 to speed of 6,005 feet per second; on June
   27, 1962, Joseph Walker flew to 6,020 feet per second, at about
   the same altitude as Major White.   However, because of differences in atmospheric temperatures for the two days, Major
   White attained higher mach number (M 6.04) than Walker
   (M 5.94).

August 4–5: Thousands of teen-agers toured NASA Lewis Research
   Center and witnessed special lecture-demonstrations during
   "Youth Days," co-sponsored by LRC and the *Cleveland Press.*

August 5: USAF launched an unidentified satellite from Vandenberg
   AFB, using an Atlas-Agena launch vehicle.

• Soviet Cosmonauts Gherman S. Titov and Yuri A. Gagarin hinted
   in a press interview that the U.S.S.R. would attempt prolonged
   manned orbital flight some time during 1962.   Colonel Gararin
   remarked: "Recently in Japan I said that new flights through
   space are not far off and expressed confidence that they would
   certainly take place this year."

   Major Titov then added: "I also am of the same opinion and
   want to add that the time already is past when the length of
   cosmic flights will be only hundreds of thousands of kilometers.
   I think that flights of future cosmonauts will be more prolonged
   and the route of their cosmic ships will measure millions of
   kilometers."

August 5: The World's Fair Corp. announced it was developing plans for a unified space exploration exhibit called "Aerospace Island," as part of the U.S. exhibit in the 1964–65 World's Fair.

- U.S.S.R. resumed nuclear tests with a high-altitude blast in the 40-megaton range, the first Soviet test since November 4, 1961.

- Michael Friedlander, associate professor of physics at Washington University, St. Louis, was preparing to send up a series of huge balloons for cosmic ray experiments, beginning about August 15, from Calvina, South Africa.

- Dr. Clarence P. Oliver of the University of Texas zoology department suggested to the House Space Committee that astronauts traveling through space "for any extended period" should refuse to have any more children. Radiation exposure exceeding normal dosage on earth would pose a "genetic risk," he testified.

August 6: Testifying before Senate Foreign Relations Committee on communications satellite bill, Secretary of State Dean Rusk stated: "Most of the discussion of the foreign policy provisions of the bill has centered on whether they delegate to the corporation a part of the President's authority to engage in international negotiations on behalf of the United States Government. Let me state most emphatically that they do not. . . .

"I can assure the committee that we in the State Department are fully aware of the broad range of questions involving foreign policy interests that may arise in connection with this satellite communications system."

- Edward R. Murrow, Director of USIA, testified before the Senate Foreign Relations Committee: "Having contributed to developing the [communications satellite] system, Government should not be on a parity of payment with other commercial users. We strongly believe that affordable rates for our agency's use is an appropriate partial repayment for that national investment." He replied, in answer to question, that he would endorse the bill as it was.

- NASA announced signing of $215,502,744 contract with Chrysler Corp. for production of Saturn S–I stages. Chrysler would deliver the 21 units (instead of 20 as originally planned) between early 1964 and late 1966, beginning with the S–I for Saturn SA–9. NASA Marshall Space Flight Center, with development responsibility for Saturn, would fabricate boosters for first eight vehicles. Chrysler would build the S–I stages at NASA Michoud Operations plant near New Orleans.

- Announced that Boeing had received $15,954,096 supplementary contract from NASA Marshall Space Flight Center for work leading to design, development, fabrication, and test of Saturn C–5 first stage (S–IC).

- Col. Daniel McKee, Project Gemini director in NASA's Office of Manned Space Flight, reported as saying that two-man Gemini spacecraft might be used for the next four or five years, and describing it as "a reliable and flexible spacecraft which has a lot of potential for scientific investigations of space. If additional applications develop, it would be used for various purposes." At present, primary goal of Project Gemini is one-week-long, two-man mission with testing of orbital rendezvous operation.

August 6: Secretary of the Air Force Eugene M. Zuckert, writing in the *General Electric Forum*, said that there were unique military problems in space that a civilian space program never could solve. "The Air Force will make orbital flights much sooner than previously planned," he wrote.

- Reported by *Aviation Week* that NASA was considering use of area near Corpus Christi, Texas, for landing two-man Gemini spacecraft after its re-entry from first orbital flight, in 1963. Landing mode for the spacecraft would include Rogallo wing and skids.

- MIT ordered a Honeywell 1800 electronic computer from Honeywell Electronic Data Processing, for work on Project Apollo navigation system. After installation in 1963, the computer would aid in circuitry design of Apollo spacecraft's guidance computer and also would simulate full operation of spaceborne computer during ground tests.

- Sweden launched U.S. Nike-Cajun sounding rocket, first of series of U.S.-Swedish cooperative program for peaceful exploration of the upper atmosphere, the rocket reaching 68-mi. altitude after launch from Kronogård range in northern Sweden. Fifteen minutes after launch, canister containing sample cloud particles was recovered; particles would be analyzed to provide information on composition and origin of noctilucent clouds.

- Gen. Bernard A. Schriever (USAF), addressing the Society of Photographic Instrumentation Engineers, declared development of the laser may prove more important to the world than "development of the ballistic missile, the discovery of the transistor, or the reality of TELSTAR. . . . As a new method for transmission of energy, the Laser has almost infinite potential. The Laser can direct a concentrated beam of light across great distances with extreme precision. It is therefore ideally suited for application in space. In addition, it appears to have many practical applications in research, industry, medicine and communications. . . .

  "In the Air Force, our current funding for Laser technology is about six million dollars. Our efforts range from instrumentation development to the search for means of using Laser beams in support of our military space objectives. Our budget for Fiscal Year 1964 should include an increase for Laser research and development.

  "The Laser is a striking example of the potential to be found in today's advanced technology. We must make full use of its possibilities for national security purposes. . . ."

- Senator John L. McClellan recommended establishment of a Senate commission on science and technology to study Government's growing scientific programs, including the desirability of a Cabinet post for science and technology.

- Dr. Melvin Calvin, 1961 Nobel Prize winner for his biochemical research on photosynthesis, said in interview that he was investigating effects of ultraviolet light on chemical elements basic to living organisms, in effort to discover the chemical evolution of life from lifeless matter. He stated that results of his experiments with those of Dr. Stanley Miller (whose earlier research produced simulated amino acids) implied that "any planet with an appropriate temperature and an atmosphere similar to that of the primitive earth will, in time, develop the very things which we now recognize as the essential molecules of living organisms. . . ."

He further indicated that, considering the numerous planets within our galaxy that have existed much longer than the earth, it was logical to conclude that some forms of intelligent life elsewhere in the galaxy are more advanced scientifically and technologically than man.

August 6: First task of ONR's 85-ft. radiotelescope at Hat Creek, Calif., would be mapping the visible galaxy, with emphasis upon studies of gaseous galactic nucleus, according to *Missiles and Rockets.*

* NAA Rocketdyne's Propulsion Field Laboratory, near Los Angeles, completed its 150,000th test in 12 years of engine development.

* Reported that Sikorsky helicopters of Okanagan Helicopters, Ltd., had been used in British Columbia, Canada, to replant telephone poles for the British Columbia Telephone Company.

During early August: USAF awarded one-year study contract to Aeronutronic Division of Ford Motor Co. to isolate, identify, define, and code tasks and skills required of members of future aerospace crews. Study would be conducted at Behavioral Science Laboratory, Wright-Patterson AFB.

August 7: President Kennedy, in ceremony at the White House, presented the President's Award for Distinguished Federal Civilian Service to Robert R. Gilruth, Director of NASA Manned Spacecraft Center, and five other recipients of the award for 1961.

* Former President Dwight D. Eisenhower, writing on the U.S. space program in *The Saturday Evening Post,* stated that he did "not see the need for continuing this effort of such a fantastically expensive crash program. . . .

"Why the great hurry to get to the moon and the planets? We have already demonstrated that in everything except the power of our booster rockets we are leading the world in scientific space exploration.

"From here on, I think we should proceed in an orderly, scientific way, building one accomplishment on another, rather than engaging in a mad effort to win a stunt race."

* House Committee on Science and Astronautics approved a bill to revise space research patent laws. Bill allowed private companies to retain title to patents earned under Government contract, with Government retaining a license to use such inventions without cost.

* House Committee on Science and Astronautics recommended that Project Anna mapping satellite program be transferred from jurisdiction of DOD to NASA. The satellite was designed to give extremely accurate measurements of the shape of the earth, its magnetic field, and distances from one point to another on earth.

* Dr. Hugh L. Dryden, Deputy Administrator of NASA, testified before the Senate Foreign Relations Committee that the U.S. "can be the leader in the establishment of a communications satellite system to serve the communications needs of the world, thus demonstrating our technical capabilities and our desire to utilize these capabilities for the benefit of all mankind. . . . I urge that the Committee report the bill favorably."

August 7: Secretary of Defense Robert S. McNamara told the Senate
Foreign Relations Committee that the Department of Defense
"strongly supports the objective of establishing a civil communications satellite system as expeditiously as practicable."

- Dr. Wernher von Braun, Director of NASA's Marshall Space Flight
Center, said in an interview in *General Electric Forum* that the
U.S. space program would be paying for itself within another 10
years. "The real payoff does not lie in mining the Moon or in
bringing gold back from the Moon, but in enriching our economy
and our science in new methods, new procedures, new knowledge,
and advanced technology in general."

- General Electric announced that control system for first Orbiting
Astronomical Observatory (OAO) had successfully completed its
first simulated space flight test.

- USN launched first flight model of advanced Polaris missile (A–3)
capable of 2,880-mi. range. Among innovations in A–3 were
bullet-shaped nose (rather than rounded shape of earlier models),
and guidance system about one third the size and weight of the
ones in earlier Polaris models and the smallest and lightest yet
developed for U.S. ballistic missiles. Launched from land pad
at Cape Canaveral, the missile fell short of its planned range
("in excess of 1,975 miles") because of second stage malfunction.
Test was termed "partially successful."

- Announced that Georgia Nuclear Laboratories' nuclear reactor
had been licensed by AEC to operate at one megawatt. Laboratories would participate with Lockheed Missiles and Space Co.
in development of NASA's Rift rocket stage.

- Testifying before the Senate Foreign Relations Committee on the
pending communications satellite bill, Washington lawyer
Joseph L. Rauh, Jr., stated that the bill would "give away not
only billions of taxpayers' money already spent to develop both
space and space communications, but also the vast unknown discoveries of the future." Another Washington lawyer, Benjamin
V. Cohen, testified that the bill was "filled with pitfalls" and advised that its enactment now would be "premature."

August 8: NASA launched Aerobee 150A sounding rocket from Wallops
Station, its 256-lb. payload lofted to 92-mi. altitude and 60-mi.
distance. Efforts to recover the payload were not successful,
but scientists were able to analyze data telemetered during
flight about performance of an attitude control system as well
as four scientific experiments: to measure solar flux in two ultraviolet spectral regions; to measure radiation emerging from the
top of the earth's atmosphere; to obtain ultraviolet photographs
of Venus from outside the earth's atmosphere; and to study distribution of atmospheric atomic hydrogen and Lyman Alpha
radiation. Also flight-tested was an experimental, transistorized
telemetry system.

- Maj. Robert Rushworth (USAF) piloted X–15 No. 2 in flight to
record pattern of aerodynamic heating at moderately low speeds,
relatively low altitudes, and moderate angle of attack, the steel-
skin craft withstanding temperatures up to 900° F. Maximum
altitude was about 90,000 ft., maximum speed about 2,898 mph
(mach 4.39) in series of maneuvers near Hidden Hills, Calif.,
to build up heat on airplane's surface. After successful 8-min.

flight, Maj. Rushworth reported observing particles of insulation floating between twin panes of X–15's windshield, while climbing at 70,000-ft. altitude.

August 8: Special subcommittee of the House Committee on Science and Astronautics conducted hearings on solid propellants. Chairman David King of Utah was quoted as saying he wanted the hearings to be detailed enough to commit DOD and NASA to a definite plan for solid-propellant rockets.

Testifying before a subcommittee of the House Committee on Science and Astronautics, NASA Deputy Associate Administrator Thomas F. Dixon said that NASA was "giving solid rockets equal consideration with liquid rockets in decisions on future launch vehicle designs. . . . We have provided the Air Force with our requirements for an advanced technology or feasibility demonstration program on large solid rocket motors pertinent to the development of solid rocket-powered launch vehicles.

"To make an informed choice of propulsion between liquid propellant engines (where the technology is fairly well in hand) and solid propellant motors (where it is not), it is necessary to advance the technology of large solid motors to the point of actual demonstration firings. The demonstration program we have proposed will allow us to make a wise choice between the two types of propulsions because a better understanding of costs and schedules as well as developmental problems will be available."

- U.S. Senate passed S. 2771 to establish a Commission on Science and Technology, which would coordinate Government research and development programs with those of business and industry and would establish coordinated systems of information storage and retrieval.

- NASA Flight Research Center released photographs of a mysterious floating object taken during Maj. Robert M. White's July 17 record-breaking flight of the X–15. Pictures, extracted from film taken by movie camera in tail of X–15, showed gray-white object of undetermined size tumbling above and behind the aircraft. NASA officials reported that it was "impossible to identify or explain the object's presence at this time."

- National Geographic Society and Lowell Observatory announced preparation of an atlas, "A Photographic Study of the Brighter Planets," containing "the most detailed and revealing pictures obtained since telescopes were perfected especially for observing the planets rather than the distant stars."

- Bidders' conference held at NASA Hq. NASA requested proposals from field centers and industry for two lunar logistics studies— one of a spacecraft bus concept which could be adapted for initial use on the Saturn C–1B and later use on Saturn C–5 launch vehicles; and one of a variety of payloads which could be soft-landed near manned Apollo missions, the latter study to determine how man's stay on the moon might be extended, how man's capability for scientific investigation of the moon might be increased, and how man's mobility on the moon might be increased. Contract proposals were requested by August 20.

- NASA had awarded $141.1 million contract to Douglas Aircraft Co. for design, development, fabrication, and test of Saturn S–IVB stage to be used as upper stage of three-stage Advanced Saturn (C–5). Contract called for eleven S–IVB units, including three for ground tests, two for inert flight, and six for powered flight. S–IVB uses NAA's J–2 liquid-hydrogen/liquid-oxygen engine generating 200,000 lbs. of thrust.

August 8: NASA Marshall Space Flight Center announced contract award to Pratt and Whitney, to investigate the feasibility of variable thrust in RL–10 rocket engine. The study was not directed toward any specific mission but would have wide application in future space flights.

- AEC and Westinghouse Electric Corp. announced a joint study program to determine technical feasibility of a 1,000,000 kw nuclear power plant using a single, closed-cycle water reactor.

- Reported that RCA had proposed unification of all U.S. international communications carriers into a single privately owned and independently operated company, to expedite development of a global communications service.

- Prime Minister Jawaharlal Nehru told Indian Parliament that India had agreed to have a rocket launching station on her territory under U.N. auspices for international use.

August 9: Third balloon in radiation-experiment series conducted by Ames Research Center was launched from Goose Bay, Labrador. In two hours, the balloon had carried its payload of two monkeys, four hamsters, and instruments to an altitude of 85,000 feet, where it was to catch prevailing winds and drift westward across Canada.

- USAF launched two Atlas D missiles in quick succession from Vandenberg AFB toward impact area 5,000 miles away. Officials said the tests, the first demonstration of a multiple countdown capability, were successful.

- Gen. Bernard A. Schriever, Commander of Air Force Systems Command, testified before a subcommittee of the House Committee on Science and Astronautics that the revised USAF solid-propellant development program would cost considerably less than the $60 million originally planned. He said that current program review ordered by Secretary of Defense and NASA Administrator had shown that solid-fuel program was larger than necessary to demonstrate the feasibility of the boosters. Once approved by the Secretary and the Administrator, the new master plan for solid-propellant boosters would limit the program to feasibility study, since neither DOD nor NASA had specific mission requirements for their application.

- Speaking on the Senate floor, Senator Estes Kefauver reiterated his belief that the communications satellite bill (H.R. 11040) "proposed the most gigantic giveaway in the history of this country. It would turn over to a governmentally created private monopoly the benefits of hundreds of millions of dollars of taxpayers' money which have been invested in the development of space and satellite communications technology. . . ."

- Establishment of NASA Industrial Applications Advisory Committee announced, with Earl P. Stevenson (former president and chairman of the board of Arthur D. Little, Inc.) as chairman. The committee would assist in transferring new scientific knowledge from NASA's research and development programs to industry.

August 9: USAF announced awarding of Distinguished Service Medals to Lt. General Thomas P. Gerrity (USAF), formerly commander of Air Force Systems Command's Ballistic Missile Division (AFSC/-AFBMD) and now Deputy Chief of Staff for Systems and Logistics; and Maj. General Osmond J. Ritland (USAF), formerly head of AFBMD and AFSSD and currently Deputy for Manned Space Flight, AFSC.

- *Congressional Record* reprinted monograph on "U.S. Space Legal Policy—Some Basic Principles," by Robert D. Crane, director of the Space Research Institute at Duke University. Mr. Crane urged U.S. initiative in the formulation of space law, which "can serve not only to promote scientific research and economic progress and to facilitate the growth of a free and peaceful world order, but to implement on a higher moral level American military and political strategies . . ."

August 10: Senate Foreign Relations Committee returned the communications satellite bill (H.R. 11040) to the Senate floor, after defeating amendments sought by opponents of the bill. Senate filibuster against the communications satellite bill was resumed.

- NASA Administrator James E. Webb, speaking before a subcommittee of the Senate Committee on Appropriations, said that 32 candidates for astronaut positions had been selected from the 200 applicants, and that by September 1 NASA would select 10 of these men to enter astronaut training.

- USAF Atlas F ICBM was destroyed by the range safety officer a few seconds after launch from Vandenberg AFB.

- Advanced model of Navy's Polaris A–2 missile was successfully test-fired from a land pad at Cape Canaveral, in 1,700-mi. flight downrange.

- Reported that NASA had selected Aerojet-General's Algol solid-propellant motor to power Little Joe II booster, which would be used to flight-test the command and service modules of Apollo spacecraft.

- NASA announced appointment of Dr. Robert L. Barre as Scientist for Social, Economic, and Political Studies in the Office of Plans and Program Evaluation. Formerly a private consultant for both Government and industry, Dr. Barre would be responsible for developing NASA's program of understanding, interpreting, and evaluating the social, economic, and political implications of NASA's long-range plans and accomplishments.

- DOD announced first flight of Army's new VTOL, turbojet airplane, the Lockheed VZ–10 "Hummingbird," at Lockheed-Georgia's plant, Marietta, Ga. After further tests to prove Hummingbird's aerodynamic and handling features in conventional flight, Lockheed would begin vertical takeoff and landing (VTOL) tests with the twin-jet research aircraft.

- British Blue Water missile development was canceled, after British Government had spent $84 million on the 75-mile-range, surface-to-surface missile. Defense Ministry said that Blue Water program was canceled because of the increasing number and yield of other tactical weapons available to NATO.

August 11: U.S.S.R. launched VOSTOK III into orbit at 11:30 AM Moscow time, the spacecraft piloted by Maj. Andrian G. Nikolayev. Initial orbital data: apogee, 156 miles; perigee, 113 miles;

period, 88.5 min.   Cosmonaut Nikolayev remained in orbit throughout the day and night.

August 11: Soviet space expert Anatoly Blagonravov, writing in *Izvestia*, said that U.S.S.R. was "leaving far behind everything that has been attained by America in carrying out manned orbital flights," and that, contrary to the U.S. program, Soviet space flights were "of an exclusively peaceful nature."

• Second joint U.S.-Sweden sounding rocket launched from Kronogård site in project to gather data on noctilucent clouds, the Nike-Cajun rocket sending instrument-packed nose cone 43.4 miles (70 km.) high before it parachuted back to earth.

• NASA Ames Research Center announced its high-altitude radiation-research balloon was flying at 135,000-ft. altitude over eastern Manitoba and that the balloon's payload capsules containing two monkeys, four hamsters, and instrumentation would be released early August 12.

• Senate Majority Leader Mike Mansfield, on behalf of himself and Minority Leader Everett M. Dirksen, filed a cloture petition intended to break the Senate filibuster against Administration-backed communications satellite bill.

• Senator Thomas Kuchel of California, speaking in the Senate, said that the bill to amend the National Aeronautics and Space Act of 1958 (H.R. 12812) would provide NASA a needed flexibility in patent regulations, including authorization to secure royalty-free license or complete title for Government-sponsored research, and he recommended that the Senate Committee on Aeronautical and Space Sciences hold hearings on the bill.

• Representative Joseph E. Karth, chairman of a subcommittee of the House Committee on Science and Astronautics, announced that hearings on four communications satellite systems would begin August 15, to determine whether duplication existed among the projects (Relay, Syncom, Advent, and Telstar).

• Evidence that atmosphere of Saturn contains molecular hydrogen was announced by astronomers Guido Munch of Mt. Wilson and Palomar observatories and Hyron Spinrad of JPL.   Previously, only the trace substances of methane and ammonia had been detected on Saturn.   Munch and Spinrad also found evidence that winds blow many hundreds of miles per hour on Saturn, stated that such winds could sweep the gases into Saturn's bands.

• Winners of the 1962 Harmon International Trophies were announced as follows: Aviator's Trophy—Lt. Col. William R. Payne (USAF), for his May 1961 non-stop, record-speed flight of B-58 from Carswell AFB, Texas, to Paris, via Washington and New York; Aviatrix Trophy—Jacqueline Cochran, for setting 8 world class records with T-38 jet trainer and for flying F-104 jet fighter at twice the speed of sound, between August 24 and October 12, 1961; Aeronaut's Trophy—Cdr. Malcolm D. Ross (USN) and the late LCdr. Victor A. Prather (USN), for their record-altitude balloon flight of May 4, 1961, over the Gulf of Mexico.

• Dr. Harry Wexler, meteorological research director of U.S. Weather Bureau since 1955, died unexpectedly.   Holder of distinguished service citations from the Air Force, Navy, Commerce Department, and National Civil Service League, Dr. Wexler had been with the Weather Bureau since 1934.   With Col. Floyd Wood as pilot, he made the first aircraft penetration of an Atlantic hurricane, in September 1944.   In June 1961, Dr. Wexler and Prof. Viktor A. Bugaev of the U.S.S.R. were principal figures in drafting plans for cooperative use of satellites to improve weather forecasting.

August 11: Reported that USAF had offered to train NASA's "second generation" astronauts at its Aerospace Research Pilot School, Edwards AFB, California.

- Reported that NASA was conducting a study to determine what effect Soviet nuclear test series might have on forthcoming orbital flight (MA–8) of Astronaut Walter M. Schirra, Jr.   Final evaluation was expected within a week.

- A modified B–17 (Flying Fortress) was being used by Denmark's Geodetic Institute to photograph northern Greenland.

August 12: Lt. Col. Pavel R. Popovich, piloting VOSTOK IV spacecraft, was launched into orbit (157-mi. apogee, 112-mi. perigee) at 11:02 AM Moscow time by U.S.S.R.   Within about an hour, Cosmonaut Popovich made radio contact with Cosmonaut Nikolayev, traveling in nearly the same orbit in VOSTOK III launched the previous day.   Nikoleyev reported shortly thereafter that he had sighted VOSTOK IV, and within three hours after VOSTOK IV's launch the two spacecraft had completed two orbits of the earth in adjacent flight.   Moscow radio announced that objective of orbiting the two spaceships close to each other was "to obtain experimental data on the possibility of establishing contacts between the two ships, coordinating the actions of the pilot-cosmonauts and to check the influence of identical conditions of space flight on human beings."   Tracking by Sohio Research Center (Cleveland, Ohio) placed the two spacecraft at 75 miles apart during VOSTOK III's twenty-second orbit and VOSTOK IV's sixth orbit, and at 385 miles during later orbital passes.   This was the first launching of two manned space flights within a 24-hour period.

- Cosmonaut Andrian G. Nikolayev, pilot of VOSTOK III, broke three records when he entered his eighteen orbit of the earth at 5:49 AM Moscow time, completing 25 hours and 19 minutes of flight time in space and more than 440,000 miles distance.   Previous records—number of orbits, time aloft, and distance covered— had been set by second Soviet cosmonaut, Maj. Gherman S. Titov, August 6–7, 1961.

   By 10:00 PM Moscow time, Nikolayev had completed 24 orbits of the earth (602,370 mi.) and Popovich had completed 8 orbits (186,300 mi.).   Tass reported orbital data of VOSTOK III: apogee, 141.3 mi.; perigee, 109.7 mi.; period, 88.2 min.; inclination, 64.59 degrees to Equator; VOSTOK IV: apogee, 145.8 mi.; perigee, 110.5 mi.; period, 88.3 min.; inclination, 64.57 degrees to Equator. Both spacecraft remained in orbit throughout the night.

- TV pictures of Soviet Cosmonauts Nikolayev and Popovich orbiting the earth in spacecraft VOSTOK III and IV were viewed by the public in Russia, Western Europe, England, and the U.S.   The pictures were put on videotape or kinescope film during London broadcast for TV viewing in America.

August 12: Second birthday of ECHO passive communication "balloon" satellite.   Since launch, the mylar balloon had orbited the earth 9,000 times and traveled 277,257,677.67 miles; it had been used for approximately 150 communication experiments.   ECHO proved that inflatable structures would survive for long periods in space, despite sensitivity to aerodynamic drag and solar radiation. Initial orbit was 1,049-mi. apogee and 945-mi. perigee; before first birthday, orbit was 1,350-mi. apogee and 580-mi. perigee; solar pressures caused orbit to become nearly circular again; then, on second birthday, orbit was 1,175-mi. apogee and 704-mi. perigee.

- NASA Ames Research Center said that its high-altitude, radiation-research balloon with cargo of four hamsters and two monkeys had been lost in a storm over Manitoba.   The balloon was last seen nearing the end of its 1,900-mi. flight as it entered a storm over God's Lake, in northeastern Manitoba.   This was the third in Ames' series of balloon flights and the first loss of payload.
- Five NASA scientists, led by Ozro M. Coverington of GSFC, arrived in Australia to inspect proposed sites for new tracking stations.
- Four German scientists, including Eugen Saenger, former director of Stuttgart Institute of Physics, West Germany, were reportedly employed by Egyptian government to direct Egypt's new rocket design and development program.
- Reported that Soviet technicians "probably have their craft in preparation" for a flight toward Venus within the next two weeks.

August 13–17: Lunar Exploration Conference of more than 250 teachers and scientists conducted at Blacksburg, Va., by Virginia Polytechnic Institute in cooperation with National Science Foundation and NASA Langley Research Center.   Major discussion topics were "The Moon as an Earth Satellite," "Studies of the Lunar Surface," and "Lunar Flight Exploration."

August 13: Cosmonaut Nikolayev, by 10:00 PM Moscow time, had orbited the earth 40 times and had passed the million-mile mark in VOSTOK III.   At this time, his companion Cosmonaut Popovich in VOSTOK IV completed 24 orbits of the earth and logged 621,000 miles of space flight.   Sohio Research Center calculated the two spacecraft were separated by 1,793 mi. at about 8:00 PM EST.

- President John F. Kennedy, in nationally televised address, commented on the "extraordinary achievement of the Soviet Union" in double manned flights of VOSTOK III and IV.   He stated that the U.S. was behind the U.S.S.R. in space exploration, and that it would "be behind for a period in the future. . . .   [But] we are making a major effort now, and this country will be heard from in space as well as in other areas in the coming months and years."
- Congressman George Miller, Chairman of the House Committee on Science and Astronautics, spoke on the House floor in acknowledgment of successful Soviet space flights of VOSTOK III and IV. Senator Alexander Wiley acknowledged the event on the Senate floor and pointed to need for advancing the U.S. space program.
- Dr. Eberhard Rees, Deputy Director of NASA Marshall Space Flight Center, was quoted as stating that Soviet dual-flights VOSTOK III and IV came "six months to a year" earlier than he had expected.

August 13: Soviet scientist Ram Bayevsky, writing in *Pravda*, said that miniaturized equipment was recording brain and eye reactions of Cosmonauts Andrian Nikolayev and Pavel Popovich. Other instruments reported were two recorders to register information when radio transmissions ceased (during re-entry) and when the spacemen left their cabins; microphones, in the form of thin rubber tubing, to register breathing; electrodes on the skin to record heart reactions and to follow eyeball muscles.

- John Hodges, director of Canada's Regina Observatory, said in an interview that VOSTOK III and IV were orbiting in space at a time when millions of meteorites were bombarding the earth's atmosphere in an unusually heavy concentration, and that scientists and astronomers were keenly interested in the outcome of meteorites colliding with the Soviet spacecraft.

- Despite Soviet announcement that U.S.S.R. had developed a "new calculation technique" for launching space vehicles as "a matter of routine," U.S. space experts were reported to agree that the U.S.S.R. had not surpassed the U.S. in application of electronic computers to space technology.

- NASA Manned Spacecraft Center officials commented that flights of Soviet spacecraft VOSTOK III and IV would not affect programing of U.S. Project Gemini two-man orbital flights, scheduled to begin in 1964.

- U.S.S.R. had declined the offer of U.S. networks to use TELSTAR satellite for relaying live telecasts of orbiting Soviet cosmonauts, said Russell Tornabene, Chairman of the U.S. joint network committee coordinating transatlantic TELSTAR programs.

- In an interview with Tass, Soviet medical scientist Dr. Georgi Arutyunov stated that Cosmonauts Nikolayev and Popovich were supplied with sandwiches, pastries, cutlets, roast veal, chicken filets, fruit, water, coffee, and various fruit juices—in addition to special foods in tubes similar to those supplied on earlier space flights.

- Filibuster on communications satellite bill (H.R. 11040) continued in Senate, with opponents filing between 75 and 100 amendments by noon. An amendment submitted by Senator Wayne Morse would permit "the enlargement of NASA so we can apply to the satellite system the contract, lease, and license system which has permitted the Department of Defense to do such a remarkable job with defense contracts. . . ."

- USAF Atlas F missile flown 5,000 miles from Vandenberg AFB carrying a special package of cameras to help locate and correct propulsion problems encountered in previous tests, the missile meeting all test objectives.

- Reported that responsibility for all U.S. space bioscience programs, including animal flights, would be given to NASA as result of informal agreement between Secretary of Defense Robert S. McNamara and NASA Administrator James E. Webb.

- India's first satellite-tracking station had been opened at Uttar Pradesh State Observatory, Nainital, one of 12 such stations established by the Smithsonian Astrophysical Observatory.

- Gen. Bernard A. Schriever (USAF) stated in interview with *Space Business Daily* that the U.S. was ahead of the U.S.S.R. in basic space sciences, but that U.S.S.R. led in space boosters and in bioastronautics.

August 13: Ten Air Force pilots emerged from simulated space cabin in which they had spent the last month, participating in psychological test to determine how long a team of astronauts could work efficiently on a prolonged mission in space. Project director Earl Alluisi said the experiment had "far exceeded our expectations."

- University of Maryland announced its Institute for Fluid Dynamics and Applied Mathematics had received $18,179 grant from NASA, for work on an interplanetary space probe to measure solar winds and study interplanetary plasma (mixture of ions and electrons).

- Reported that USAF had contracted with Westinghouse Research Laboratories for modification for a solid-state fuel cell to determine its technical feasibility for aerospace applications and to evaluate the cell in comparison with other space power systems.

- Construction of a new institute of medical radiology in the U.S.S.R. was nearing completion, it was reported. Institute's purpose was to study the biological effect of ionizing radiation, the potentially poisonous effect of radioactive substances, possible means of protecting human beings and animals against radiation of all types, ways of using radioactive isotopes and other radiation sources for diagnosis and treatment of various diseases, and effects of radioactive fallout.

- Speaking before the Lunar Exploration Conference in Blacksburg, Va., Nobel physicist Dr. Harold C. Urey said that modern space science indicated the moon and the earth originated independently. It appeared likely that the moon was drawn into earth orbit early in the history of the solar system, some time after both bodies had been essentially formed in their present sizes and masses. He said that science now tended to disprove two widely accepted theories on the moon's origin: (1) that the moon had been broken from the earth by tidal action, and (2) that the moon and earth had originated together in one gaseous cloud.

August 14: Soviet Cosmonauts Nikolayev (VOSTOK III) and Popovich (VOSTOK IV) continued to orbit the earth, their flight paths reported by Sohio Research Center as ranging from 1,382 mi. to 1,793 mi. apart. Radio Moscow reported the cosmonauts were maintaining radio contact with earth and between themselves. By 4:10 PM Moscow time VOSTOK III completed 52 orbits; VOSTOK IV, 36 orbits. Soviet government remained silent on plans to land the pair. Cosmonaut Popovich, orbiting the earth in VOSTOK IV, passed the million-mile mark shortly after retiring, about 9:00 PM Moscow time.

- X–15 No. 3 flown by NASA pilot Joseph A. Walker to 197,000-ft. altitude and 3,784 mph speed (mach 5.13), in 9-min. flight near Delamar Lake, Nev., to evaluate new re-entry technique. Automatic roll damper ceased operating at 100,000-ft. during descent, causing X–15 to go into a simultaneous rolling and yawing motion. Pilot Walker regained stability with manual control and landed safely.

- Senate voted (63–27) to invoke cloture rule on communications satellite debate, cutting off the filibuster against the bill and limiting each Senator to an hour of speaking time.

August 14: Radiotelescope operators at Jodrell Bank Experimental Station, England, reported VOSTOK III and IV were orbiting within about 1,600 mi. of each other, as they continued their extended flight in space.

- Reported from Moscow and confirmed by Goddard Space Flight Center that pilots of VOSTOK III and IV were using radio frequencies of 19,990 megacycles, 20,006 megacycles, and 143,625 megacycles.

- Announced that U.S.S.R. Ministry of Communications had already issued postage stamp honoring double space flight of VOSTOK III and IV.

- Five live telecasts of Cosmonauts Nikolayev (VOSTOK III) and Popovich (VOSTOK IV) reportedly were received and relayed by Moscow television when the spacecraft passed within range. Soviet heart specialist Prof. Alexander Myasnikov, speaking on Moscow television, said heart activity data received from Cosmonauts Nikolayev and Popovich gave no cause for apprehension. Pulse and breathing rates, accelerated during the launchings, were normal, and the cosmonauts had not complained of any disturbing cardiovascular activity.

- British astronomer Zdenek Kopal of the University of Manchester said the gravitational pull of the earth might have heated the interior of the moon to nearly 2,000° C.   He ascribed the moon's growing diameter and its rising temperature to constant push-and-pull action of the earth's gravitational force, causing tidal friction within the lunar interior.   The friction generates heat faster than the heat can escape from the moon, thus contributing to the addition of tens of thousands of sq. mi. to the moon's surface and 1,832° of temperature over the last 4.5 billion years.   Kopal was speaking before the Conference on Lunar Exploration, Blacksburg, Va.

- Announced that 1962 Daniel Guggenheim Medal for achievement in aviation would be awarded posthumously to James H. Kindleberger, late board chairman of North American Aviation, Inc.

- Soviet scientist A. Prokhorov, writing in Soviet labor organization's publication *Trud,* stated that study of structure of living organisms would provide great opportunities for their utilization in space exploration. He stated that the "general reliability of modern automation systems is less than the reliability of each of their components taken individually.   Living organisms, on the other hand, having developed over millions of years, are much more reliable than their component organs."

August 14–17: Construction projects at NASA Marshall Space Flight and Army Missile Command, Huntsville, Ala., halted when 1,500 workers refused to cross picket lines of Local 588, International Brotherhood of Electrical Workers, protesting employment of nonunion electricians by Baroco Electrical Construction Co. District Court injunction ordered 5-day stop of strike, effective August 18, and set hearing for August 22.

August 15: In early morning broadcast, Radio Moscow announced Cosmonaut Nikolayev (VOSTOK III) had completed 61 earth orbits and Cosmonaut Popovich (VOSTOK IV) had finished 45 orbits.

August 15: Landing safely in Kazakhstan, U.S.S.R., Cosmonauts Andrian G. Nikolayev and Pavel Popovich had broken all records for number of orbits, miles traveled, and time in flight. Maj. Nikolayev in VOSTOK III had completed 64 earth orbits; 1,645,000 miles; and 95 hours, 25 minutes flying time. Lt. Col. Popovich in VOSTOK IV had completed 48 earth orbits; 1,242,000 miles; and 70 hours, 29 minutes flying time.

• President John F. Kennedy, in message to U.S.S.R. Premier Nikita Khrushchev, said: "I send to you and to the Soviet people the heartiest congratulations of the people and the Government of the United States on the outstanding joint flights of Maj. Nikolayev and Col. Popovich. This new accomplishment is an important forward step in the great human adventure of the peaceful exploitation of space. America's astronauts join with me in sending our salute to Maj. Nikolayev and Col. Popovich."

• NASA press conference at NASA Hq. on the tandem Russian orbital flights of VOSTOK III and IV. Participants: NASA Administrator James E. Webb; Deputy Administrator Dr. Hugh L. Dryden; Associate Administrator Dr. Robert C. Seamans, Jr., and Director, Office of Manned Space Flight, D. Brainerd Holmes. Mr. Webb paid tribute to the Russian accomplishment as "demonstrations of a very real technological capacity, an ability to plan and engineer and build and fly vehicles that can carry man for extended periods of time. . . . They do have significance." Dr. Dryden felt that the flights essentially conformed to the stated Russian objective, "to subject two men to identical space exposures, and weightlessness . . . plus . . . longer duration, which is needed to do this." Dr. Seamans noted that "as far as guidance accuracy is concerned, what they achieved is roughly comparable to what we achieve in our Mercury flights. To me the significant element here is that on the second flight they were able to take off within a limited period of time . . . countdowns can be protracted for one reason or another." Mr. Holmes said he was not surprised at the Russian flights: "I think we would be selling the Russians pretty short if we didn't feel that a year after they launched a Vostok on a booster that could lift that kind of a weight, something of the order of 10,000 pounds, that they couldn't indeed perform in this fashion."

Asked whether the NASA manned space flight program could be accelerated, Mr. Webb answered that the booster program could be accelerated with a crash program spending another $1 to $2 billion a year. "You can get more done. It will be done inefficiently. We believe we have a program [now] that marries all of the considerations in an effective way . . ."

To the question of who would first land on the moon, Mr. Webb replied, "I think we will make the manned lunar landing and return before they do."

As to future Russian capability, Dr. Dryden said: "there is a possibility that the Soviets can do circumlunar flight before we can. I once said there is a fifty-fifty chance, certainly no better than that, that we could do that as early as they can. It is for this reason, in fact, that our national goal was set at the lunar flight because this does require another booster on the part of the Russians as well as ourselves.

"Ever since the Space Agency has been formed we have been waiting for this other shoe to be dropped. People have told us every month, the Russians are going to produce this big booster in the next few months. Now four years have gone by and they have not yet shown us this big booster. To the best of our knowledge they have developed lighter-weight nuclear weapons and ballistic missiles rather than bigger space boosters. This is not to say that they may not be doing this. All I am saying is that they will require a bigger booster to land men on the moon."

August 15: Former President Dwight D. Eisenhower, in London press conference, said he did not believe the U.S.S.R. was leading the U.S. in space exploration.

"We have been putting all kinds of satellites in the air for meteorological readings and other scientific information, . . ." but the U.S. has not been indulging in "the same kind of spectacular" as the U.S.S.R.

"I think we should develop achievement upon achievement until it should be almost a matter of course for us to go to the moon, rather than indulge in the spectacular."

• Dr. Wernher von Braun, Director of NASA Marshall Space Flight Center, said in Blacksburg, Va., that dual space flight of VOSTOK III and IV "was impressive from the standpoint of the whole operation, . . . [but] it does not look like the Russians used any new equipment. I don't think there was a technical breakthrough. . . ."

When asked to comment on U.S. prospects of being first to reach the moon, Dr. von Braun said that manned lunar flights would require rockets bigger than any used either by the U.S. or U.S.S.R. so far. "We have such rockets under development, and the Russians also have to develop a new rocket to do this job . . . . Therefore I don't think we have any serious handicaps to overcome so far as the lunar program is involved."

• Sir Bernard Lovell, director of Jodrell Bank Experimental Station, England, said that flights of VOSTOK III and IV indicated "the Russians have a clear space superiority in the military, if not the scientific, sense." He urged that the U.S. and U.S.S.R. join each other in a cooperative attempt to explore the solar system with manned space ships.

• Congratulating the U.S.S.R. for success of its VOSTOK III and IV space flights, Prof. Hermann Oberth, leading German space scientist, added: "I am convinced that the Americans will soon come up with something similar."

• NASA announced that a second attempt to send a Mariner probe towards Venus would be made within the next few days. The spacecraft and its mission would be identical to that of the first Mariner probe, which had to be destroyed shortly after launch on July 22, 1962, because of a deviation from planned flight path.

• Soviet press agency Tass said that the cosmonauts traveling in weightless state could not depend on their sensory organs for correct signals. The cosmonauts had to rely on instruments to determine whether they were upside down or not, Tass added.

• Prof. G. V. Petrovich, Soviet rocket specialist, wrote in *Komsolskaya Pravda* that within a few decades the solar system will seem no more forbidding than once-formidable Antarctica does today. When the solar system is conquered, man will be able to undertake interstellar flights. Noting that the rockets boosting VOSTOK III and VOSTOK IV into orbit were 20,000,000 horsepower, he predicted that during the Twentieth Century rocket liners would be built with engine capacity greater than 1,000,000,000 horsepower.

August 15: Walter L. Lingle, Jr., was appointed Assistant Administrator for Management Development, NASA.

- NASA announced that its Goddard Space Flight Center had awarded three three-month study contracts for design of an Advanced Oso (orbiting solar observatory) satellite, to be launched into polar orbit during 1965. Advanced Oso would aid development of method of predicting solar flares. Contractors were Republic Aviation Corp., Space Technology Laboratories, and Ball Brothers Research Corp.

- Announced that NASA had awarded contract to Rocketdyne Division of NAA for two-year continuation of H-1 engine research and development. Preliminary letter contract of $700,000 was signed toward estimated $9,000,000 total cost. H-1 engine is used in clusters of eight to power S-I stage of Saturn C-1 launch vehicle.

During mid-August: Chandler Ross, Aerojet-General Corp. Vice President, told House Committee on Sciences and Astronautics that two Aerojet concepts of recoverable/reusable boosters compared favorably to cost and capability of Saturn C-5. One concept was Sea Dragon two-stage unmanned vehicle with payload capability "well in excess" of a million pounds; and the other was Astroplane single-stage manned vehicle with payload capability of about 550,000 pounds. "In each case the incremental cost savings is of sufficient magnitude to warrant continued feasibility studies of these systems," Ross said.

August 16: Adlai E. Stevenson, U.S. Ambassador to the U.N., stated in letter to the Senate that the communications satellite bill "provides the President and the executive branch of this Government with adequate control and influence to ensure that the instrument proposed here can be fitted or adapted to an international system when we learn enough to design one."

- D. Brainerd Holmes, NASA Director of Manned Space Flight, said that Astronaut Walter M. Schirra, Jr., probably would make his orbital flight (MA-8) the middle or end of September. Speaking at the Lunar Exploration Conference in Blacksburg, Va., Mr. Holmes made it clear that the success of recent Soviet prolonged manned space flights would not change the six-orbit mission of Commander Schirra.

- H.R. 12812, bill to amend the National Aeronautics and Space Act of 1958 regarding property rights in inventions by private companies under NASA contract, was reported from the House Committee on Science and Astronautics.

- DOD activated NRL's huge radar system near Chesapeake Beach, Md. Central feature of system was 150-ft.-diameter antenna, capable of obtaining signals from small aircraft or satellites from as far as 1,200 miles away and of receiving very strong signals from the moon (240,000 miles away). Radar would be used solely for research purposes.

August 16: Soviet Defense Minister, Marshal Rodin Malinovsky, declared in message to Cosmonauts Nikolayev and Popovich published in *Krasnaya Zvezda:* "Let our enemies know what techniques and what soldiers our Soviet power disposes of."

- S–IV stage of Saturn C–1 launch vehicle was successfully static-fired for first time in 10-sec. test at Sacramento, Calif., facility of Douglas Aircraft Co. First flight test of the liquid-hydrogen/liquid-oxygen-powered upper stage would be made in 1963.

August 17: U.S. Senate passed Administration communications satellite bill (66 to 11 vote), ending 20-day debate and liberal bloc opposition which had invoked seldom-used anti-filibuster rules. Bill was returned to the House, which had previously passed a slightly different version.

- NASA selected Dalmo-Victor Co. and Amelco, Inc., to negotiate for R&D and production services for initial systems of Satan (satellite automatic tracking antenna). New antenna system would orient itself automatically to orbiting spacecraft and would be a major improvement in NASA's worldwide Minitrack system.

- NASA Administrator James E. Webb announced appointment of Dr. Howard S. Turner to the Industrial Applications Advisory Committee. Dr. Turner was vice president for R&D of Jones & Laughlin Steel Corp. and formerly president and a director of Industrial Research Institute.

- DOD spokesmen considered there was nothing in the space flights of VOSTOK III and IV that DOD had not anticipated, and that the double flight appeared to have no immediate military significance.

- In joint interview of Cosmonauts Nikolayev and Popovich published in Soviet newspaper *Izvestia*, Major Nikolayev stated: "In an unstrapped condition in the cabin I carried out communications and ate. In a condition of weightlessness one can live and work completely. . . . In accordance with the schedule, we got up, conducted scientific work, maintained communications with the earth, made new entries in the flight log. We also had a special time for rest. In general, we worked and lived just as we do at home."

   Lt. Col. Popovich said that there had been a lot of work to do in collecting the maximum amount of scientific data. . . . "We were not idle."

- Reported that launch of NASA's second Mariner Venus probe would be delayed until August 26 or later, because of "unusual" failures in rocket's guidance equipment revealed during prelaunch checkout at Cape Canaveral.

- Navy's Polaris A–2 missile, testing components for A–3 version, exploded a few seconds after launching at Cape Canaveral.

- Seven newspaper pages were transmitted by facsimile from Westrex Corp. laboratories, New York; to Andover, Maine; to TELSTAR satellite; and back in successful 12-minute communications experiment.

- Announced that American Cable & Radio Corp., subsidiary of IT&T, would conduct the first experiments in sending teletype messages between the U.S. and Great Britain via TELSTAR communications satellite. Tests would be made between August 20 and September 7, in cooperation with AT&T and British Post Office Department.

August 17: British Air Ministry announced that three U.S. Lockheed U-2 aircraft would be stationed at RAF station at Upper Heyford, England, to conduct high-altitude weather research and atmospheric sampling for radioactive dust in flights over international waters in the North Atlantic area. Announcement said that results of studies would be published and would contribute to the work of technical committees of the United Nations.

August 18: Massive celebration in the Kremlin on Soviet Air Force Day as the new Russian space heroes, Cosmonauts Major Andrian Nikolayev and Lt. Colonel Pavel Popovich, were feted. In round of speech making, Premier Nikita Khrushchev welcomed the cosmonauts and their families, awarded them "Hero of the Soviet Union" medals, and said: "We are gathering new strength to surprise the world with our discoveries and victories." He also demanded that the Western Allies get out of Berlin.

- U.S.S.R. launched COSMOS VIII into orbit. Reported preliminary orbital data: perigee, 159 mi.; apogee, 375 mi.; inclination, 49° to Equator; period, 92.9 min.

- Norway, in cooperation with Denmark and the U.S., launched its first ionosphere probe from Andoeya, Norway, the probe reaching altitude of about 100 km. to measure electron density and collision frequency in the ionosphere. Under arrangements between the Norwegian Committee for Space Research, NASA, and the Royal Technical University of Copenhagen, Denmark, the Norwegian Defense Research Establishment designed and fabricated the instrument equipment. For this and future launchings, Norway was assembling and launching Nike-Cajun rockets, purchased in U.S. under NASA sponsorship. NASA personnel and equipment supported telemetry operations. Results of flight would be compared with data obtained from past and future probes launched at NASA Wallops Station in Norway-U.S. cooperative program.

- Reported that Secretary of Defense McNamara was not surprised by the Soviet double-orbiting of VOSTOK III and IV, and that he did not consider the flights as cause for changing the U.S. military space program.

- Senator Wiley of Wisconsin urged that Congress before adjournment conduct hearings on the military implications of U.S.S.R. two-man orbital space flights.

- Dr. Carl Sagan, University of California astronomer, warned scientists at the Lunar Exploration Conference, Blacksburg, Va., of the need for sterilization of lunar spacecraft and decontamination of Apollo crewmen, pointing out that LUNIK II and RANGER IV probably had deposited a million terrestrial micro-organisms on the moon, possibly contaminating areas around the impact points. Even more serious, he said, was the possibility that lunar micro-organisms might be brought to earth where they could multiply explosively.

- Washington *Evening Star* reported that USAF had orbited or attempted to orbit 24 unmanned, unidentified satellites since autumn 1961 when DOD adopted its policy of not issuing details of satellite launchings.

August 19: National Aviation Day, commemorating the birth in 1871 of Orville Wright in Dayton, Ohio.

- NASA launched Scout vehicle from Wallops Island in flight experiment to make direct measurements of radiative heating during atmospheric entry.
- Project to beam U.S. messages from New York to Soviet Union via TELSTAR failed when relay station at Pleumeur-Bodou, France, malfunctioned.
- In NBC TV interview, Dr. F. J. Krieger, specialist of the RAND Corporation on Soviet space affairs, pointed out that the 50th anniversary of the Communist Revolution in Russia was in 1967, and that that year may be the political target date for the accomplishment of a Soviet lunar landing mission.

August 20: Senator Howard W. Cannon spoke on the Senate floor on the need for an enlarged military space program. ". . . the major emphasis being placed by present U.S. space efforts has been on the purely scientific exploration of space and on its civilian applications, in priority above and indeed to the detriment of development of vitally needed military capabilities in space. If continued, this could prove to be a fatal mistake . . . . Unfortunately the present direction of our national efforts in space gives little or no assurance that attention is truly directed to the development of our military capabilities . . . . Clearly, we ought to have an energetic development program underway, adequately supported with funds, to find means to defend ourselves against attacks that could come from hostile orbiting space vehicles."

Senator Cannon listed several "minimum essential required space capabilities:" (1) near-space operations using manned maneuverable vehicles "capable of self-defense and having the capability of conducting offensive, defensive, and passive support missions;" (2) a standard military-civilian space launch system; (3) unmanned satellites for military communications missions; (4) improved tracking, control, and detection facilities (5) space-located facilities for R&D; (6) space-located bioastronautics research by means of a manned orbiting satellite.

- X–15 No. 2 flown to 3,433 mph speed (mach 5.22) and 87,000-ft. altitude by Maj. Robert Rushworth (USAF), the roll damper failing three times and forcing pilot Rushworth to rely on the backup system. Electronic malfunction did not interfere with flight's primary mission: to chart aerodynamic heating rates at relatively high airspeeds and moderate angles of attack.
- DOD-AEC jointly announced they were making a detailed study with NASA to determine possible effects of new radiation belt created by U.S. high-altitude nuclear blast on July 8, 1962, on manned orbital space flights. Preliminary study showed new belt was 400 mi. wide and 4,000 mi. deep, stretching around the earth above the geomagnetic equator. Government scientists stated the belt had lost half its radioactivity over a 2–3 week period and predicted the decay would be more rapid at lower altitudes than at higher ones.
- D. Brainerd Holmes, NASA Director of Manned Space Flight, stated in press interview that the recently-created radiation belt had not altered plans for Project Mercury manned flight MA–8. "We

don't have enough data on this. We are looking at this thing, but until we have looked at it a lot more completely, I would not like to see us change our plans."

August 20: DOD announced plans to develop Titan III launch vehicle powered by both solid and liquid rocket fuels and capable of thrust greater than 2.5 million pounds. To become operational by 1965, Titan III would be used primarily to launch X–20 (Dyna Soar) manned spacecraft as well as heavy military satellites. Martin-Marietta Corp. was selected as prime contractor for Titan III, to cost between $500 million and $1 billion.

- Soviet scientist A. Mikhailov, writing in *Pravda*, reported twin space flights of Maj. Andrian Nikolayev and Lt. Col. Pavel Popovich apparently proved that "normal and corpuscular radiation" presented no danger to space travelers unless it was "catastrophically intensified" and that the chances of harm from large meteorites were "rather slight."

August 21: Secretary of Defense Robert S. McNamara, in news conference, stated that U.S.S.R. was "substantially ahead" in some phases of military space development but the U.S. was acting to overtake the lead. Confirming reports that DOD space budget would not be increased, McNamara added: "Lest you interpret that as an indication that the Russian flight was not a great one, or lest you interpret it as my belief that the Russians didn't do anything we couldn't do, let me hasten to add that we have not accomplished what they have accomplished to date.

"We are behind in certain space developments, particularly those associated with large booster capabilities."

He cited development of Titan III booster as a major step toward overtaking the Soviet Union. Titan III would have between two and three times the thrust of the Soviet boosters used to launch manned Vostok satellites. Secretary of Defense McNamara stated that DOD FY 1964 funding would include increases for development of 120-in., 156-in., and 260-in. solid-propellant rocket motors. 120-in. motors would be used in Titan III launch vehicles.

- Secretary of Labor Arthur Goldberg, in telegram message, told unions involved in strike at NASA Marshall Space Flight Center it was "imperative" that union members "return to work immediately and remain at work."

- Astronaut M. Scott Carpenter, at NASA Manned Spacecraft Center conference on Project Mercury flight MA–7, stated that his May 24 orbital space flight demonstrated there was no problem associated with prolonged drifting flight, a procedure necessary in future long-duration space flights. Reporting on re-entry and landing, he said that his preoccupation with floating particles in space consumed time available for stowing loose gear, so that his attention to last-minute stowing and to difficulties in horizon pitch scanner circuit, coupled with shift to manual control of retrofire, resulted in 3-sec. delay of retrofire and 250-mi. overshoot of target landing point.

- Soviet cosmonauts held news conference in Moscow, revealing they had parachuted to earth from their space capsules, landing about 124 mi. apart. Maj. Nikolayev reported that he and Col. Popovich had unstrapped themselves and floated weightless in their cabins for an hour, maintaining communications with each other at the same time. He also stated that the flight program had not called for rendezvous of VOSTOK III and IV, and that the closest the two capsules had come to each other was 5 kilometers (a little over 3 mi.).

Col. Popovich reported they both had observed the glowing particles seen by Cosmonauts Gagarin and Titov and by U.S. Astronaut John Glenn. Popovich said they believed the particles had been "merely the exhaust of the rocket motors."

It was stated that the results of the cosmonauts' flights would be published after the collected data had been processed.

August 21: In response to statement by Col. Gen. Pyetr Braiko, Soviet Air Force chief of staff, regarding U.S.S.R. lead over U.S. in aircraft records, National Aeronautic Association president Martin M. Decker announced that Soviet claim was based on January 1, 1961, figures. At that time, U.S.S.R. held 176 world records— 113 aircraft, and 63 parachute; U.S. held 106 world records—all aircraft. By July 20, 1962, U.S.S.R. still led U.S. in total world records (including parachute), but U.S. held more aircraft records than U.S.S.R. Parachute records: 83 U.S.S.R. to 8 U.S.; aircraft records: 112 U.S.S.R. to 161 U.S.

August 22: In regular press conference, President Kennedy announced that the U.S.S. *Skate* and U.S.S. *Seadragon*, nuclear-powered submarines based in different oceans, had made underwater rendezvous beneath the ice at the North Pole.

In answer to questions on the present and future significance of the flights of VOSTOK III and IV, President Kennedy said: "Well, we are second to the Soviet Union in long-range boosters. I have said from the beginning—we started late; we've been behind. It's a tremendous job to build a booster of the size that the Soviet Union is talking about, and also of the much larger size which we're presently engaged in in the Saturn Program.

"So we are behind. We're going to be behind for a while. But I believe that before the end of this decade is out, that the United States will be ahead. But it's costing us a tremendous amount of money. We're putting a tremendous effort in research and development; but we just might as well realize that when we started—we started late.

"Last year, as you know, we made a decision to go to the moon, with bipartisan support, and it's going to take us quite a while to catch up with a very advanced program which the Soviets are directing . . . there's no indication the Soviets are going to quit . . . .

"But we're in for some further periods when we're going to be behind. And anybody who attempts to suggest that we're not behind misleads the American people—we're well behind—but we are making a tremendous effort.

"We increased, after I took office, after four months, we increased the budget for space by 50 per cent over that of my predecessor. The fact of the matter is that this year we submitted a space budget which was greater than the combined eight space budgets of the previous eight years . . . .

". . . we are spending for military purposes in space three times what we were in 1960—about $1,500,000,000. The two, at least at present, the two important points that should be kept in mind are the ability to build a large booster which can put a large satellite into atmosphere. That is being done. NASA is doing that, though there has been, of course, under Titan III contract, a booster program for the military.

"In addition, the guidance, navigation, etc., that's extremely important. That we are making a major effort in. And so that I recognize that there are those who oppose this program and then suddenly a month later say we ought to suddenly go ahead on a different basis.

"The fact of the matter is that 40 per cent of the R and D funds in this country are being spent for space. That's a tremendous amount of money and a tremendous concentration of our scientific effort.

"I'm not saying that we can't always do better. But I think the American people ought to understand it's billions of dollars we're talking about which I believe a month ago a prominent— mentioned as a great boondoggle.

"I think it's important, vital and there is a great interrelationship between military and peaceful uses of space. But we're concentrating on the peaceful use of space which will also help us protect our security if that becomes essential. . . ."

August 22: U.S. District Judge Clarence W. Allgood ordered striking electricians back to work at Marshall Space Flight Center, Huntsville, Ala. President of the AFL-CIO International Brotherhood of Electrical Workers, Gordon Freeman, also directed the workers to return to their jobs.

- French government announced first satellite, weighing 150 lbs., would be launched in March 1965 and would be followed by others three and four times as large.
- National Aeronautic Association announced that U.S. Astronaut Alan B. Shepard would receive FAI medal on September 28 for his suborbital space-flight records May 5, 1961. Also to receive medals: U.S.S.R. Cosmonauts Yuri A. Gagarin and Gherman S. Titov, for their record-setting orbital flights in 1961. Another American, Lt. Col. R. G. Robinson (USMC), would receive award for closed-course, aircraft speed record of 1,606.5 mph in F4H over Edwards AFB, Calif., November 22, 1961.
- IAS announced the 1962 Guggenheim International Astronautics Award would be made to Dr. James A. Van Allen, head of State Univ. of Iowa physics and astronomy department and discoverer of earth's radiation belts.

August 23: Nike-Apache launch vehicle carried 63-lb. payload to approximately 80-mi.-altitude from NASA Wallops Station, in experiment to measure ion concentration and composition in the upper atmosphere. Impact occurred approximately 67 mi. from the launch site, and no attempt was made to recover the payload. Experiment was conducted by NASA Goddard Space Flight Center in cooperation with Lockheed Missile and Space Co.

- USAF launched an unidentified satellite with a Blue Scout booster from Point Arguello, Calif.
- Electricians began to return to work at NASA Marshall Space Flight Center, Huntsville, Ala., after a ten-day walkout. Estimated 70% of total construction work-force, including workers of other

trades who had refused to cross electricians' picket lines, had returned to work.   Hearing on the strike by President's Missile Sites Labor Commission was scheduled for August 27.

August 23: International Edition of *New York Times* contained articles (about 5,000 words) transmitted from New York to Paris via TELSTAR satellite.

- First attempted live radio broadcast between U.S. and Europe via TELSTAR satellite was unsuccessful because of failing radio connections between Boston broadcasting station and Andover, Maine, relay station.   Boston station received, but could not understand, Swedish message; Sweden station did not receive message from Boston.   TELSTAR performed perfectly, however.

- Two new sources of radiation in the Milky Way had been discovered by USAF Aerobee probe on June 18, physicist Riccardo Giacconi announced at Third International Symposium on X-Ray Optics and X-Ray Microanalysis, Stanford University.   Giacconi said that knowledge of x-ray emissions was valuable because this kind of radiation was "intimately connected with the creation and behavior of cosmic rays and the properties of a galaxy. . . ." Discovery and investigation of such x-rays would "further understanding of the origin and dynamics of the universe . . . and may have a bearing on the study of communication of matter between galaxies."   A second probe was planned for launch on October 2, to investigate the discoveries.

- Soviet Cosmonaut Pavel R. Popovich, in an interview on Moscow television, was asked if he had seen God in space during his recent orbital flight in VOSTOK IV, and he replied: "Yes, I can confirm this.   I have seen God.   I asked his surname and he replies, 'Nikolayev, Andrian Grigorevich.' "

August 24: Speaking on the floor of the Senate, Senator Thomas Dodd supported statements by Senator Cannon on August 20, and added: "The primary goals of our foreign policy, as I see it, are first, the protection of our Nation; second, the maintenance and extension of freedom; and third, the preservation of peace. . . . Our ability to protect our freedoms and preserve the peace depends . . . on our military power in-being—on land, on sea, in the air and, tomorrow, in space. . . .

"I earnestly hope that it will prove possible to maintain our scientific space program and our moon program intact while instituting an immediate program to bolster our flagging military effort in space. . . .

"The reorganization of our space effort will require a number of changes.   First, and preceding everything else, there must be a change in policy, which clearly and publicly assigns a greater emphasis to the military potentialities of space. . . .

"Second, there must be a specific directive to the Defense Department to proceed as a matter of the utmost urgency with the development of military space systems, in particular of manned space vehicles . . .

"Third, I believe that while proceeding with the development of specialized military spacecraft, our Military Establishment should be authorized to embark immediately on a launching program of its own for manned space vehicles, perhaps employing modified versions of the equipment now under development by NASA. . . .

"Fourth, to support the establishment of military launching and vehicle development programs, our astronaut training program will have to be stepped up, and military personnel will have to be trained in the use of the complex monitoring and control systems that constitute the brain of our entire space vehicle effort.

"Fifth, our defense program in space will have to be provided with whatever money may be necessary to achieve its objectives, and to overtake the Soviet lead in manned space vehicles.

"Sixth, the machinery of cooperation between NASA and Defense will have to be overhauled in a manner which gives due consideration to the requirements of defense as well as to the requirements of science. . . ."

August 24: NASA issued proposal requests to 23 industrial firms for research, development, delivery, and installation of two separate shock-heating, free-flight wind tunnel structures and gun development facilities. Proposals were due at NASA Ames Research Center by September 17.

- Balloonist Don Piccard set new altitude record in ascent of 17,000 ft. during two-hr. and two-min. balloon flight from Sioux City to Kennebec, Iowa.

- Laser (narrow and intense beam of light) had been sent through 28 miles of haze over the Chesapeake Bay, from Tilghman's Island to Patuxent, Md., in recent nighttime experiment by USN scientists.

August 25: Launch of NASA's Mariner Venus probe, scheduled for August 26, postponed until August 27 because of technical difficulty encountered during prelaunch countdown on Atlas-Agena launch vehicle.

- Astronomers released photographs of the Humason comet in recent collision with solar wind of magnetic particles. Jesse L. Greenstein of Mt. Wilson and Palomar observatories said: "When comets are close to the sun, the disintegrating effects of solar radiation can be observed on them. This is the first time such effects have been observed at anywhere near this distance from the sun." Humason comet was about 240 million miles from the sun, traveling in a solar orbit billions of miles long.

- Congressman George P. Miller, Chairman of House Committee on Science and Astronautics, said in press interview that the greatest accomplishment of Soviet VOSTOK III and IV flight was putting two men in orbit and bringing them close together. "They're well ahead in that respect. But we can meet that in a couple of years. We'll be ahead when Titan 3 is ready."

- AEC announced two U.S.S.R. nuclear tests in the atmosphere had been detected, one with a yield of several megatons in the Novaya Zemlya area and the other test of low yield at the Semipalatinsk test site in central Siberia, the sixth and seventh Soviet tests reported by AEC in the current series.

- U.S.S.R. made unsuccessful attempt to launch Venus probe, the launch vehicle failing to achieve escape trajectory and remaining in parking orbit, NASA Administrator James E. Webb reported in Sept. 5, 1962, letter to Congress.

August 27: Atlas–Agena B vehicle boosting MARINER II Venus probe launched from Cape Canaveral at 2:53 AM EDT, the Atlas boosting its load to about 115 mi. and then the Agena B igniting to project it into parking orbit. About half an hour after launch the Agena B engines reignited, accelerating the probe's speed from 18,000 mph to 25,551 mph, when it escaped earth's gravitational pull and flew into deep space.

Shortly after launch, scientists reported that "normal dispersions" in launch vehicle engines would cause the probe to fly not more than 600,000 mi. from the planned point of intersection with orbit of planet Venus. Later calculations pinpointed the deviation at 250,000 mi., a distance that could be corrected by radio signals within a few days. After 109-day flight covering 180,200,000 mi., MARINER II would come within 10,000 mi. of Venus on fly-by mission. Instrumentation aboard the 447-lb. craft included a microwave radiometer to determine the surface temperature of Venus and details of its atmosphere; an infrared radiometer to determine fine structure of cloud layers; a magnetometer to measure changes in Venusian and interplanetary magnetic fields; ion chamber and particle flux detector to measure charged-particle intensity and distribution in interplanetary space and vicinity of Venus; cosmic dust detector to measure density and direction of cosmic dust; and solar plasma spectrometer to measure intensity of low energy protons from the sun.

- House approved Senate version of communications satellite bill (H.R. 11040) in vote of 371–10, under suspension of House rules to permit neither further changes nor conference on the measure. The bill, which would establish a corporation responsible for U.S. commercial communications satellites, was sent to the White House.

- Senate Committee on Appropriations reported H.R. 12711, Independent Offices Appropriation Bill for FY 1963, restoring $77.5 million to NASA authorization for FY 1963. Committee recommendations: $2,917,878,000 for research, development, and operation; $786,237,000 for construction of facilities; and $35,000 for general provisions. (Total $3,704,150,000.)

- Sir Bernard Lovell, Director of England's Jodrell Bank Experimental Station which was tracking MARINER II, reportedly said that the Venus probe appeared destined to become the world's most successful space effort to date.

- Subcommittee on construction of President Kennedy's Missile Sites Labor Commission conducted closed hearing on recent 10-day work stoppage at NASA Marshall Space Flight Center, Huntsville, Ala. Although electricians and other construction workers had returned to work, the basic labor dispute—union members' stopping work on vital U.S. missile and space projects because of employment of nonunion members—remained to be settled.

- Announced that Anna 1–B geodetic satellite would not be launched until October because recently-created radiation belt might adversely affect Anna's solar cells. Previously scheduled for August 14 launch, satellite would orbit in 600-mi. circular path where radiation of new belt reached peak intensity. Also reportedly postponed was launching of Canada's S–27 Alouette satellite, planned for 625-mi. circular polar orbit. Center of peak electron intensity in the new belt had shifted from 16,000 mi. above the earth to 700 mi. Since its creation, the belt had decreased in intensity by about one half; scientists expected the radiation would soon be dissipated.

August 27: DOD announced that TELSTAR was being used to synchronize master clocks in England and the U.S. Accuracy of 10 millionths of a second was obtained in first demonstration when stations at Andover, Maine, and Goonhilly Downs, England, simultaneously sent time-check signals via TELSTAR, on August 25.

- Soviet Cosmonaut Andrian G. Nikolayev, interviewed on Moscow television, indicated that although he and Cosmonaut Pavel R. Popovich had parachuted to earth from VOSTOK III and VOSTOK IV, "if we had been ordered to land in the ship we would have done this. It is very easy to land in the ship." When asked why they and Cosmonaut Gherman Titov (VOSTOK II) had all landed by parachute, Maj. Nikolayev replied that they all were very fond of parachute jumping.

- NASA Goddard Space Flight Center (GSFC) announced it was training Italian scientists and engineers for the launching of Italy's first satellite. The 165-lb. satellite would be launched by 1965 from platform in Indian Ocean off eastern coast of Africa.

- Reported that DOD was "taking steps toward developing the Nike-Zeus antimissile missile into an antisatellite weapon capable of shooting down any hostile space craft."

- USN test-fired Polaris missile in 1,400-mi. flight from Cape Canaveral.

August 28: USAF announced launch of an unidentified satellite with Thor-Agena D vehicle from Vandenberg AFB.

- NASA reported that power output of orbiting satellites ARIEL I, TRANSIT IV–B, and TRAAC dropped after U.S. high-altitude nuclear blast on July 8, the energetic particles created by explosion damaging the satellites' solar cells and thereby reducing their ability to convert sunlight into electricity. Many satellites, including TELSTAR, showed no signs of damage from radiation. TELSTAR's solar cells survived because they were specially designed to withstand radiation in the Van Allen belts.

- Speaking at meeting of American Institute of Biological Sciences in Corvallis, Ore., James C. Finn, Jr., of North American Aviation, predicted a permanent moon base would be established before 1990. He said the base would be dug at least 40 ft. underground to protect crew from bombardment by meteors and radiation and from temperature extremes between day and night.

August 29: Radio signals from Johannesburg, South Africa, activated four observation instruments on the MARINER II space probe. Transmitted data would be evaluated by Cal Tech's Jet Propulsion Laboratory (JPL).

- X–15 No. 2, piloted by Maj. Robert Rushworth (USAF), reached 3,443 mph (mach 5.21) and 97,000-ft. altitude in 9-min. flight near Hidden Hills, Calif., to obtain data on heat transfer rates at moderate airspeeds and high angles of attack

August 29: British scientist Sir John Cockcroft, addressing the British Association for the Advancement of Science in Manchester, said the British had a "good deal to learn from some American organizations who have a consistent record of success in developing new products by objective basic research and applied research." He cited three discoveries of Bell Laboratories of New Jersey as "among the principal promoters of economic growth today": extraction of pure crystals of the metal germanium; creation of strong magnetic fields in tin-and-niobium alloy; and development of the maser (molecular amplification of stimulated emission of radiation).

• Dr. Fred Singer, Director of National Satellite Weather Center, testifying before subcommittee of House Committee on Science and Astronautics, confirmed reports that Nimbus meteorological satellite system was 6 to 12 mo. behind schedule. Dr. Singer recommended the U.S. keep at least two Tiros satellites in orbit until Nimbus was ready.

• Official Yugoslav news agency Tanjug announced that U.S. Astronauts John Glenn and Alan Shepard would join U.S.S.R. Cosmonauts Yuri Gagarin, Gherman Titov, Andrian Nikolayev, and Pavel Popovich at International Astronautical Congress in Bulgaria, September 23 to 29.

August 30: Dr. Brian O'Brian, State University of Iowa physicist, predicted in interview that the new radiation belt created by July 8 high-altitude nuclear blast could last for many months—possibly as long as a year. He added that only a major sun-spot storm would eradicate all traces of the band before mid-1963, and such a storm was unlikely. An associate of Dr. James A. Van Allen, Dr. O'Brian based his prediction on data from orbiting Injun satellite which he had been studying for the past month.

• Dr. Morris Tepper, NASA Director of Meteorological Systems, told subcommittee of House Committee on Science and Astronautics that he doubted the launch schedule for first Nimbus satellite could be accelerated and stated that the Nimbus program had never been considered "urgent."

• National Aeronautic Association announced John Stack, former NASA Director of Aeronautical Research, would receive 1962 Wright Brothers Memorial Trophy. Stack was cited for his major contributions to aeronautical research and for "his leadership and vision" during 34 years with NACA and NASA.

August 31: NASA launched four-stage Scout from Wallops Station with re-entry heat-shield experiment, two stages carrying payload to near 80-mi. altitude while other two stages fired payload back to earth near speeds of 18,000 mph.

• President Kennedy signed into law the communications satellite bill, which would establish a private corporation in charge of U.S. portion of future global communications satellite network. Satellite corporation board of directors would include six named by communications industry, six elected by public stockholders, and three selected by the President.

August 31: James E. Webb, NASA Administrator, was awarded the McCurdy Medal, honoring NASA for "outstanding contributions to the world in the field of science," at the Canadian International Air Show, Toronto.

- Senate passed FY 1963 appropriations bill (H.R. 12711) as amended, giving NASA $3,704,115,000 for FY 1963. This amount represented $60,115,000 restoration of funds cut by House in its appropriation of $3,644,115,000 for NASA in FY 1963.
- Soviet Cosmonauts Andrian G. Nikolayev and Pavel R. Popovich described their re-entry in VOSTOK III and IV, in *Pravda* article. Their experience was "probably one of the most tremendous impressions in life. . . . Our antennas charred, the [radio] connection between the earth and the spaceships stopped. We did not close the shields over the portholes and watched with curiosity, from behind the heat-proof glass, flames that raged in different colors alternating from blue to dark red."
- NASA-supplied Nike-Cajun sounding rocket launched by Swedish Committee of Space Research at Kronogård, Sweden, in experiment to study auroral event in progress, the rocket performing as predicted but payload failing to separate.
- Levitt Luzern Custer, pioneer balloonist and member of the Early Bird Club, died at 74 in Dayton, Ohio. In 1909, Custer's first balloon flight set record for longest flight of free balloon in an hour's time, from Dayton to Middletown, Ohio.
- Vincent Johnson, NASA Centaur program manager, reportedly said NASA would spend an additional $33 million to accelerate slipping Centaur launch vehicle development. NASA would divert funds from other programs rather than request supplement from Congress.

During August: First full-scale model of 3-man Apollo spacecraft underwent preliminary ocean drop-testing off California coast and was shipped to NASA Manned Spacecraft Center for further testing. J. Thomas Markley, Apollo project engineer for NASA, described all spacecraft structural tests thus far as "successful."

- Dr. Robert R. Gilruth, Director of NASA Manned Spacecraft Center, presented details of Apollo spacecraft at Institute of the Aerospace Sciences (IAS) meeting in Seattle. During launch and re-entry, 3-man crew would be seated in adjacent couches; during other phases of flight, center couch would be stowed to permit more freedom of movement. Apollo command module cabin would have 365-cu. ft. volume, with 22-cu. ft. free area available to crew. "The small end of the command module may contain an airlock; when the lunar excursion module is not attached, the airlock would permit a pressure-suited crewman to exit to free space without decompressing the cabin. Crew ingress and egress while on Earth will be through a hatch in the side of the command module."

During August: Under Secretary of the Air Force Joseph V. Charyk
told Seventh Symposium on Ballistic Missile and Space Technology that USAF had flight-tested two Agena D upper-stage
vehicles, the first firing occurring in June. Agena D would
accommodate a variety of payloads, whereas earlier models
Agena A and B had integrated payloads.

• Congressman Bob Wilson of House Armed Services Committee,
addressing the Institute of World Affairs, recommended establishment of a U.S. Military Space Academy. He said he was "deeply
concerned over the tendency of our space program to emphasize
the experimental and scientific and de-emphasize the military
aspects of space. . . . The Air Force has started a limited
program to train spacemen for the future. It should be expanded
into a full-fledged space manpower program. Perhaps a 'fourth
service' should be set up to train men in astronautics, hypersonic
aerodynamics, the physiological and psychological aspects of
flight."

• American Rocket Society gave Congressman George Miller of
House Committee for Science and Astronautics a special award
for his "outstanding leadership in space."

SEPTEMBER 1962

September 1: NASA Venus probe MARINER II passed the million-mile mark on its interplanetary journey. JPL scientists said that they would send signals Sept. 3 to alter the trajectory and swing MARINER II on path that would come within 10,000 mi. of the planet Venus.

- U.S.S.R. made another unsuccessful attempt to launch Venus probe, the launch vehicle failing to achieve escape trajectory and remaining in parking orbit. (See Sept. 5).
- U.S. sources disclosed that U.S.S.R. had unsuccessfully attempted to launch a Venus probe on August 25, and that three fragments of the spacecraft were orbiting the earth. In accordance with its policy, U.S.S.R. made no announcement of unsuccessful launch.
- USAF launched an unidentified satellite with a Thor-Agena B booster from Vandenberg AFB.
- AEC-DOD jointly announced that the man-made radiation belt created by U.S. high-altitude nuclear explosion was stronger than anticipated and might last for many years. The blast also intensified natural Van Allen radiation belt more than anticipated. AEC-DOD added that the new radiation belt had completely knocked out communications from satellites TRANSIT IV–B, TRAAC, and ARIEL, but it would "not constitute any hazard to manned satellite launchings that we have planned in the near future." The belt extended from 200 mi. above the earth to 700 mi., where it merged into the Van Allen belts.
- INJUN satellite was apparently unaffected by artificial radiation belt, reported Dr. Brian J. O'Brian, State University of Iowa physicist. INJUN orbits earth at about 500-mi. altitude, near center of the new radiation belt which had destroyed transmission from satellites ARIEL, TRANSIT IV–B, AND TRAAC. Dr. O'Brian speculated that INJUN transmission survived because satellite used low-efficiency solar cells, not as rapidly affected by radiation as high-efficiency cells used in ARIEL.
- Reginald Lascelles, personal assistant and spokesman for Sir Bernard Lovell, director of Jodrell Bank Experimental Station, said Lovell was afraid the new radiation belt would "seriously affect space exploration by radio astronomy."

September 2: MARINER II, 1.2 million mi. from earth on its voyage to vicinity of Venus, successfully swung a yard-wide antenna around and focused it on earth, sending radio beam back to scientists at Jet Propulsion Laboratory. The antenna maneuver was performed on command of on-board timer and was first of two in-flight maneuvers crucial to correcting the probe's path to within 10,000 mi. of Venus; after scientists evaluated results of antenna orientation, NASA Goldstone Tracking Station would send radio command to fire rocket aboard the 447-lb. craft.

September 2: In editorial noting addition to FCC of 19 specialists to plan enforcement of communications satellite bill's provisions for public interest protection, *New York Times* said: "the policy-making decisions confronting the FCC will become increasingly complex. How well and how wisely it discharges its duties will be of crucial importance in shaping the future of this mechanism [communications satellite] for bettering life through man's conquest of space. . . . The march of science is sure to create many more situations in which the partnership of private and public efforts will entail a need for dependable instruments of Government regulation. This makes doubly urgent the successful exercise of the commission's first test."

- More than 1,600 U.S. military aircraft participated in simulated attack and defense maneuvers during NORAD's Operation Sky Shield III, with all civilian aircraft grounded during the 5½-hr. exercise.

- Tenth anniversary of Livermore Branch of Lawrence Radiation Laboratory, University of California. Established to "help cope with problems involved in maintaining American superiority in nuclear deterrence," the laboratory devoted about half of its effort to weapons research and the remainder to peaceful uses of atomic energy.

September 3: MARINER II flight-path correction was postponed for at least 24 hours, because scientists could not determine if the Venus probe was stabilized on the moon or the earth. JPL scientists received moon-reference indication from MARINER II's earth-sensing system and earth-reference indication from its high-gain antenna. If mid-course maneuver were executed with moon-lock instead of earth-lock, the probe probably would come within 12,000 mi. of Venus instead of 10,000 mi.

- Dr. Hugh L. Dryden, NASA Deputy Administrator, interviewed in *The National Observer*, said: "Some people have drawn the conclusion that the Russians are ahead of the United States in the endeavor to land men on the moon and return them safely. This is based on a misconception that going to the moon is something like progress down a one-way street, passing successive street-corners.

   "That is far from being the case. The enterprise requires mobilization of a large task force with many tasks to be done and brought to fruition simultaneously. In some aspects, it makes small difference in what order the tasks are performed. The fact that the Russians have done one part of the work of going to the moon does not mean that they are ahead in the entire enterprise. . . ."

   When asked about U.S. participation in international space programs, Dr. Dryden said: "The United States now has arrangements with more than 55 countries for space activity. . . . We hope to increase the number of countries with which we are co-operating. We have been engaged in negotiations with the Soviet Union toward that end.

   "Recently, Academician Anatoli Blagonravov and I concluded the first phase of the negotiations with a joint recommendation to our governments that they undertake co-operative efforts in meteorology, communications, and the investigation of the earth's magnetic field. . . ."

September 3: Congressman George P. Miller, chairman of House Committee on Science and Astronautics, told the press that "since the orbiting of the Soviet twin Cosmonauts, the agitation [among members of the Senate, professional military societies, and the trade press] borders on panic and constitutes a threat to a program which is not only very important but complex and carefully planned. . . . I have been a consistent advocate of a military capability in space for the United States second to none, but I see no reason why our military requirements cannot be met without hamstringing or jeopardizing the civil space program. . . . There has been so much misinformation and misimpression generated about our space program, both the civil and military aspects, that I can no longer remain silent. The record has to be set straight and I intend to do it." He said he would deliver a major speech on the House floor later this week, when he would place the space program into "balanced perspective."

- Report outlining NASA's manned space flight projects was published by Senate Committee on Aeronautical and Space Sciences, with cooperation of NASA. Project Gemini two-man flights, planned for 1964 and aimed at orbital rendezvous, might be as much as a week long. Three-man Apollo flights would follow, circumlunar flights to be followed by lunar landing missions. Advance-design Gemini spacecraft would be first designed specifically by U.S. to come down on dry land, using Rogallo wing to control descent.

- Preparations for Operation Harp (High Altitude Research Project) announced by Dr. D. L. Mordell, Faculty of Engineering Dean at Canada's McGill University. In the McGill-initiated project, McGill's 66-in., 470-lb. Martlet missiles would be fired to approximately 200-mi. altitude from two 16-in., 140-ton guns provided by U.S. Navy. With the first launching in October from Barbados, B.W.I., McGill would become the first university to conduct its own space research program. At least 6 firings would be made before Christmas. Also involved in Harp were U.S. Army, which transported naval guns from Hampton Roads, Va., to Barbados; Florida State University meteorologists; and USN weather-observation aircraft.

- P–1127 VTOL jet fighter airplane was publicly demonstrated for first time at Farnsborough, England, Air Show. Developed by Hawker Siddeley Aviation, the P–1127 uses single Bristol Siddeley Pegasus jet engine whose thrust is directed vertically for takeoff, then directed backward for horizontal flight at supersonic speeds.

- U.S.S.R. conducted atmospheric nuclear test explosion of intermediate-range yield, between 20,000 tons and one megaton. The firing took place near Novaya Zemlya and was the ninth in current Soviet series to be announced by AEC.

- Prof. Alexander A. Mikhailov, chairman of U.S.S.R. Academy of Sciences Council for Astronomy, was reported by U.S. magazine as saying that the U.S.S.R. had 36 astronomical observatories in operation and 6 new facilities under construction.

- Prof. Bruce Patton of London's Institute of Education told British Association for the Advancement of Science that English should be adopted as the universal language. He gave as one reason the fact that English was the accepted language of civil aviation (except in Communist countries) and as another the fact that approximately half the world's scientific journals were in English.

September 3–8: Tenth Conference on Science and World Affairs held in London, with more than 200 scientists from 37 countries attending. President Kennedy urged the conference to "explore fully and objectively the basic reasons for our failure thus far to reach agreement" on a nuclear test ban.

September 4: Radio signals to Venus probe MARINER II, nearly 1.5 million mi. from earth, repositioned the craft and fired an onboard rocket to send the probe on desired trajectory toward Venus. Launched August 27, the 447-lb. interplanetary probe would have missed Venus by 233,000 mi. if the mid-course maneuver had not corrected its path. Signals from JPL's Goldstone Tracking Station were sent after 24-hr. delay for scientists to determine that the spacecraft's antenna was pointing to earth and not the moon. Assuming that its corrected course was nominal, it was estimated that MARINER II would come within 9,000 mi. of Venus December 14, covering about 180,200,000 mi. through space and relaying scientific data on interplanetary space and Venusian atmosphere.

- In joint letter to NASA Administrator James E. Webb, Chairman of Senate Committee on Aeronautical and Space Sciences Robert S. Kerr and Chairman of House Committee on Science and Astronautics George P. Miller said: "The world must of necessity admire the remarkable achievements of the Soviet Union in the field of space. A shadow is thrown over the entire space effort through their refusal to admit to failures. . . .

  "We feel it is important that if the U.S. Government possesses any information relative to unsuccessful attempts by the Soviet Union to launch a spacecraft to Venus, or other planetary probes, that this information should be made available to our committees and to the American people."

- John Rubel, Director of DOD Office of Research and Development, testified before subcommittee of House Committee on Science and Astronautics that it was "highly doubtful that DOD would undertake a new weather satellite program at this time," and stated the military interest in weather satellite development was adequately protected by existing NASA and Weather Bureau programs.

- Dr. S. Fred Singer, director of National Weather Satellite Center, said he was recommending an operational Tiros weather satellite system by next June because of Tiros' "quite astounding" ability. TIROS V was first to spot five of the ten major tropical storms around the world in August. He added that scheduled launch of sixth Tiros satellite would not be delayed by the man-made radiation belt.

- U.S.S.R. formally protested "provocative flight" of a U.S. U–2 reconnaissance airplane over Sakhalin Island in the Pacific. Replying to the U.S.S.R., U.S. stated that the pilot of the August 31 flight "has reported he was flying a directed course well outside the Soviet territorial limits but encountered severe winds during this nighttime flight and may, therefore, have unintentionally overflown the southern tip of Sakhalin." The note reaffirmed

U.S. policy of not permitting reconnaissance flights over Soviet territory, established after U-2 piloted by Francis Gary Powers was brought down in U.S.S.R. May 1, 1960.

September 4: British physicist Dr. Eric Mendoza of University of Manchester explained superconductivity principle at British Association for the Advancement of Science meeting. He stated the problem as why several superconducting metals (such as lead and aluminum) at extremely low temperatures suddenly lose all traces of electrical resistance and are able to conduct currents without producing heat. Normal electrical resistance is produced when electrons are knocked out of their paths by collision with heavy metallic atoms. In superconductors, he said, the electrons get past the atoms without being deflected, because "electrons, instead of moving as individuals, move in pairs." Vibrations of the metallic atoms in superconductors are so slowed down by extremely low temperature (minus 273° C) that they are unable to break the pairs, so current flows without friction.

- Reported that U.S.S.R. had invited Dr. Bernard M. Wagner of the U.S. to visit U.S.S.R. for cooperative exchange of space medicine data, including information obtained from VOSTOK III and IV manned orbital flights. Dr. Wagner, chairman of Department of Pathology, New York Medical College, would leave for Moscow October 5.

September 5: NASA Administrator Webb reported that the U.S.S.R. had tried to send four probes to Venus and two to Mars since October 10, 1958. In letter to chairmen of the Senate Committee on Aeronautical and Space Sciences and the House Committee on Science and Astronautics, Mr. Webb said Venus attempts were made on February 4, 1961; August 25, 1962; and September 1, 1962; and each time the launch vehicle achieved parking orbit but failed to send probes on escape trajectory. The one partially successful Venus probe, launched February 12, 1961, achieved proper flight path but its radio transmission failed when probe was 4.5 million miles from earth. Mars attempts were made October 10 and 14, 1960, and each time the launch vehicle failed to achieve parking orbit.

- Agreement establishing U.S.-Italy cooperative space program, signed in May, was confirmed in Rome by Vice President Lyndon B. Johnson and Italian Foreign Minister Attilio Piccioni. The Memorandum of Understanding between NASA and the Italian Space Commission provided for three-phase program, expected to culminate in launching of a scientific satellite into equatorial orbit. Generally, NASA would provide the Scout rockets and personnel training; Italians would launch the vehicle with its Italian payload and would be responsible for data acquisition as well as for towable launch platform located in equatorial waters.

- NASA announced it would negotiate with three companies to conduct three-month studies of a lunar logistics system. Negotiations with Space Technology Laboratories, Inc., related to $150,000 study of types of spacecraft which could carry supplies to manned Apollo landing site on the moon; negotiations with Northrop Space Laboratories and Grumman Aircraft Engineering Corp. related to $75,000 studies and operational analyses of possible cargoes for the lunar logistics spacecraft. Various NASA centers simultaneously would study lunar logistics system trajectories, launch vehicle adaptation, scheduling, alternate spacecraft propulsion concepts, lunar landing touchdown dynamics, and use of roving vehicles on the lunar surface.

September 5: White House announced President Kennedy would visit military and civilian missile and rocket installations at Cape Canaveral, Fla.; Huntsville, Ala.; Houston, Tex.; and St. Louis, Mo., on September 11–12. President Kennedy would be accompanied by Vice President Johnson, Secretary of Defense McNamara, NASA Administrator Webb, BOB director David E. Bell, and several Congressmen. Purpose of trip was "to study the work being done in this most important area and in connection with the preparation of the fiscal year 1964 budget. . . ."

- 13-min. radio program relayed from New York via TELSTAR satellite to Europe, where it was broadcast by Radio Free Europe and Radio Liberty to audiences in U.S.S.R. and Communist-bloc countries. Program featured message on U.S.-U.S.S.R. nuclear disarmament by Senator Hubert H. Humphrey.

- Deputy Secretary of Defense Roswell L. Gilpatric, addressing representatives of Midwest industries and universities in South Bend, Ind., said:

  "The United States believes that it is highly desirable for its own security and for the security of the world that the arms race should not be extended into outer space, and we are seeking in every feasible way to achieve this purpose. Today there is no doubt that either the United States or the Soviet Union could place thermonuclear weapons in orbit. . . .

  "We have no program to place any weapons of mass destruction into orbit. An arms race in space will not contribute to our security. . . ."

  He added that U.S. military space program has two objectives:

  "First, as part of our overall Defense effort, we have continuing programs to ensure that the United States will be able to cope with any military challenge in outer space. Our programs in this area are under constant review, and this review indicates that our present rate of effort is entirely adequate.

  "Second, as a part of our national space program, we in the Defense Department, along with NASA, are actively exploring the potentialities of outer space as a useful part of our expanding universe. We are developing through activities in space and observations from space our ability to improve our capabilities in fields such as communications, navigation, meteorology, mapping and geodesy. Many branches of industry are contributing to this endeavor by improving propulsion, electronic, photographic, communications and other components of systems for space research and utilization. These programs have great significance not only for our military forces but for the economic and scientific advance of the United States and of the whole world. The progress that they represent, like all scientific advances, is neutral in its political and moral content. . . ."

- NASA announced that nine companies had submitted proposals to develop lunar excursion module (LEM), one of three modules comprising the Apollo spacecraft. After evaluating proposals, NASA would award contract "within six to eight weeks," according to the *Wall Street Journal.*

September 5: NASA Marshall Space Flight Center announced award of $4,673,327-contract to Ets-Hokin and Galvin, Inc., of San Francisco, for construction of main structural portion of Advanced Saturn test stand at MSFC, Huntsville, Ala. When completed, the 405-ft.-high test tower would be used for static-firing the 7.5-million-lb.-thrust first stage of Advanced Saturn.

- Re-entry and disintegration of SPUTNIK IV reported by Edward A. Hallbach, director of Milwaukee Astronomical Society. Hallbach and others observed the satellite as it broke into about 24 pieces, most of which fell from orbit toward Green Bay, Wisc., area and Lake Michigan. Law officers in a wide area of northern Wisconsin reported seeing flaming objects at about the same time. SPUTNIK IV was launched by U.S.S.R. on May 15, 1960.
- Chunk of metal, too hot to touch, was discovered in Manitowoc, Wisc., street by two policemen. Considered as possibly part of SPUTNIK IV, the 20-lb. object was sent by local members of nationwide Moonwatch tracking network to Moonwatch hq. at Smithsonian Astrophysical Observatory, Cambridge, Mass., where fragment would be analysed.
- Hughes Research Laboratories announced development of high-power amplifier tube that could eliminate spacecraft radio blackouts such as those experienced by astronauts during atmospheric re-entry. Tubes produce about 10 times the continuous power output of any previous tubes and could pierce heat-induced ion shields to transmit messages.
- General Mills launched balloon, trailing 315-ft.-long reflectorized polyethelene tube in test to improve long-range communications, from GM's research center at New Brighton, Minn. The cylinder, 23 ft. in diameter, was designed as a relay device to receive commercial TV signals from WKBT–TV, La Crosse, Wisc., and relay them 260 mi. to USAF base, Wadena, Minn.

September 6: NASA Aerobee 100 sounding rocket launched at White Sands Missile Range (WSMR) by Jet Propulsion Laboratory (JPL), the rocket reaching 46-mi. altitude in successful experiment to measure ultraviolet ray airglow spectrum.

- JPL scientists reported that Venus-bound MARINER II was successfully transmitting data from nearly two million mi. in space. John W. Thatcher, staff engineer of JPL's Deep Space Instrumentation Facility (DSIF), said, "We fully expect that we will be able to track it and follow it well beyond Venus." Signals were reported coming in relatively "loud and clear" at receiving stations in Goldstone, Calif.; Johannesburg, So. Africa; and Woomera, Australia.
- International Telephone & Telegraph Corp. (IT&T), announced plans for NASA Project Relay satellite communication experiment to link North and South America. To be launched into orbit from Cape Canaveral this year, the satellite would relay 12 telephone calls at once, from IT&T ground relay station in Nutley, N.J., to portable ground station near Rio de Janeiro. The satellite, built by RCA for NASA, would also transmit TV from North America to Europe via ground stations at Andover, Me.; Goonhilly Downs, England; and Pleumeur-Bodou, France. In another set of experiments, Relay would establish telephone communications between Brazil and England. Relay would radiate microwave signal of 10 watts, compared to TELSTAR's 2-watt signal.

September 6: Congressman George P. Miller, chairman of the House Committee on Science and Astronautics, defended the U.S. space program in speech on floor of the House. Arguing that Soviet VOSTOK III and IV manned orbital flights did not justify changes in the U.S. program, he said:

"Certainly I will not stand here and say that more and better military space programs should not be undertaken. What I do say is that a considerable effort in this field has long been underway and that the Monday-morning quarterbacks whose teeth start chattering after every Russian 'spectacular' might exhibit a bit more faith in those who have the actual responsibility for the defense of the country. . . .

"[We should not overlook] . . . the fact that the Soviets have now demonstrated two important capabilities: First, that they have sufficient launch facilities and rocket reliability to launch two manned spacecraft within a short time, and second, that they have the ability to time launches with great precision. These capabilities are an important step toward the development of the rendezvous and docking technique which will be of great value in achieving many advanced objectives in space exploration, some of which may have potential military value.

"These, I think, are the formidable implications of their recent feat. But these implications are no reason for us to put our program in a constant state of flux, with projects starting, stopping, and shifting in response to each new Soviet development. Our undertaking is gigantic, immensely complex. It cannot be assembled and disassembled and redesigned and reassembled without losing its direction and momentum, and the space contest itself. . . .

"There is a suggestion that the military services should duplicate work now in progress under the civilian agency. There is nothing particularly military about solutions to problems involved in the effects of long-term weightlessness, radiation, and isolation. The capabilities of both civilian and military research agencies should be brought to bear on these problems in a coordinated manner.

"On the record, then, it appears that the critics of our space program are not asking that we do twice as much—but that we do everything twice. I do not think the economy will take that, but even if it would—it just does not make sense.

"And how about the reciprocal use of the knowledge we are gaining? Science knows no exclusive applications technologically speaking, what is developed by civilians can be used militarily and vice versa. . . .

"It seems farily well established that the real cause of all this squabble about the military-in-space stems from an inhouse difference of opinion within the military establishment. The problem is not that our civil space program is retarding the military. On the contrary, it is enhancing it and will continue to do so in the future. The problem is that the military space enthusiasts have not been able to obtain all the green lights they want from their bosses. . . . I am confident the controversy will be resolved in time—and expeditiously, I hope."

September 6: USIA Voice of America began broadcasting reports of U.S.S.R. unsuccessful interplanetary probes and sent the announcement to newspapers all over the world via its wireless press network. Public release of information on Soviet space failures was made September 5 in form of letter from NASA Administrator James E. Webb to Congressmen.

- TELSTAR satellite was used to relay exchange of reports of action on New York and Paris stock exchanges, in 10-min. test transmission of telephone call between Michael W. McCarthy, chairman of Merrill Lynch, Pierce, Fenner, and Smith, Inc., to Frederick J. Sears, manager of Merrill Lynch's Paris office.

- Weather Bureau announced weather data gathered by Tiros satellite was relayed to France via TELSTAR satellite.

- Ernest Brackett, NASA procurement officer, told American Management Association that about 80% of NASA's contracts are made for research and development, and outlined NASA procedures of awarding R&D contracts.

- USN launched second Polaris A–3 in partially successful test from Cape Canaveral, a malfunction during second-stage flight causing the missile to fall far short of its intended 1,950-mi. range.

- In response to U.S. explanation that U–2 reconnaissance plane may have unintentionally overflown Soviet territory, Soviet government newspaper *Izvestia* said U.S. Government spokesmen in the U–2 situation were "making clumsy attempts to exonerate themselves, attempts unworthy of responsible politicians." Soviet Communist Party newspaper *Pravda* claimed U.S. was renewing "provocative" reconnaissance flights and said that "such activities only intensify the threat of war, heat up the international situation and stir up suspicion and hostility between countries."

- British Minister of Aviation Julian Amery told London audience that Aviation Ministry was studying an "aerospace plane" that could be operated either in the atmosphere or in space. London newspapers reported that the craft was a "space fighter" that takes off in conventional manner but can fly into orbit and back at speeds up to 18,000 mph. Space fighter was reportedly based on concept of sustained flight rather than boost-glide principle used in U.S. Dyna Soar.

- U.S. and Australian scientists reported in *Science* that phases of the moon affect rainfall on earth. Their studies in both northern and southern hemispheres disclosed that heavy rains fall most frequently in the first and third weeks of the synodical month— the period from one new moon to the next.

September 7: AEC-DOD-NASA announced that results of joint study of artificial radiation belt indicated Project Mercury flight MA–8 could be made with no change of plans and with no fear of harm to Astronaut Walter M. Schirra, Jr. (Cdr., USN). Study showed that radiation on outside of capsule during Cdr. Schirra's 6-orbit flight would be about 500 roentgen (R). Shielding, vehicle structures, and flight suit would cut this to 8 R on the astronaut's skin. "This exposure is well below the mission limit previously established by NASA for the manned flight program," the report said.

Other results of the investigation: Soviet Cosmonauts Nikolayev and Popovich (VOSTOK III and IV orbital flights in August) received very small radiation doses on their double-mission flight; radio interference caused by new radiation belt constituted "a significant problem to radio astronomy"; radiation doses caused rapid deterioration of solar cells aboard ARIEL, TRAAC, TRANSIT IV–B satellites as previously reported, but TELSTAR solar cells were apparently unaffected; future space missions were being reviewed "in light of the present information concerning the new belt." It appeared likely that "a substantial number" of future space experiments would be changed to avoid possible damage from radiation.

September 7: Replying to queries of whether Administrator Webb's public letter of Sept. 5 on Soviet failures marked a reversal of NASA policy, NASA said no change had been made in policy of not releasing information on unsuccessful Soviet space attempts. The September 5 letter to Congressmen was "an exception to a continuing policy."

• Aerospace Industrial Life Sciences Assn., established to "assist the industrial aerospace life scientist through exchange of information and to increase his contribution to the nation's space efforts through better professional communication in the industrial life sciences field," with Dr. Charles M. Gell of Ling-Temco-Vought as president. AILS was accepted as full affiliate of U.S. Aerospace Medical Association.

• D. Brainerd Holmes, NASA Director of Manned Space Flight, addressed National Advanced Technology Management Council in Seattle on the U.S. lunar program. "Although we are still in a very early phase of the greatest engineering project that man has ever undertaken, the Nation has made four crucial management decisions which clear the way for action. First, the goal has been identified. We are to land American explorers on the moon. Second, the timetable has been laid down. We are to carry out the mission and return the U.S. explorers safely to earth within this decade. Third, we have settled on the organization to meet the national goal on schedule. Finally, we have selected the method which we believe can accomplish the lunar expedition on the shortest possible time schedule, with the greatest assurance of success, and at the least expense. You do not need to be a specialist in management to realize that decisions such as these are central to the success of any job—what to do, when to do it, who is to be in charge, and how it is to be done. In the United States space program, these fundamental decisions have been made. And we have proceeded with the work to carry them out at the swiftest pace consonant with other processing needs of the country and with the resources available . . . .

"The achievement which stands out [in the manned lunar landing program] . . ., in my opinion, has been the launching by the people of the United States of a truly national effort to demonstrate our determination and our capacity to sail on the new ocean of space and to master the technology of the space age in all its aspects—for the security of the Free World and the good of all mankind.

"The lunar landing is no stunt. The most important accomplishment will be the development by this Nation of the ability to make the landing, and not the landing itself. Achievement of such ability is worth the great investment in brains and industrial capacity and technological advance that we are making. The lunar landing is a yardstick—a measure of this country's technological leadership, a measure of the ability of this democracy to 'manage,' if you will, a great engineering and technological undertaking in the national interest."

September 7: In private audience with Pope John XXIII, Vice President Lyndon B. Johnson presented the Pope with a small model of TELSTAR communications satellite.

September 8: RL–10 rocket engine successfully static-fired in second test at NASA Marshall Space Flight Center. Cluster of the liquid-hydrogen engines would power second stage (S–IV) of the Saturn C-1.

- Atlas rocket for manned orbital flight MA–8 was static-fired successfully for 11 seconds at Cape Canaveral.

- USAF announced Deputy Chief of Staff for Research and Development would be focal point for space projects within the Air Staff. Lt. Gen. James Ferguson, DC/S for R&D, would be responsible for such projects as Midas, Titan III, Dyna Soar, communications satellites, and NASA liaison and support.

- Soviet Cosmonaut Yuri Gagarin, speaking at Danish-Soviet Friendship Advancement Society in Copenhagen, attacked U.S. space projects. He charged U.S. launched satellites which spied on other countries, endangered space travel by its atmospheric nuclear tests, and put millions of copper needles in space, making space research difficult. He added: "I also know a certain country is considering plans to set up bases on the moon from which attacks on other parts of the world can be made."

September 8–22: First NASA exhibit in Europe was displayed in pavilion of Swiss National Trade Fair, in Lausanne.

September 9: Three-man crew "survived" a simulated week-long, round-trip to the moon, in simulation chamber at Martin Co. plant, near Baltimore. Crew in the simulator, which resembled Apollo spacecraft in size and shape, consisted of 3 NASA civilian test pilots: Donald L. Mallick and Harold E. Ream of Langley Research Center, and Glenn W. Stinnett of Ames Research Center. During the simulated flight, made in July, the men took turns as pilot, co-pilot, and navigator, and each made a simulated landing on the moon's surface in the special one-man capsule designed by Langley scientists.

- Plan calling for international sounding-rocket launching site to be set up under U.N. auspices would be presented to U.N. General Assembly by 28-nation Committee on Outer Space, *New York Times* reported. Originally proposed by U.S. last June, plan provided that all member countries would be given access to facility for collection of meteorological and other scientific data.

- U.N. Scientific Committee on the Effects of Atomic Radiation released detailed report warning that any radiation dose, however small, can cause biological and hereditary effects. "The Committee therefore emphasizes the need that all forms of unnecessary exposure be minimized or avoided entirely." Adopted unanimously by the 15-nation committee, the 442-page report was the most comprehensive in the 7-year history of the Committee.

- 25-franc postage stamp picturing TELSTAR communications center at Pleumeur-Bodou was issued by France, first nation to so honor TELSTAR and its international communications achievements.

September 10: NASA Project Mercury officials postponed for three days the 6-orbit space flight of Astronaut Walter M. Schirra (MA–8), now scheduled for September 28. Delay was to permit further time for flight preparation.

- AEC–DOD announced that a "few" atmospheric tests would be conducted with nuclear devices dropped from airplanes, beginning late this month. Extension of atmospheric series, originally planned to end in July, was expected to end in 60 days. In addition to the atmospheric tests, plans called for three more high-altitude explosions, to be conducted with rocket-launched nuclear devices and at altitudes below 100 mi. to avoid creation of further radiation belts.

- NASA indicated it would begin series of flights with an inflatable paraglider in mid-1963. Flights would provide information on frequency and size of micrometeoroid particles in space and allow post-flight laboratory study of impacts caused by such particles. Paraglider would be launched by Aerobee research rocket from White Sands Missile Range (WSMR) to height of 700,000 ft., and would descend to earth at speed of mach 5. Its re-entry path would be almost straight down—at an angle of 82°. Inflated paraglider would be 14-ft. long with span of 18 ft.; it would weigh 170 lbs. during launch and 85 lbs. during re-entry.

- Fire in simulated space cabin at SAM, Brooks AFB, Texas, caused serious smoke-inhalation injuries to Capt. Carl C. Fletcher, Jr. (USAF), one of two men spending 13th consecutive day in cabin; his partner, Capt. Dean B. Smith (USAF), escaped serious injury. Atmosphere in the sealed cabin was almost pure oxygen when fire occurred. Airman 1/C Henry W. Hall broke the cabin's seal, quickly depressurized it, and brought out the two men. Cause of fire, which badly damaged the cabin, was unknown.

- Soviet delegate Platon D. Morozov demanded that U.N. Committee on Peaceful Uses of Outer Space reach agreement on legal aspects of space exploration before considering scientific and technical program, at opening meeting of Committee's autumn 1962 session. Morozov argued that U.S. high-altitude nuclear test in July had caused atmospheric radiation that imperiled the lives of Soviet cosmonauts; he also commented that international space cooperation would be possible only "in a disarmed world." When U.S. delegate Francis T. P. Plimpton said he hoped Mr. Morozov was not trying to "block progress" on technical issues, Morozov said U.S.S.R. had accepted "in principle" the recommendations of the scientific and technical committee and later added that his Government would not make agreement on international technical cooperation dependent upon acceptance of Soviet views on legal issues. Committee finally agreed to consider the two subcommittee reports jointly. Subcommittee on international space law had met in Geneva last summer but had failed to reach agreement.

- David J. Mann of Thiokol Chemical Corp. told American Chemical Society that Thiokol was developing new rocket fuel, oxygen difluoride, 15% more powerful than current rocket fuels. Used as oxidizer with such substances as diborane or monomethylhydrazine, oxygen difluoride could save 7 ft. in length and 350 lbs. in weight on an Apollo-type launch vehicle. Other claimed advantages were improved storability and capacity to remain in liquid form under wider range of conditions than possible with liquid hydrogen/liquid oxygen. New fuel's density would permit large amounts to be stored in smaller volume.

September 10: Communist China announced that U.S.-built "U–2 plane of the Chiang bandit group intruded into the airspace of East China on the morning of September 9 and was shot down by the air force of the Chinese People's Liberation Army." The U.S. confirmed that the U–2 missing over East China was one of two sold to Nationalist China by Lockheed Aircraft Corp. in 1960.

• Tass announced death of Maj. Gen. Vladimir Y. Klimov of Soviet Corps of Engineers. General Klimov developed V.K. jet engine used in MiG–15 and MiG–17 fighter planes.

September 11: Venus probe MARINER II passed the 2.5-million-mile mark on its 180-million-mile journey to vicinity of Venus. Speed relative to the earth was 6,512 mph. Four of the six scientific experiments aboard were sending data back to earth; the other two experiments would not be activated until spacecraft approached Venus.

• President Kennedy toured NASA Marshall Space Flight Center, Huntsville, Ala., and Launch Operations Center, Cape Canaveral, Fla. At MSFC, he inspected a mock-up of the F–1 engine and a Saturn C–1 launch vehicle, and later witnessed 30-sec. static-firing of 1.3-million-lb.-thrust S–1 stage for Saturn SA–4 launch vehicle. At LOC, he inspected launch complexes for Mercury-Atlas vehicle, Titan rocket, and Saturn C–1. He then flew to Houston, Tex., to visit NASA Manned Spacecraft Center on September 12. Addressing workers at Cape Canaveral, the President said: "I don't think we can exaggerate the great advantage which the Soviet Union secured in the '50s by being first in space. They were able to give prestige to their system, they were able to give force to their argument that they were an advancing society and that we were on the decline.

"But I believe that we are an advancing society, and I believe that we are on the rise, and I believe that their system is as old as time. . . ."

He said that, as long as the American system was to be judged in at least one degree by its achievements in space, the United States "might as well be first, and therefore, this country, both political parties, have determined that the United States shall be first."

• Reported in the *New York Times* that new radiation belt created by July 200-mi.-high nuclear explosion in the Pacific was expected to cause cancellation of U.S. planned 500-mi.-high test in the Johnston Island series. Within this new belt, electron intensity was about 100 times greater than usual radiation peak in the outer part of the Van Allen belt; at higher altitudes the electrons could persist for 5 years or more. Before the July test, U.S. had predicted any trapped electrons would disappear "within a few weeks" of the explosion.

• U.S. delegate to U.N. Committee on the Peaceful Uses of Outer Space, Francis T. P. Plimpton, replying to Soviet criticism of U.S. high-altitude nuclear explosion, charged that U.S.S.R. had secretly conducted at least one high-altitude nuclear explosion within the last year. "Although the Soviet Union has never announced it, it is a fact that the Soviet Union itself carried out high-altitude nuclear tests during its series which began a year ago this month."

September 11: U.S.S.R. delegate to the U.N. Committee on the Peaceful Uses of Outer Space expressed approval of a subcommittee's recommendation to establish an international sounding rocket launching base on the geomagnetic equator. U.S. delegate Plimpton welcomed the Soviet position.

• National Research Corp. had determined that certain micro-organisms may be able to survive such space environment effects as temperature extremes and ultrahigh vacuum, in first phase of study program sponsored by NASA. Five known laboratory strains of micro-organisms were subjected to simulated space conditions for five days, and certain species were found to be "highly resistant" to temperature and ultrahigh vacuum effects. Future studies would consider heat and ultraviolet solar effects and gamma irradiation.

• Congressman Victor Anfuso (N.Y.), speaking before the Pan American Management Club at Cape Canaveral, said:

"Recognizing the very significant contrast between our mission in space and that of the Russians, we must do three things:

"1. Convince the world that our venture into space is a peaceful one—to bring back benefits which will create a world of abundance, making war and strife among nations unnecessary. . . .

"2. We must, at all costs, achieve dominance in all phases of space exploration for our own security and that of all peoples of the earth.

"3. We must establish freedom in space and invite all nations of the world, having a contribution to make, to join our efforts.

"To accomplish all these three things requires an all-out effort on our part—a mobilization of all our resources—sacrifices on the part of Government, business, industry, and labor. . . ."

• AEC reported that KIWI B-1B reactor had been damaged recently during its experimental ground test series with a static-power run at the Nevada Test Site. Further analysis would be required to determine exact cause and extent of the malfunction.

• United Auto Workers and International Association of Machinists unions in Los Angeles announced they had set September 22 as strike date against 4 aerospace companies—North American Aviation, Inc., Lockheed Aircraft Corp., General Dynamics Corp., and Ryan Aircraft Co. President Kennedy warned that such a strike, involving nearly 100,000 workers, "would seriously set us back in space exploration and would imperil the Nation's defense."

• In interview in Yugoslav Government newspaper *Borba*, Soviet Cosmonaut Gherman Titov said U.S.S.R. was planning flight to the moon in 1965. "If some problem arises and a solution is not found immediately, the time is extended."

• NORAD announced Soviet manned spacecraft VOSTOK III and IV crossed the North American continent 70 times during their double orbital trips August 11–14. NORAD space detection and tracking system tracked the flights for 112 earth orbits.

September 12: President Kennedy toured NASA's Manned Space Center, after speaking at Rice University, Houston, where he said:

". . . Man, in his quest for knowledge and progress, is determined and cannot be deterred. The exploration of space will go ahead, whether we join in it or not, and it is one of the great adventures of all time, and no nation which expects to be the leader of other nations can expect to stay behind in this race for space.

"Those who came before us made certain that this country rode the first waves of the industrial revolutions, the first waves of modern invention, and the first wave of nuclear power, and this generation does not intend to founder in the backwash of the coming age of space.

"We mean to be a part of it. We mean to lead it, for the eyes of the world now look into space, to the moon and to the planets beyond, and we have vowed that we shall not see it governed by a hostile flag of conquest, but by a banner of freedom and peace.

"We have vowed that we shall not see space filled with weapons of mass destruction, but with instruments of knowledge and understanding.

"Yet the vows of this Nation can only be fulfilled if we in this Nation are first, and, therefore, we intend to be first. In short, our leadership in science and in industry, our hopes for peace and security, our obligations to ourselves as well as others, all require us to make this effort, to solve these mysteries, to solve them for the good of all men, and to become the world's leading spacefaring nation.

"We sail on this new sea because there is new knowledge to be gained, and new rights to be won, and they must be won and used for the progress of all people. For space science, like nuclear science and all technology, has no conscience of its own.

"Whether it will become a force for good or ill depends on man, and only if the United States occupies a position of preeminence can we help decide whether this new ocean will be a sea of peace or a new, terrifying theater of war.

"I do not say that we should or will go unprotected against the hostile misuse of space any more than we go unprotected against the hostile land or sea, but I do say that space can be explored and mastered without feeding the fires of war, without repeating the mistakes that man has made in extending his writ around this globe of ours.

"There is no strife, no prejudice, no national conflict in outer space as yet. Its hazards are hostile to us all. Its conquest deserves the best of all mankind, and its opportunity for peaceful cooperation may never come again. . . . .

"We choose to go to the moon . . . in this decade and do the other things, not because they are easy, but because they are hard, because that goal will serve to organize and measure the best of our energies and skills, because that challenge is one that we are willing to accept, one we are unwilling to postpone, and one which we intend to win, and the others, too.

"It is for these reasons that I regard the decision last year to shift our efforts in space from low to high gear as among the most important decisions that will be made during my incumbency in the office of the Presidency. . . .

"We have had our failures, but so have others, even if they do not admit them.   And they may be less public.

"To be sure, we are behind, and will be behind for some time in manned space flight.   But we do not intend to stay behind. In this decade we shall make up and move ahead. . . ."

September 12: NASA announced it would launch a special satellite before the end of the year to "obtain information on possible effects of radiation on future statellites and to give the world's scientific community additional data on the artificial environment created by the [radiation] belt."   The 100-lb. satellite would be launched from Cape Canaveral into an elliptical orbit ranging from about 170-mi. perigee to 10,350-mi. apogee.

- USAF Titan II rocket flew 5,000 miles in its third success of five test launchings.   All test objectives were met on the 30-min. flight from Cape Canaveral to target area near Ascension Island in the South Atlantic Ocean.

- U.K. and Canada supported U.S. request that U.N. Committee on Peaceful Uses of Outer Space concentrate on international cooperation on space programs rather than on legal matters that contained elements of "a highly political and controversial nature."

- NASA management meeting held to brief key officials on a manned space station program and to discuss possible FY 1964 funding.

- Senate passed bill (S. 3138) to authorize FAA to conduct research to "determine what means and criteria can be employed to reduce and, hopefully, eliminate objectionable aircraft noise."   Bill was referred to the House.

- Walter Reed Army Institute of Research dedicated a new, 55,000-watt nuclear reactor, to be used in research on "creating a medication to ward off radiation."

- First "mystery" satellite in history of space exploration was launched, according to British magazine *Flight International.* Magazine said satellite orbited at height of 113 mi. and re-entered earth's atmosphere 12 days later.   Satellite was listed as belonging to USAF, but spokesman said this was a "scientific guess based on our assessment of previous satellite launchings." Launching was not confirmed, and no official U.S. listing included such a satellite.

- Balloon-borne Cassegrain telescope was launched from Artesia, N.M., to altitude of about 86,000 ft. by USAF Cambridge Research Laboratories.   Besides the telescope, the 850-lb. payload included interferometer spectrometer to take spectra of moon and planet Venus.   Other information obtained on this initial Project Skytop flight concerned amount of water vapor present in upper atmosphere.   Third objective of flight was to obtain information required to measure amount of different gases in upper atmosphere of Venus.   Instrumentation failure prevented the telescope from orienting on Venus for sufficient length of time.

- USAF named first two test pilots of the mach three RS–70: Lt. Col. Joseph F. Cotton, of Rushville, Ind.; and Maj. Fitzhugh Fulton, of Talladega, Ala.

- Two USN pilots, in separate flights, officially claimed two world flying records held by U.S.S.R. since 1940.   LCdr. Fred A. W.

Franke reached 27,380-ft. altitude in UF–2G Albatross amphibian with 4,410-lb. load; LCdr. Donald E. Moore reached 29,460-ft. altitude in same aircraft with 2,205-lb. load.   Both flights were made from New York NAS, Floyd Bennett Field.

September 13: Back in Washington after two-day tour of U.S. space facilities, President Kennedy held news conference.   Asked why decision was made to extend U.S. atmospheric nuclear tests, the President said:

"There are two reasons.   One is that, as you know, because of the blow-up in the pad at Johnston Island and because of the earlier failures of the communication system in the missile, we were not able to carry out these tests which were . . . among the most important, if not the most important of our series.

"So we are going to finish those.   In addition, as a result of the earlier tests of this Dominic series, there were certain things learned which we would like to prove out.

"So we have agreed to a limited number of tests in concluding the Dominic series, and also we have taken some steps to prevent a repetition of the incident which caused an increase in the number of electrons in the atmosphere, by lowering the altitude and the yield so that lunar flights will not be further endangered."

Asked whether U.S. position of refraining from competition with U.S.S.R. military space vehicles would condemn U.S. to second place in the military field, President Kennedy replied:

"No, Mr. Alexander.   As I said last week, in the first place we are spending $1.5 billion a year on our military space program. What is the key for the success both of peaceful exploration of space as well as the military mastery of space are large boosters, effective control of the capsule, and the ability to rendezvous, and all of the rest, so that there is an obvious usefulness if the situation should require, military usefulness for our efforts, peaceful efforts, in space.

"In addition, as you know, very recently we determined to go ahead with the Titan III, which gives the United States Air Force a very strong weapon if that should become necessary.   So that the work that NASA is doing on Saturn and the work the Air Force is doing on Titan and the work being done on the Apollo program and Gemini and the others, all have a national security factor as well as a peaceful factor. . . .

"I think the United States is attempting, and this Administration, as you know, is making a very massive effort in space.   As I said, we are spending three times what we spent last year in space, and more in this year's budget than the eight previous years, so that this is a tremendous effort, $5.5 billion as well as the money that we are spending for the military use of space.

"As I say, the size of the booster and the capsule and the control all would have, if the situation required it, a military use. We hope it does not; we hope that space will be used for peaceful purposes.   That is the policy of the United States Government, but we shall be prepared if it does not; and in addition, as I said from the beginning, both the Soviet Union and the United States have a capacity to send a missile to each other's country with a nuclear warhead on it, so that we must keep some perspective as to where the danger may lie.   But the United States, in the effort it is making both in the peaceful program and the military program, all of this will increase our security if the Soviet Union should attempt to use space for military purposes."

Asked if he were hopeful that U.S.-U.S.S.R. cooperative ventures in space exploration were likely in the near future, he replied:

"No. As you know, Dr. Dryden had some conversations in Geneva with regard to the matter, and some progress was made, but it is limited in its scope and we would hope more could be done and more, perhaps, could be done if the atmosphere between the two countries should be improved."

September 13: Paresev (Paraglider Research Vehicle) flown at NASA Flight Research Center. First manned vehicle designed to be towed aloft and released like a conventional glider, Paresev was towed to 6,000-ft. altitude by Stearman biplane. Released, the craft was flown back to Rogers Dry Lake bed in 3 min. 44 sec. by NASA project pilot, Milton O. Thompson. Paresev flight program provided data on (1) flare and landing capabilities; (2) stability and controllability; and (3) pilot training. Possible uses of the paraglider concept included recovery of future spacecraft and rocket boosters.

- NASA–DOD jointly announced undertaking of a joint study to improve field operations involved in contract management for both DOD and NASA contracts. Under direction of committee chaired by Assistant Secretary of Defense (I&L) Thomas D. Morris, the study was to improve effectiveness of field contract management, to improve responsiveness to buying offices and systems project offices, to assist NASA's increasing contract management requirements with minimum of additional personnel, and to ensure continued contract management efficiency in times of national emergency. Combined value of DOD and NASA contracts during FY 1963 would be approximately $30 billion.

- Hearings on U.S. program for atomic propulsion held by Joint (Congressional) Committee on Atomic Energy, with NASA Administrator James E. Webb, AEC Commissioner Leland Haworth, MSFC Director Dr. Wernher von Braun, Los Alamos Scientific Laboratory Director Norris Bradbury, and SNPO Director Harold Finger all agreeing on need for living facilities near atomic test site in southern Nevada. NASA-AEC Nerva (nuclear engine) development would require between 1,500 and 2,000 employees at the Nevada site. However, the prospect of commuting 180 mi. daily to and from the nearest town (Las Vegas) was scaring away prospective workers, according to testimony.

- Dr. Harold B. Finger, Director of joint AEC-NASA Space Nuclear Propulsion Office (SNPO), told Joint Committee on Atomic Energy that first flight date in nuclear-powered space vehicle program (Project Rover) had been moved from 1966 to 1967. Over-optimism, technical difficulties, organizational problems, and a situation in which construction workers have to spend 4 hours a day commuting to the Nevada test site have all contributed to the delay, he testified.

- Testimony before the Joint (Congressional) Committee on Atomic Energy revealed management difficulties on Project Pluto nuclear-powered vehicle, to fly at 2,000 mph within the atmosphere. Dr. Leland Haworth, member of AEC, testified that AEC was "awaiting a firm decision" from DOD on whether the project would be continued. Reactor for nuclear ramjet rocket, under development since 1956, had been successfully ground-tested in Nevada. Dr. Harold Brown, director of DOD research and engineering, said DOD hoped that a decision could be reached "within a month." An estimated $500 million would be required to continue the project through flight-test phase.

September 13: President Kennedy urged four major aerospace manufacturers (North American Aviation, Lockheed, General Dynamics, and Ryan) to accept terms proposed by a special Presidential board for settlement of their labor contract dispute with two unions (United Auto Workers and International Association of Machinists). He also named another special board to try to mediate a dispute between the International Association of Machinists and the Boeing Co., asking the union not to strike but to continue work until November 15 so the board could try to work out a solution.

- Astronaut Walter Schirra, Jr., interviewed on nationally-televised CBS program, criticized delayed reporting of Astronaut M. Scott Carpenter's landing during flight MA–7: "We had a lot of information there [during the 35-min. period from blackout of communications when spacecraft re-entered until public announcement of reception of signals from spacecraft] and yet it never came out and it was disgusting, you know."

  Schirra also commented on the heavy demands on Astronaut John H. Glenn for "outside" personal appearances. Interview had been taped several weeks before its broadcast.

- India offered to provide launching base for U.N. international rocket probes in session of U.N. Committee for Peaceful Uses of Outer Space. India's proposal was related to a proposal that 1964 be designated International Year of the Quiet Sun (IQSY), and include equatorial sounding rockets to gather data above the earth's Equator. Southern part of India lies within 10° of the Equator and would provide suitable launching site for the project.

- Congressman George Meader (Mich.) introduced H.R. 13130, bill to establish a Commission on Research and Development. The new bill was a "perfected" version of his earlier bill. He called attention to S. 2771, the bill which had passed the Senate, and urged that "speedy action be taken on the legislation."

- Analysis of the metal chunk believed to have been part of Soviet satellite SPUTNIK IV showed the object had greater radiation level than any other object previously recovered from space. AEC scientists at Los Alamos Scientific Laboratory (N. Mex.) said one reason for high radiation was that the chunk was analyzed soon after being found, whereas previous objects had not been analyzed until at least three weeks after being found. Radiation expert Dr. Ernest C. Anderson of Los Alamos reported that the object had same sort of radioactivity shown by natural iron meteorites, evidence that the object was part of a spacecraft that had been orbiting for long time above the atmosphere. SPUTNIK IV went into orbit May 15, 1960, and disintegrated September 5, 1962, over the U.S. Los Alamos forwarded 14-lb. chunk to U.S. delegation to U.N. Committee on Peaceful Uses of Outer Space;

Smithsonian Astrophysical Observatory had kept a 6-lb. slice of the original object for further examination.

September 13: First operational Atlas-F launching silos turned over to Strategic Air Command. Ceremonies were held at one of the underground launching complexes near Salina, Kans.

- Reported that Hughes Aircraft Co. had developed special devices to determine radiation dosage an astronaut may absorb in space flight. Called "tissue-equivalent ionization chambers" and developed under USAF contract, the instruments would be installed in three USAF "plastinauts," man-sized dummies of astronauts made of plastic simulating human tissue. The plastinauts were designed "to fly in any space vehicle capable of carrying a man."

- USAF Skybolt missile was intentionally destroyed when missile veered off course, third failure in three flight attempts of the air-to-surface Skybolt.

- International Air Transport Association (IATA) technical committee reported it was too early to state with certainty whether commercial supersonic aircraft were practical. Committee said there was a growing possibility that a supersonic airliner could fly within the next few years, but that "considerable research into such matters as sonic boom and cosmic radiation is still required before a final conclusion can be reached."

- President Kennedy, replying to question at regular news conference, said the U.S. had no plans to permit sale of U–2 high-altitude reconnaissance planes to Nationalist China. He added that the U.S. had not sold and would not sell U–2's to any third nation. Sale of two U–2 aircraft to Nationalist China was arranged in July 1960, during Eisenhower Administration, President Kennedy pointed out.

- Senator Hubert Humphrey announced public hearing on improvement of Government management of information and coordination of reporting would be conducted by Senate Government Operations Subcommittee on Reorganization and International Organizations, September 20. Among officials testifying would be Melvin Day, Director of NASA Office of Technical Information.

September 14: MARINER II, U.S. Venus probe, continued to transmit signals from interplanetary space to earth on its nineteenth day of flight, thus surpassing record of U.S.S.R. Venus probe which stopped transmitting on its eighteenth day of flight in February 1961. At 8:00 PM EDT the spacecraft was estimated to be 3,067,471 mi. from earth and traveling at a speed of 6,463 mph relative to the earth. JPL scientists said the probe appeared to be continuing "right on course."

- In meeting of U.N. Committee for the Peaceful Uses of Outer Space, U.S. offered to U.S.S.R. a 14-lb. metal chunk, believed to be part of SPUTNIK IV, that had landed in Wisconsin September 5. Soviet delegate to the U.N. Platon D. Morozov rejected the offer, saying "We do not know to whom it belongs." He said it was not necessary to make dramatic gestures to show that falling satellite fragments could do harm, and charged the U.S. with "dramatic staging" to divert the Committee from Soviet proposals on legal principles for use of outer space. Displaying the chunk following the meeting, U.S. delegate Francis T. P. Plimpton said the object weighed 20 lbs. when it fell from the sky, but U.S.

scientists of Smithsonian Astrophysical Observatory had retained a six-lb. piece for further examination.

September 14: House and Senate Conference Committee on FY 1963 appropriations agreed to one half the restoration of NASA funds recommended by Senate, allowing total of $3,674,115,000 for NASA FY 1963 appropriation.

- NASA announced that nine new astronauts for Projects Gemini and Apollo would be named at Manned Spacecraft Center, Houston, on September 17.

- Dr. Wernher von Braun, Director of NASA Marshall Space Flight Center, said in published interview: "People talk too much about crash programs and too little about sustained support. These space programs take time. A decision you make today on some engineering or scientific aspects of our space program may take anywhere from four to five years before it takes the form of hardware.

    "When you make a wrong decision now, the payoff, good or bad, could come five years later, not sooner than that. We can't work miracles overnight. . . ."

- In prepared press statement, Astronaut John H. Glenn commented on September 13 televised remarks by Astronaut Walter Schirra, and said: "I will continue to support his flight [Astronaut Schirra's orbital flight MA–8] just as well as I possibly can. . . . I don't want to add any more statements to a situation that has all the aspects of a tempest in a teapot. Scheduling of my time for maximum benefit from a technical and national space program standpoint has been continually reviewed by both Dr. Gilruth and the management of NASA." Dr. Robert R. Gilruth, Director of NASA Manned Spacecraft Center, stated there was no real controversy over Col. Glenn's position in the space program or on other questions raised by Cdr. Schirra.

- AEC announced 50th U.S. nuclear test in current series, an underground low-yield explosion at Nevada test area.

- Dr. Leland J. Haworth of AEC led delegation of scientists and Government officials in dedication of $12 million electron accelerator at Cambridge, Mass. To be operated by MIT and Harvard Univ., the facility would be supported by AEC, which financed its construction.

- DOD announced International Telephone and Telegraph Corp. (IT&T) had leased to USAF two transportable ground communications stations for operational training of communications personnel. The units would be used in new program to provide data on problems of using satellites for communications support.

- Communist China stated the U–2 shot down September 9 was fourth U.S. military-type aircraft China had downed since October 1959.

September 15: Signals from ARIEL satellite had been received during past week by tracking stations. ARIEL had stopped transmitting after U.S. high-altitude nuclear test because of radiation damage to the satellite's solar cells. Although resumed transmission was not continuous, it did demonstrate ARIEL's regained capability to return scientific data from space. Transmission apparently occurred whenever the satellite stayed in sunlight long enough for its damaged solar cells to absorb sufficient solar energy to power its electronic equipment.

September 15: NASA announced the sixth Tiros weather satellite would be launched into orbit from Cape Canaveral on September 18, at the earliest. Launch date was moved two months ahead to provide backup of TIROS V cloud-cover photography during last half of current hurricane season and to provide weather forecasting support for Astronaut Walter M. Schirra's orbital space flight September 28. Wide-angle TV camera in TIROS V continued to operate, but medium-angle Tegea lens stopped functioning on July 2 because of "random electrical failure in the camera's system."

• Magnetic fields had been discovered in outer space by two scientists working in Australia, the Commonwealth Scientific Research Organization announced. Brian Cooper, of Sydney, and Marcus Price, of the U.S., made their discovery with new radiotelescope at Parkes, New South Wales. Charting of the magnetic field near galaxy Centaurus A was termed "radio astronomy's biggest discovery in 10 years." Centaurus A is 20 million light years away from the earth.

• Announced that 140-million-candlepower lights would burn near Durban, So. Africa, for three minutes during fifth and sixth orbital passes of Astronaut Walter M. Schirra's Mercury space flight (MA–8).

• Piloting an Albatross amphibian, LCdr Richard A. Hoffman (USN) claimed speed record of 151.4 mph average over 3,100-mi. flight course for amphibians carrying 2,205-lb. loads. Flight was made from New York NAS, Floyd Bennett Field.

September 16: NASA announced the advanced version of liquid-hydrogen/liquid-oxygen rocket engine had completed its preliminary flight rating tests. In the test program, the 15,000-lb.-thrust engine underwent 20 static-firings for a total of 2,820 sec. of firing; it also passed humidity and vibration tests.

• Dr. Albert Kelley, NASA Director of Electronics and Control (OART), told Electronics Industry Association that between 90 and 95% of U.S. space failures could be traced to failures in electronic components. "We are trying to build a reliable spacecraft with unreliable components." He added that NASA had begun an electronic components' reliability program separate from that of DOD—a program providing a "great challenge" with prospects for rewarding technological payoffs.

• NASA announced contract award for research in communications between man and dolphins, results of which could apply if man encounters other species on distant planets. The $80,800, one-year contract was awarded to Communication Institute of St. Thomas, V.I., directed by Dr. John C. Lilly. Dr. Lilly's research had already determined that dolphins communicate with each other and that they could be taught to mimic human speech.

• NASA announced new policy under which astronauts would be permitted to negotiate individually for the sale of their stories, the policy being "aimed at assuring equal access by all news media to the Astronauts' stories of their flight missions." Notable provisions of new policy were: (1) requirement for a second post-flight news conference in which representatives of news media could question the astronaut "in depth" about his flight; and

(2) prohibiting of any publication's advertising "exclusive" stories purchased from astronauts, with requirement that the contracting publication provide a method to avoid this inference and with all such agreements subject to approval of the NASA Administrator. Inherent in the policy was that "all information reported by the Astronauts in the course of their official duties, which is not classified to protect the national security, will be promptly made available to the public." New guidelines applied to all NASA astronauts, civilian and military.

September 16: U.S.S.R. exploded second nuclear device of "several megatons" in two days. AEC announced the atmospheric test had been conducted in vicinity of the island Novaya Zemlya.

September 17: NASA's nine new astronauts were named in Houston by Dr. Robert R. Gilruth, Director of NASA Manned Spacecraft Center (MSC). Chosen from 253 applicants, the former test pilots who would join original seven Mercury astronauts in training for Projects Gemini and Apollo were: Neil A. Armstrong, NASA civilian test pilot; Maj. Frank Borman (USAF); Lt. Charles Conrad, Jr. (USN); LCdr. James A. Lovell, Jr. (USN); Capt. James A. McDivitt (USAF); Elliot M. See, Jr., civilian test pilot for General Electric Co.; Capt. Thomas P. Stafford (USAF); Capt. Edward H. White, II (USAF); and LCdr. John W. Young (USN). Dr. Gilruth stressed that they would not all necessarily make actual space flights. "Assignment to flight crews will depend upon the continuing physical and technical status of the individuals concerned and upon the future flight schedule requirements."

- USAF launched unidentified satellite with Thor-Agena B booster from Vandenberg AFB.
- NASA Manned Spacecraft Center announced Mercury Astronaut Donald K. Slayton had been named "Coordinator of Astronaut Activities." Slayton would be responsible for assignment of training and engineering duties of all the astronauts.
- Reported that NASA would build facilities worth $15 to $18 million at White Sands Missile Range (WSMR), N.M., for testing Apollo spacecraft's propulsion and abort systems. Facilities for flight tests of the abort system and the lunar excursion module would be ready in January 1963; static-firing sites for testing propulsion systems would be operational in mid-1963.
- DOD established conflict-of-interest ruling on industry's role in DOD research contracts. New policy specified that a company serving as technical adviser in a research program was ineligible to compete for contracts in the hardware production phase. Ruling was first established in Project Advent communications satellite project.
- First Snap–8 nuclear reactor was operated successfully at Santa Susana, Calif., AEC announced. The experimental reactor was of type designed to ultimately propel spacecraft through deep space.
- Reported that Astronaut John H. Glenn, Jr., had received less than half the heavy primary cosmic radiation dosage expected during his 4.5-hr. orbital flight MA–6. According to Dr. Hermann J. Schaefer of USN Biophysics Branch, Pensacola, the MA–6 spacecraft walls and capsule instrumentation probably absorbed many of the rays and kept the total dosage lower than expected.

September 17: Lockheed Aircraft Corp. Board Chairman, Courtlandt S. Gross, wrote to 56,000 employees that the union-shop proposal by President Kennedy's fact-finding panel was "just plain wrong." In response to President Kennedy's plea that Lockheed and three other aerospace companies accede to labor demand for union shop, Gross protested that the "freedom to join or not to join an organization is a basic individual right. . . ." The other three companies—North American Aviation, General Dynamics/Convair Div., and Ryan Aeronautical Co.—reportedly were willing to withdraw their opposition to union shop, and the two unions—United Auto Workers and International Association of Machinists—indicated they would accept the proposed settlement.

- Federal Radiation Council (Chairman: Anthony J. Celebrezze, Secretary of HEW) announced that radiation exposure greatly above the safety guide levels set two years ago "would not result in a detectable increase in the incidence of disease. . . . The radiation protection guides are not a dividing line between safety and danger in actual radiation situations; nor are they alone intended to set a limit at which protective action should be taken or to indicate what kind of action should be taken." Individual situations require individual evaluation before action is taken, the council continued.

- M. G. O'Neil, president of General Tire & Rubber Co., told national convention of tire dealers that in 10 years astronautics industry may exceed size of combined automotive industries of the world. He compared the impact of the space age to that of the discovery of America in 1492.

- Reported that telemetry equipment developed for missiles and earth-orbiting satellites was available to automobile engineers to help make cars quieter. To check noise and vibration accurately, car manufacturers could use two-channel electronic telemetry system that picked up sounds made by test cars and transmitted signals to receiver in another vehicle.

September 18: NASA's TIROS VI weather satellite was placed in orbit by three-stage Delta vehicle launched from Cape Canaveral at 4:45 PM EDT (apogee, 442 mi.; perigee, 425 mi.; period, 98.7 min.; inclination, ranging between 58.3° north and south latitude). Orbit was believed too low for satellite's 9,120 solar cells to be unduly damaged by artificial radiation zone created by U.S. high-altitude nuclear explosion in July. By midafternoon, pictures taken by TIROS VI's two television cameras were being incorporated with ground data into conventional weather forecasts. Weather Bureau reported quality of the photographs was "as good, if not better than that of any taken by the five previous Tiros satellites." The launching marked the sixth straight success in Tiros program and the eleventh straight successful satellite launching by the Delta rocket.

- MARINER II entered gravitational field of the sun; the Venus probe was 3,608,857 miles from earth. The velocity of MARINER II relative to the earth now stopped decreasing and began increasing due to the effect of the gravitational field of the sun.

- House adopted conference report on Independent Offices Appropriations Bill, allowing $3,674,115,000 for NASA in FY 1963.

September 18: First photographs of Gemini two-man spacecraft mockup were released by NASA and McDonnell Aircraft Corp. Three-section capsule would be used to train pilots to maneuver in space, particularly to rendezvous and to dock with another orbiting space vehicle.

- DOD announced award of $12.2 million contract to IT&T for systems engineering and technical advice in development of communications satellite systems.

- Editorial in the *New York Times* said:

 "Since the National Aeronautics and Space Administration is a civilian agency, it might be wiser to make all the astronauts civilians so that no questions of inequality or discrimination arise among them and also so that they might receive more adequate pay than is provided by the low military pay scales. . . .

 "In permitting the astronauts to cash in on their exploits, the Kennedy Administration is following an unwise precedent set by the Eisenhower Administration.

 "While the practice of profiting from memoirs of Government service is an old one, such memoirs are normally written by persons who have already left Federal employment. . . .

 "The government would be far wiser if it paid its astronauts a sufficiently generous salary so that it could in good conscience ask them to observe the same practices of discretion and modesty which have hitherto been considered normal for all other Government employees."

- Dr. Ivan A. Getting, President of Aerospace Corp., told National Rocket Club in Washington that the U.S. space program had been plagued from the beginning by an artificial "dichotomy" that assumed peaceful activities in space were "pure" and military activities were "evil."

 "We know, all the world knows, that we have no intention of exploiting space for reasons of aggression. But now the Russians are demonstrating to the world that their space exploits are straightforward demonstrations of raw military power. . . .

 "We in the United States need to reaffirm our traditional position—but with pride instead of seeming shame—that our presently great military strength is the most potent force in the world and that it is working 24 hours a day to help keep the peace. . . .

 "If our strength is to be maintained we must have the military tools that are best suited to help keep the peace . . . operational however and whenever and wherever necessary—in space, in the atmosphere, under the seas, or even underground.

 "We need to restate the historic peacetime military role of sharing in exploring the frontiers . . . now in space . . . and that this sharing be on a basis of both cooperation and also some healthy competition. . . .

 "We must recognize that many practical space missions have both military and civilian uses; that the exploitation of space for these missions necessarily involves both the development and continuing operation of these systems, and that as a consequence, each should be evaluated on the basis of how it can be realized most effectively from the standpoint of overall national benefit.

"It will take our best efforts in an overall unified plan. But to assure success there must be no dichotomy in space any more than has been in any other worthwhile national undertaking . . . ."

September 18: Three fliers were credited with breaking two Soviet world records by flying their B–58 Hustler supersonic bomber to 85,360-ft. altitude with cargo of 11,023 lbs., DOD announced. Fliers were: Maj. Fitzhugh L. Fulton (USAF), pilot; Charles R. Haines, civilian flight test engineer; and Capt. William R. Payne (USAF), navigator.

• Dr.Fred P. Adler, Director of Space Systems Div. of Hughes Aircraft Co., told House subcommittee of Committee on Science and Astronautics that the U.S. should not make a "major investment" in the Telstar communications satellite system until the Syncom system was proven or disproven. Dr. Adler said Hughes' Syncom could perform better tha AT&T's Telstar and at one third the cost.

September 18-20: First national conference on the problem of Technical Manpower in the Space Age was held in New York City by Institute of the Aerospace Sciences, at request and under auspices of Executive Office of President Kennedy.

September 19: President Kennedy announced labor agreements had been reached with two aerospace industry companies, North American Aviation and General Dynamics/Convair, ending threat of strike. Agreement was based on recommendations of the President's Aerospace Board.

• Mercury capsule for orbital space flight MA–8 was named "Sigma 7" by Astronaut Walter M. Schirra. Sigma, the eighteenth letter of the Greek alphabet and an engineering symbol for "many," symbolized the teamwork involved in the manned space flight; "7" referred to the seven-man Project Mercury astronaut team.

• Third Saturn launch vehicle (SA–3) arrived at Cape Canaveral after 10-day trip by barge from NASA Marshall Space Flight Center, Huntsville, Ala. Flight test was scheduled for November.

• Leonard Jaffe, NASA Director of Communications Systems, told House subcommittee of Committee on Science and Astronautics that NASA was "endeavoring to assist in determining as rapidly as possible which of the various system designs . . . should be used in the establishment of operational communications satellite systems." He described various "orbital configurations proposed for communications satellites" and reviewed passive satellites (Echo) and active-repeater satellites (Telstar, Relay, Syncom). "From these experiments and others which will follow, we will obtain engineering data upon which to base operational system designs. . . .

"Satellites of the near future must be designed to exhibit reliability and dependability, unattended in the space environment for many years, if we are to have economically viable communications systems."

• Senate began consideration of conference report of H.R. 12711, FY 1963 Independent Offices Appropriations bill. NASA appropriations items were not in dispute.

September 19: Reported in *New York Times* that U.S. had notified the U.S.S.R. of its willingness to sign an agreement to cooperate in the peaceful exploration and utilization of space. Within the past 3 weeks, U.S. State Department had sent note to Moscow calling for formal signature of a bilateral agreement on joint weather and communications satellites and study of the earth's magnetic fields.

- USAF Atlas missile was successfully flown 5,000 mi. downrange from Cape Canaveral. As the Atlas rose, two cameras attached to provide pictures of blast-off were ejected and landed on the Cape. Two additional camera capsules were ejected and recovered near Grand Turk Island. This was the 100th Atlas missile launching at Cape Canaveral.

- USAF Minuteman ICBM successfully fired from silo at Cape Canaveral, the second such firing in as many days.

- Air Force Association (AFA) convention in Las Vegas unanimously adopted resolution calling for immediate clarification of U.S. policy on the military uses of space. AFA policy statement said: "Soviet space achievements, with their military implications, make it clear that we cannot satisfy the national security requirement in space with by-products from our civilian space program. . . . Space must be used to press our deterrent capability, to protect the future against the agonies and miseries of war, and thus to provide the climate required for the growth of freedom."

- USAF named six men selected to pilot X–20 (Dyna Soar) orbital spacecraft. The pilots, all from Edwards AFB, included five Air Force officers and one NASA civilian: Capt. Albert H. Crews, Jr.; Maj. Russell L. Rogers; Maj. James W. Wood; Maj. Henry C. Gordon; Capt. William J. Knight; and Milton O. Thompson (NASA). USAF also displayed full-scale mock-up of X–20 at Air Force Association convention, first public showing of the space glider.

- 14,000 International Association of Machinists (IAM) workers at Lockheed Missiles & Space Co. plants threatened walkout effective September 24.

- Announced that the opening ceremony of the Roman Catholic Church ecumenical council on October 11 would be televised in the Vatican and relayed to the U.S. via TELSTAR satellite.

September 20: World's largest movable radiotelescope, located at National Radio Astronomy Observatory (NRAO), Green Bank, W. Va., was turned on at midnight for the first time. The 300-ft.-diameter telescope, built for $800,000 in about one year, was expected to be trained on Venus as its first operational target.

- AEC Chairman Glenn T. Seaborg told panel discussion of International Atomic Energy Agency in Vienna that a U.S. nuclear-powered spacecraft now in the initial-test phase could take two men on a round trip to Mars in the 1970's. He said that the round trip, including one-month visit on Mars, would take about a year.

September 21: Dr. Edward C. Welsh, Executive Secretary of NASC, told Air Force Association convention in Las Vegas that "the objective of our [national space] policy is to obtain and maintain leadership in space activities for the benefit of man's freedom, man's well-being, man's understanding, and man's scientific progress. The details of the policy are not quite so clear, as those who have studied our budgets and public documents have reason to know. It is clear, however, that our policy includes going to the moon during this decade. It includes developing an operational communications satellite system as well as navigation and meteorological systems on a world-wide basis. The specifics are less clear as to the roles of man in space and what can and should be done to maintain peace in outer space. This lack of clarity, however, is partly due to the difficulty of knowing what can be done and what cannot be done. Continuing efforts will be made to clarify policy, while keeping it necessarily flexible. . . ."

September 21: Replying to September 20 statement by AEC Chairman Glenn T. Seaborg regarding U.S. nuclear propulsion capabilities for a 1970 Mars mission, NASA spokesman said: "There is no such approved program for such a flight at this time, but the joint AEC-NASA program is aimed at giving such a capability, and at this time it appears it may be feasible to conduct such a flight."

• Announced that International Association Machinist workers and Lockheed Corp. had agreed to extend their contract indefinitely, thus averting threat of strike in this aerospace industry.

• Astronaut John H. Glenn, Jr. (Lt. Col., USMC), was chosen as Marine Aviator of the Year for 1962–63. This was the first annual award of the Alfred A. Cunningham Trophy.

• NASA contracted with Armour Research Foundation for an investigation of conditions likely to be found on the lunar surface. Initial research would concentrate on evaluation of effects of velocity of landing, size of the landing area, and shape of the landing object with regard to properties of the lunar soils. Earlier studies by Armour had indicated the lunar surface may be composed of very strong material.

• USN announced tests conducted at U.S. Naval Ordnance Test Station, China Lake, Calif., had included successful free-flight of a new type of rocket propulsion system called "Hybrid" because of its use of combination liquid and solid propellant in single motor.

• Two leading U.S. space scientists born in Germany were voted honorary members of the German Rocket Society: Dr. Kurt H. Debus, Director of NASA Launch Operations Center, and Prof. K. O. Lange, Director of University of Kentucky's Aeronautical Research Laboratory.

• Secretary of the Air Force Eugene M. Zuckert, addressing the Air Force Association in Las Vegas, said: "The United States, in keeping with our motivations in freedom and peace, has embarked upon a space program aimed at peace building, constructive exploitation of space—second to none.

"In the National Aeronautics and Space Administration, the nation has mobilized an effort unprecedented in history. NASA is making progress, because it has imaginative leadership, a competent staff, extensive facilities, and wholehearted support from all elements of industry and government which can contribute, especially the Air Force.

"In addition to our contribution to the success of the NASA program, the Air Force is required by its own mission to put its energies into a different kind of space effort. We are reaching a clearer definition of that mission as it relates to space. . . .

"The United States is dedicated as a matter of national policy to the peaceful exploitation of the space medium. The United States does not intend to extend the arms race into space.

"We are, however, taking the step to enable us to protect ourselves in the event the Soviets or any other nation were to undertake missions in space that would endanger our security. . . .

"The dual-vehicle orbital experiment of the Soviets, when coupled with their previous claims, seems to indicate that the need for protection against possible threats to our security will be in the near orbital stage of space, rather than farther out. One possible instrument of security that might be useful at this level in space may be found in extending the X–15 and X–20 technology, leading to craft which could operate from surface to orbit and back, and perform defense missions at the edge of the atmosphere. Another might be the permanently manned orbital space station designed for military purposes. Progress toward utilization of such a vehicle for security and protection would be speeded by capitalizing upon the NASA program to acquire knowledge and competence in manned orbital flight. . . .

"These are the principles on which Air Force space program must be based: (a) Ample preparation to utilize the space medium; (b) Ample preparation to defend ourselves in case others choose to extend the threat of aggression to outer space; (c) Concentration on the really important tasks, including today's job and not just tomorrow's, and finally, (d) An organizationally disciplined space program—well planned and specific—which fits into overall national plans and objectives.

"We in America have no choice but to extend our defenses as far as they need to be extended to save freedom on earth. . . .''

September 22: Four-stage Journeyman rocket carried 145-lb. payload to 1,058-mi. altitude, from NASA Wallops Station. Primary experiment was measurement of the intensity of RF energy at medium frequencies; secondary experiment was measurement of electron densities in the upper ionosphere and investigation of possibility of using ionosphere as a focusing medium in future radio astronomy mapping experiments. Payload landed about 1,323 mi. from launch site in the Atlantic Ocean.

• NASA launched three-stage, solid-fuel Ram-B (Radio Attenuation Measurement) rocket from Wallops Station, Va., all contact with the vehicle lost 11 sec. after second-stage ignition. Project Ram was designed to provide information on communications blackout during space vehicle re-entry.

• Aerobee 150A launched from NASA Wallops Station, the rocket reaching 117-mi. altitude in experiment to measure absolute intensity of the spectrum of stars with 50 angstrom resolution and to measure ultraviolet fluxes. No usable data were received because rocket failed to despin as planned.

• Faulty valve in MA–8 spacecraft was replaced by technicians, a repair which required removal of Mercury capsule from the Atlas booster. Astronaut Walter M. Schirra's orbital space flight was planned for October 3.

• NASA announced it would launch Canadian spacecraft Alouette (S–27) from Point Arguello, Calif., no earlier than September 26. Named for high-flying songbird of Canada, Alouette was first satellite to be both designed and built by a nation other than U.S. or U.S.S.R. The launching would mark NASA's first orbiting attempt from Point Arguello and its first use of a Thor-Agena B launch vehicle.

September 22: Unnamed NASA official said that a four-stage rocket probe would be launched in October or November to check the strength of the radiation belt created by U.S. high-altitude nuclear explosion over the Pacific on July 9, 1962.

• NASA announced award of facilities grants to five universities, first such grants awarded by NASA. Worth a total of $6,410,000, the grants were to provide research facilities for activities in space-related sciences and technology to universities making "substantial contributions" to the U.S. space program: University of California, Berkeley; University of Chicago; Rensselaer Polytechnic Institute; State University of Iowa; and Stanford University.

• According to the *Washington Post,* USAF second attempt to orbit belt of copper filaments in Project West Ford experiment was apparently inadvertently revealed by Eugene C. Fubini, Deputy Director of Defense Research and Engineering, testifying before the House Committee on Science and Astronautics. The attempted orbiting was inferred to have been made sometime this summer, failed because of launch vehicle malfunction (first failure and only attempt publicized had been made on Oct. 21, 1961). Of this report, Assistant Secretary of Defense for Public Affairs, Arthur Sylvester, said to the press: "The facts are that only one attempt has been made [to orbit tiny metal filaments] and the U.S. Air Force will coordinate with the National Academy of Sciences on any future attempt." Confusion had arisen when the *Washington Post* had reported that the USAF had made an unsuccessful attempt to orbit such a payload during the summer.

• Review in *Red Star* by General Pavel Kurochkin, commenting on new Soviet textbook on strategy, stated: "The Soviet people are engaged in the peaceful conquest of space. But it is perfectly clear that if the imperialists continue to conduct research for means of using cosmic space for military goals, then the interests of guaranteeing the security of the Soviet state demand definite measures from our side." Doctrinal textbook was written by several of Russia's top military leaders under the direction of Marshal V. D. Sokolovsky, and was said to be the first serious work on military strategy published in Russia since 1926.

• Israel Finance Minister, Levi Eshkol, speaking in Washington, D.C., said that the "missile race has entered the Middle East" and Israel must be prepared to defend itself "no matter what the cost." He said: "When Colonel Nasser stands by Egyptian rocket launchers and boasts they are trained in our direction, we dare not mistake this for an idle propaganda boast."

September 23: Dr. Charles M. Herzfield, Washington physicist, reported that the luminous particles sighted in space by both Col. John Glenn and Maj. Scott Carpenter probably were particles of solid nitrogen trapped in the extreme upper atmosphere. "As the temperature rises, the luminosity disappears. No other systems seem to have all the requisite properties. These particles may exist independently at the altitudes involved.

They may be produced from leaks of gas from the capsule or from gas trapped in various portions of the vehicle. . . ." Dr. Herzfield presented his evaluation in report to American Association for the Advancement of Science.

September 23: Soviet Academician Nikolai N. Semenov, Nobel Prize-winning chemist, was quoted as suggesting the moon be made into a major power source for earth. If the moon were covered with semiconductors and photo-elements of high efficiency solar energy falling on the moon could be converted to electrical power far exceeding that produced on earth. He also proposed the moon be made the site for all atomic and thermonuclear power stations, to avoid overheating and irradiating the earth. Research may find a way, he continued, to transmit the power to earth—perhaps by lasers or masers.

September 23–29: Thirteenth International Astronautical Congress was held in Varna, Bulgaria, sponsored by the International Astronautical Federation (IAF). Opening session was marked with pleas by western delegates that U.S.S.R. remove veils of secrecy surrounding her space program.

September 24: Six-engine S–IV stage for Saturn space vehicle was successfully static-fired in 60-sec. test at Douglas Missile and Space Systems, Sacramento, Calif.

• Announced that Astronaut Walter M. Schirra, Jr., would study powerful light sources in Australia and South Africa, during his MA–8 orbital flight. NASA said experiment was to determine how well a space traveler could see light sources of known density on earth. Three high-intensity flares would burn at Woomera, Australia, and electric lamps at Durban, So. Africa. Schirra would try to determine effects of atmosphere on light observed from orbital altitude and to establish which light source was more readily visible. Other scientific experiments of the MA–8 flight would be studies of cosmic radiation, search for rare particles in space, and effects of re-entry heating on new materials.

• Reported that, if Astronaut Schirra's six-orbit flight (MA–8) is fully successful, NASA would conclude Project Mercury with one 18-orbit, 24-hour flight. Present plans called for four 24-hour flights with the one-man Mercury capsule; new plan under consideration would cancel three of these and assign their mission to two-man Gemini capsule.

• FCC Chairman Newton N. Minow invited U.S.S.R. to cooperate in an international television system using high-altitude communications satellites. A high-altitude system would have advantages over low-altitude system such as Telstar, he said in *Look* magazine, but "we now lack the launching power to make it go." He proposed that Soviet rockets be used to launch U.S. communications satellites into high-altitude orbits, and added that "this would be a very dramatic area of cooperation for peaceful purposes in space."

• Congressman John E. Moss, Chairman of the Government Information Subcommittee, wrote to Secretary of Defense Robert S. McNamara asking why 1959 Army report "Project Horizon" was not yet declassified. Originally classified "Secret," the study proposing manned lunar program culminating in 1965 lunar landing had 10 separate security reviews by 4 agencies, but it was still classified "official use only." Congressman Moss observed that the House Committee on Science and Astronautics had requested declassified version of "Project Horizon" on Aug. 3, 1961, but had never received it.

September 24: Space Technology Center of General Electric Co. said it was designing an escape system for astronauts to use if orbiting capsule malfunctioned and could not return safely to earth. Called "Moose" (Man Out Of Space Easiest), the system involved use of a "prepackaged spacecraft" consisting of plastic sack, folding heat shield, retrorocket pack, containers of foaming plastic, and a standard parachute. Astronaut, zipped into sack, would step out of faulty spacecraft, aim himself with special scope, and fire the retrorocket. Then his heat shield would unfold and sack would fill with rapid-hardening plastic foam, protecting him from re-entry heat. At 30,000-ft. altitude, he would open parachute and descend slowly to earth.

- Communications lag between basic researchers and those who apply the research was cited by Assistant Secretary of Commerce J. Herbert Holloman at testimonial dinner for NAS president Dr. Frederick Seitz, in Chicago. Mr. Holloman said that communication "throughout the international scientific community" was more rapid than communication between scientists and people working in application of scientific discoveries.

- AFSC announced invention of Bio-telescanner—compact and portable instrument to determine biological properties of other planetary soil—by Dr. William G. Glenn, SAM research immunobiologist, and Wesley E. Prather, electronics expert. Designed to answer basic questions about extraterrestrial life, the instrument would analyze samples gathered by travelers to Mars or other planets and would telemeter its findings back to earth. It could be adapted to operate automatically, to gather and analyze soil samples before man set foot on the planet. Use of the Bio-telescanner would preclude necessity to bring samples of other planetary crust back to earth—an operation which could be dangerous if alien micro-organisms were present in the sample.

September 25: Senate adopted conference report on H.R. 12711, Independent Offices Appropriations Bill containing NASA appropriations for FY 1963. Bill would be sent to White House.

- Bernard Strassberg was appointed director of the FCC Office of Satellite Communications.

- In test near Florida coast, air-launched Skybolt missile flew most successful flight to date, achieving second-stage ignition for the first time. The USAF missile fell short of its 900-mi. target when malfunction caused second-stage engines to burn out after only 15 sec., several sec. sooner than programed.

September 26: Jet Propulsion Laboratory scientists reported that MARINER II Venus probe was disoriented on Sept. 8, by either a micrometeorite impact or a solar pressure wave, but that automatic devices onboard the craft had restored it to proper orientation.

- NASA announced it had completed preliminary plans for development of $500 million Mississippi Test Facility. First phase of three-phase construction program would begin in 1962 and would include four test stands for static-firing Advanced Saturn

S–IC and S–II stages; about 20 support and service buildings would be built in the first phase.   Water transportation system had been selected, the system calling for improvement of about 15 mi. of river channel and construction of about 15 mi. of canals at the facility.   In planning the facility, Sverdrup and Parcel Co. of St. Louis was preparing design criteria; Army Corps of Engineers was acquiring land for NASA in cooperation with Lands Div. of the Justice Dept.   The 13,500-acre facility in southwestern Mississippi is 35 mi. from NASA Michoud Operations, where Saturn stages are fabricated.

September 26: Liquid hydrogen-liquid oxygen engine RL–10A3 completed its preliminary flight rating test, MSFC announced.   RL–10A3 was advanced version of RL–10A1, first U.S. engine using liquid hydrogen.   Six A3 engines will be clustered in Saturn S–IV stage and will be used in advanced Centaur vehicle.   The five-day PFRT included 20 static firings for an accumulated 2,820 sec. of firing time, plus humidity, vibration, and other tests.

- Federal District Judge Joseph Lieb signed injunction against International Union of Operating Engineers, Local 267, ordering the union to stop allegedly unfair labor practices—refusing to operate cranes unloading heavy equipment at Cape Canaveral missile launch complex.

- President Kennedy signed legislation to authorize production of electricity at Hanford, Wash., atomic plant by using byproduct steam from a plutonium reactor.   He commented that such production would give the U.S. a clear margin of superiority in the peaceful use of atomic energy.

September 27: NASA announced that Venus-bound MARINER II had been in flight one month, during which time it had traveled more than five million mi. from earth and had transmitted scientific data on interplanetary space from a greater distance than any space probe except PIONEER V (launched March 11, 1960).

- U.S.S.R. placed scientific satellite COSMOS IX into orbit (apogee, 219 mi.; perigee, 187 mi.; period, 90.9 min.), in series to study ionosphere, radiation belts, and effect of meteorites on spaceships.

- NASA postponed launch of first Canadian satellite, Alouette, at Pt. Arguello.

- Astronaut Walter M. Schirra, Jr., would photograph cloud formations during his forthcoming orbital flight MA–8, it was announced.   Cdr. Schirra's special 2½-lb. hand camera would be used to test techniques for the advanced Nimbus weather satellite.

- Because of unnamed technical difficulties, flight test of X–15 rocket research aircraft was postponed minutes before X–15 was to be dropped from its mother ship.

- Results of a USIA poll in Britain showed that 82% of people questioned had identified TELSTAR communications satellite by name, compared to 78% able to identify SPUTNIK I in 1957.   Testifying before a subcommittee of the House Committee on Science and Astronautics, Robert Mayer Evans, special assistant to USIA Director Edward R. Murrow, reported the survey's results and said they showed "an extraordinary level of awareness" of the communication satellite in its first weeks of existence.

- AEC announced Soviet atmospheric nuclear explosion of nearly 30 megation yield.

September 28: Canadian satellite ALOUETTE was placed in polar orbit by Thor-Agena B vehicle launched by NASA from Vandenberg AFB. Initial orbital data: apogee, 619.2 mi.; perigee, 597 mi.; inclination, 80.84° to the equator; period, 105.4 min. Named for Canadian high-flying songbird, ALOUETTE was first satellite both designed and built by a nation other than U.S. or U.S.S.R. Launching was NASA's first into polar orbit, first from PMR, and first use of Thor-Agena B combination. To sustain its experiments in study of the ionosphere, the 320-lb. ALOUETTE had two crossed antennae, one 150 ft. long and the other 75 ft. long.

- X–15 No. 2 was flown by NASA pilot John B. McKay in longest engine run to date (127 sec.), the craft reaching 67,000-ft. altitude and 2,693-mph speed (mach 4.08) in successful 9-min. flight near Edwards AFB, Calif. Flight tested heat transfer rates over X–15's skin, craft's stability without its ventral fin, and newly designed windshield's resistance to heat.

- Astronaut Walter M. Schirra, Jr., made 6½-hr. simulated flight in Sigma 7 spacecraft for Project Mercury flight MA–8. Worldwide tracking network of 21 ground stations and ships also participated in exercise, in which Cdr. Schirra and the tracking stations practiced solving emergency situations that could arise in actual orbital flight.

- First successful firing of a Centaur flight stage at full thrust of 30,000 lbs. for ten seconds was achieved by General Dynamics/Astronautics at Sycamore Canyon, Calif.

- NASA announced plans to launch two Project Echo balloons during October. To be filled with helium while on the ground near White Sands Missile Range, N.M., one ballon would be Echo I-type measuring 100 ft. in diameter and the other would be Advanced Echo type measuring 135-ft. in diameter. They would reach respective altitudes of approximately 128,000 ft. and 115,000 ft. in short flights over the range. Purpose of launches was to determine skin smoothness for the Advanced Echo balloon satellite, to be orbited in 1963.

- USAF Minuteman ICBM fired in combat-ready test from Vandenberg AFB, Calif., range safety officer destroying missile shortly after launch.

- Spokesman for American Telephone and Telegraph Co. (AT&T) said TELSTAR communications satellite was still operating despite almost three months of intense radiation exposure.

- Cdr. Alan B. Shepard, Jr. (USN), the first U.S. astronaut, and Lt. Col. R. G. Robinson (USMC) were awarded the FAI DeLaVaulx Medal—Shepard for achieving 615,300-ft. altitude in his suborbital Mercury flight (MR–3) and Robinson for setting closed-course speed mark of 1,606.5 mph in McDonnell F4C aircraft.

- Congressman George P. Miller, Chairman of the House Committee on Science and Astronautics, summarized the year's activities of the committee in speech on House floor: "It has been another busy year for the committee, highlighted by action on the fourth annual budget of the National Aeronautics and Space Administration, now rapidly approaching the $4 billion mark. In addition to the intensive study the space budget underwent, the committee also conducted 26 investigations, issued 16 reports, held 124 days of hearings, and heard 236 witnesses.

"Indicative of the committee's activity is the more than 2,175,000 words of testimony taken this year incorporated into approximately 3,860 pages of printed hearings. . . ."

September 29: NASA Administrator James E. Webb announced that the project management of the liquid hydrogen-fueled Centaur launch vehicle and the hydrogen-fueled M-1 rocket engine would be transferred from Marshall Space Flight Center to the Lewis Research Center. Studies were being initiated to adapt Centaur as a stage to other boosters. Mr. Webb said: "The transfer will allow the Marshall Center to concentrate its efforts on the Saturn vehicles for the manned lunar landing program. . . . It will permit the Lewis Research Center to use its experience in liquid hydrogen to further the work already done on one of the most promising high-energy rocket fuels and its application to Centaur and the M-1."

• USAF launched satellite from PMR with Thor-Agena booster.

• At the 13th Congress of the International Astronautical Federation (IAF), meeting in Bulgaria, the 1962 Guggenheim International Astronautics Award was awarded to Prof. James A. Van Allen.

September 30: NASA launched Aerobee sounding rocket from Wallops Station, the 259-lb. instrumented payload reaching 106-mi. altitude in test to map sources of photons in specific wavelengths in the nighttime sky. Payload landed about 61 mi. downrange, with recovery not attempted.

• Tropical storm developed 420 mi. east of Puerto Rico, traveling on path that would cross impact area for Astronaut Schirra's third orbit. NASA flight officials were closely watching the storm to determine whether it would necessitate postponing Schirra's MA-8 flight scheduled for Oct. 3.

During September: Conclusion that wheel-shaped space station was technically feasible and identification of problems that still exist were revealed in NASA Technical Note D-1504, compilation of 11 papers by NASA Langley Research Center staff. Selection of this 150-ft.-diameter, 171,000-lb. structure was result of studies by Langley for more than 2 years and detailed analyses by North American Aviation, Inc., for past 6 months. The report indicated the space station primarily would provide a means of learning to live in space, where zero-gravity and variable-gravity experiments could be performed, closed life-support systems could be qualified, and rendezvous techniques and systems developed. Chosen model could support up to 38-man crew.

• Issue of ARS' *Astronautics* contained series of articles by NASA Langley Research Center scientists and others, reviewing LaRC research program on manned space-station technology since early 1960.

• Search for two huge meteorites believed to have fallen in western Virginia near Covington and West Virginia near Clarksburg was conducted by Dr. Frank Drake, Director of National Radio Astronomy Observatory, Green Bank, W. Va. Reports by witnesses indicated that the meteorites weighed about one ton each and that they fell in sparsely populated, national forest areas.

During September: Dr. Joshua Lederberg, of Stanford University, and Dr. Carl Sagan, former University of California professor, wrote in September *Proceedings of the National Academy of Sciences* that large amounts of moisture may be frozen into the subsoil of the planet Mars. From openings in the planet's crust, steam and gas may escape and create hot spots that would be "favorable microenvironments for life." They said that recurrent clouds detected by telescope "may be symptomatic of this local outgassing of water vapor."

- Dr. John W. Findlay, deputy director of the National Radio Astronomy Observatory at Green Bank, W. Va., said Echo-type reflective orbiting balloon and West Ford-type copper-needle belts were potential threats to science of radio astronomy. Writing in *Science* magazine, Dr. Findlay said these objects would contribute to "noise" in radiotelescope reception and "would cut down sharply on an antenna's ability to receive faint signals."

- Army Corps of Engineers announced that Launch Complex 12 at Cape Canaveral would be modified to service larger and more sophisticated space vehicles. Complex 12 was used to launch Mariner and Ranger space probes.

- World's first regularly scheduled Hovercraft transportation service ended. Operating since July with 8 round trips daily from Rhyl, Wales, to Wallasey, England, the experiment proved Vickers Armstrong 24-passenger Hovercraft could successfully negotiate soft sand and shallows as well as deeper water. Hovercraft rides 8 in. above surface (land or sea) on cushion of air.

- Secret of radio emissions from planet Jupiter was proposed by physicists Dr. Leon Landovitz of Yeshiva Univ. and Dr. Leona Marshall, New York Univ. They suggest that perhaps Jupiter acts as a gigantic maser, using energy from the sun as its "pumping" energy, to produce the decameter waves. They proposed that Jupiter's decameter radiation was based on principle that the change of state of an electron with respect to static magnetic field results in either emission or absorption of electromagnetic energy, such as light or radio waves, at frequency that depends upon the magnetic field. There may be a "cascading" of such stimulated emissions from the electrons as the disturbance propagates through the electrically charged layer of Jupiter's atmosphere, Dr. Marshall stated. Jupiter's emission of radio waves was discovered in 1955 by Dr. Bernard Burke and Dr. Kenneth L. Franklin, then of Carnegie Institution of Washington.

- Seven X–15 pilots and Paul F. Bickle, Director of NASA Flight Research Center, were awarded the first John J. Montgomery Award for aerospace achievement by the National Society of Aerospace Professionals and the Aerospace Museum of San Diego. The X–15 pilots were : Scott Crossfield, NAA; Maj. Robert M.

White, USAF; Neil A. Armstrong, NASA; John B. McKay, NASA; Joseph A. Walker, NASA; Cdr. Forrest S. Petersen, USN: and Maj. Robert A. Rushworth, USAF.

During September: Ground rehearsals for ONR-NSF-NASA Stratoscope II balloon-carried telescope experiment took place in Norwalk, Conn., the 36-in. telescope focusing on the planet Jupiter. Later in month, telescope was disassembled and transported to Palestine, Tex., where it would remain in hangar until first flight, in February or March 1963. To be lifted above most of the earth's atmosphere by two balloons, telescope would be stabilized at 80,000-ft. altitude throughout each observing night, where it will make a variety of astronomical and planetary observations.

OCTOBER 1962

October 1: Fourth official anniversary of NASA, charged by the National Aeronautics and Space Act of 1958 to organize and conduct U.S. space exploration for peaceful purposes. No new agency of the Executive Branch of the Federal Government was created by the transfer of so many units and programs of other departments or agencies as NASA.

- Project Mercury flight officials studied tropical storm "Daisy," which continued moving westnorthwest about 300 mi. east of Puerto Rico, the general landing area for third orbit of Astronaut Walter M. Schirra's six-orbit MA–8 flight planned Oct. 3. Mercury Operations Director Walter C. Williams, indicating continuing surveillance of "Daisy," ordered preparations for the launch attempt to continue.

- NASA Administrator James E. Webb, addressing the Greater Hartford (Conn.) Chamber of Commerce, said that the "habit of oversimplifying has carried over into the evaluation and understanding of our objectives in space. . . . Many of our citizens believe our major, and in some cases, our only purpose in space, to be one of winning, in a contest with the Soviet Union, 'a race to the moon.' . . . [This concept] overlooks the significant and seldom appreciated fact that learning how to get to the moon, developing the technology which will be required to get there, and employing this technology for many purposes in space, is more important than the lunar landing itself. . . .

  "Characterization of our effort solely as a 'space race to the moon' also leads to the mistaken notion that the nation first to conduct a lunar exploration will have assured itself of ultimate and enduring superiority in space. The fact is overlooked that landing men on the moon, although a great and challenging goal, is only one of many goals in space. . . .

  "With the resources which are being applied, and knowledge of how those resources are being utilized to mobilize our nation's scientific, technical and industrial strength, I am confident that we will achieve a space posture satisfactory to the United States given a continuation of our present efforts and support for those planned for future years. On the basis of what we know now of the scope of the Soviet Union's efforts, and of what is required to do the job, this country has a better than even chance to conduct the first manned exploration of the moon. For these reasons, I have said I believe we will be there first.

  "But in saying this I do not want to fall into the same trap which has snared too many of us—that of again inviting a characterization of this broad and vital national effort as a spectacular 'race to the moon.'"

  He pointed out that President Kennedy did not set the national goal of "landing the first man on the moon." Rather, the President "stated our determination to attain 'a position of pre-eminence in space' and to become the world's leading spacefaring nation . . . .

"That is our objective—to develop superior competence in space which will be available for any national purpose which may be required, whether it be the peaceful utilization of space for the benefit of mankind, or to keep the peace and forestall its exploitation by any nation for aggressive purposes . . . ."

October 1: Nova launch facilities study contract was awarded to Martin-Marietta Corp., Denver, by NASA Marshall Space Flight Center. The study assumes Nova will be launched from Cape Canaveral.

October 2: EXPLORER XIV satellite was launched into orbit by Thor-Delta rocket from Cape Canaveral on mission to study natural and man-made radiation in space. NASA spokesmen said satellite appeared to be following the planned orbital path (53,000-mi. apogee, 185-mi. perigee, 31-hr. period), but scientists could not confirm the exact orbital figures for several days (later confirmed as 61,000-mi. apogee, 175-mi. perigee, 36.1-hr. period). The 89-lb. satellite was injected into orbit at nearly 24,000-mph speed by the three-stage Thor-Delta. This was Delta's 12th straight successful satellite launching.

- 20th anniversary of first military jet-powered flight in the U.S., a Bell P–59A piloted by USAF pilot Laurence C. Craigie at Muroc, Calif., attained estimated speed of 450 mph. P–59A was powered by two GE I-A engines, each developing 1,250-lb. thrust and weighing 780 lbs.

- Project Mercury operations director Walter C. Williams ordered launch countdown to continue toward MA–8 launch at Cape Canaveral the following morning. Weather conditions which had been under surveillance were considered generally favorable, and all elements of the six-orbit manned space mission were in a "go condition."

- Nuclear physicist Dr. Edward Teller told 1962 conference of UPI editors and publishers that recent U.S. nuclear tests in the Pacific had "not been quite satisfactory" and that it was his guess "that the Russians are ahead of us in the nuclear race today."

- NASA Administrator James E. Webb, addressing the Northeast Commerce and Industry Exposition in Boston, said:
  "The achievement of . . . [U.S. pre-eminence in space] requires that we learn to travel in space as we have learned to travel on the sea or in the air. We must develop what might be termed *space power*—the capability to utilize space for every purpose which our national interest may require. And, to fulfill the directives established by the Congress, we must develop that competence in space for the benefit of our own people, and of all mankind . . . .
  "Although creative individual effort is as important today as it has ever been, and the imagination and initiative of the individual remain the primary source of scientific and technical progress, the development and perfection of the complex equipment, the systems and subsystems which are taking men into space are rarely, if ever, the work of single individuals. Today, teams of talented and imaginative technicians are joined in each major technological achievement . . . ."

October 2: Thomas F. Dixon, NASA Deputy Associate Administrator, addressed the Southern Governors' Conference, Hollywood-by-the-Sea, Fla., on the role of the South in the Space Age, and remarked: ". . . our space activities are in no sense stunts—spectaculars, if you will—nor will the landing of U.S. astronauts on the moon be our ultimate objective. Rather, the national effort is broadly organized to accomplish rapid advances in science and technology that will make it possible for us to perform any space assignment—military or civilian—that the national interest might require. Furthermore, from lunar exploration, we will go on to wider exploration of the solar system, ventures whose end no man can foresee. . . ."

• International Association of Machinists rejected contract offer of Aerojet-General Corp. and began a strike. In Washington, a Presidential committee opened talks on labor dispute between IAM and the Boeing Co. United Auto Workers employed by North American Aviation, Inc., had accepted a three-year contract, thus averting strike.

• Lewis L. Strauss, former AEC chairman, criticized U.S. manned lunar flight program as seeming "perilously like a stunt." Speaking at United Republican Fund dinner in Chicago, Strauss urged the lunar program be coordinated with military necessity.

• Science Service announced that a total solar eclipse would be visible July 20, 1963, in Alaska, Canada, and Maine.

• Secretary of Commerce Luther Hodges told news conference that U.S. emphasis on defense- and space-related research was "one of the great dangers" to the productivity of U.S. economy because such research programs attract scientists and engineers away from the civilian economy.

• Monetary awards under $1,000 were made to NASA employees for patentable inventions:

   NASA Ames Research Center—Vernon L. Rogallo; Joseph R. Smith, Jr.; John V. Foster; and Albert E. Clark, Jr.;

   NASA Goddard Space Flight Center—Harold J. Peake, Stephen Paull, and William A. Leavy;

   NASA Langley Research Center—William J. O'Sullivan, Jr.; George F. Pezdirtz; George F. Look; Frank M. Ballentine, Jr.; Virgil S. Ritchie; Howard F. Ogden; Eldon E. Kordes; Donald H. Trussell; Deene J. Weidman; George E. Griffith; Blake W. Corson, Jr.; and Emanuel Schnitzer;

   NASA Lewis Research Center—Paul F. Sikora;

   NASA Manned Spacecraft Center—Warren Gillespie, Jr.;

   NASA Marshall Space Flight Center—Donelson B. Horton.

• AEC announced first U.S. atmospheric nuclear test in the Pacific since July had been conducted near Johnston Island, with intermediate-range yield.

October 2–4: Third Symposium on Advanced Propulsion Concepts held in Cincinnati, cosponsored by USAF Office of Scientific Research and GE Flight Propulsion Division.

October 3: SIGMA 7 spacecraft with Astronaut Walter M. Schirra, Jr., as pilot was launched into orbit by Mercury-Atlas vehicle at Cape Canaveral. Orbital data: apogee, 176 mi.; perigee, 100 mi.; period, 88.5 min. Schirra traveled 160,000 mi., nearly six orbits, returning to earth at predetermined point in Pacific Ocean about

9 hrs., 14 min. after blastoff.   Within 40 min. after landing he and his spacecraft were safely aboard aircraft carrier U.S.S. *Kearsarge*.

Highly successful MA–8 flight proved the feasibility of (1) prolonged weightless flight in space and (2) drifting in orbit without consumption of hydrogen-peroxide attitude control fuel and without physically endangering the astronaut.   Schirra, who spent a total of 2½ hrs. in drifting flight, reported there were no unusual attitude control problems during drifting.

Various other scientific experiments during MA–8 flight included astronaut's sighting of luminous particles in space, also reported by Astronauts John H. Glenn and M. Scott Carpenter; and photographing cloud formations with a special hand camera.

SIGMA 7 was sighted by observers on Indian Ocean tracking ship for five min. as the spacecraft made its third orbit more than 100 mi. high.   The observers said SIGMA 7 appeared almost as bright as Venus.   This was the first reported visual sighting of a manned spacecraft during orbital flight.

Only technical difficulty of the entire operation was attaining proper adjustment of Schirra's spacesuit coolant supply and failure of body temperature measurements in first orbit.   However, desired suit temperatures were achieved early in second orbit.

October 3: In press conference following MA–8, D. Brainerd Holmes, NASA Director of Manned Space Flight, said: "This was a highly successful flight, magnificently performed.   It proved there is no substitute for sound engineering and thorough training."

Walter C. Williams, MSC Director of Mercury Operations, said that "so far as I am concerned, the mission was perfect."   He added that the next Mercury mission would be a 24-hour flight in early 1963.

• Soviet Foreign Ministry told U.S. Embassy in Moscow that the "Soviet Union will not undertake any action that might hinder the flight of the American spaceman Walter Schirra."   The Soviets were replying to U.S. note requesting the U.S.S.R. to refrain from nuclear tests during the MA–8 flight.

• President Kennedy appointed Dr. Frederick Seitz to his Science Advisory Committee.   Dr. Seitz, president of the National Academy of Sciences, succeeded Dr. Detlev W. Bronk on the committee.

• SIGMA 7 (MA–8) launching was relayed to Western Europe via TELSTAR satellite.   European TV network Eurovision broadcast the launching sequence to 17 countries.   Televised launch was also viewed by millions of Americans.

October 4: Fifth anniversary of the Space Age, inaugurated with the orbiting of first manmade satellite, SPUTNIK I, by the U.S.S.R. According to the *Goddard Satellite Situation Report* of October 10, 1962, the first half decade of the Space Age saw the orbiting of 134 satellites, lunar probes, and space probes.   U.S.S.R. orbited 26 (6 still in orbit), the U.S. 108 (48 still in orbit).   Of U.S. total, NASA orbited 36 (24 still in orbit), DOD 72 (24 still in orbit). Two of the NASA launches must also be credited to the nations contributing the experiments or the payload—U.K. for ARIEL and Canada for ALOUETTE.   Of the U.S. total, three were manned orbital spacecraft, as were four of the U.S.S.R.'s.   Totals for manned space flight—U.S.: 12 orbits, with 19 hrs., 3 min. flight time; U.S.S.R.: 130 orbits, with 192 hrs., 41 min. flight time.

October 4: Maj. Robert A. Rushworth (USAF) flew X–15 No. 3 to altitude of 106,000 ft. and speed of 3,375 mph (mach 4.91), in 10-min. flight to check out the craft's adaptive control system. A secondary objective was to further evaluate X–15 performance with ventral fin removed. All flight objectives were achieved.

• Astronaut Walter M. Schirra, Jr., on board aircraft carrier U.S.S. *Kearsarge*, began debriefing and medical check. Schirra would arrive in Honolulu Oct. 6 and then fly to NASA Manned Spacecraft Center, Houston.

• President Kennedy nominated 13 men to form satellite communications corporation, under terms of communications satellite law passed in August. The names of the nominees, who would serve only until establishment of corporation's operating procedures and arrangement for stock sale, were sent to the Senate for confirmation.

• International Association of Machinists reached preliminary agreement with Aerojet-General Corp., ending strike which began Oct. 2 at the Sacramento and Azusa plants in California and extended to Cape Canaveral this morning.

• Speaking on Radio Moscow, Prof. N. P. Arabashov said: "All the magnificent achievements of Soviet science convince us that flights by men to the moon and the planets are not only possible, but are also near. It is now clear that automatic interplanetary space ships, at first unmanned, will be sent to the moon and then to Venus and Mars."

October 4–5: Seventh NASA Management Conference held at Langley Research Center.

October 5: EXPLORER XIV was relaying "excellent quality preliminary data," NASA announced. The satellite, launched Oct. 2, was traveling in highly eccentric earth orbit: apogee, 61,000 mi.; perigee, 175 mi.; period, 36.1 hrs. Speed at perigee was 23,700 mph; at apogee, 1,500 mph. NASA said it would be some time before analysis of telemetry data on the radiation belts could be correlated.

• Walter M. Schirra, aboard the recovery vessel U.S.S. *Kearsarge*, reportedly suggested to Project Mercury engineers that the ground communications stations should give the orbiting astronaut "more quiet time" in space. Extraneous conversation could interrupt the astronaut's concentration on technical duties.

• Dr. Charles A. Berry, Chief of Aerospace Medical Operations Office, NASA Manned Spacecraft Center, reported that preliminary readings of a dosimeter indicated Astronaut Schirra had received much smaller radiation dosage than expected. Dr. Berry said post-flight examination of the astronaut determined that he was "in excellent condition and detected no physiological effects from the space flight."

• NASA announced progress of MARINER II Venus probe launched August 27: distance from earth—6,268,440 mi.; distance from Venus—34,997,199 mi.; radio signals—good.

• Cluster of six RL–10 engines for the Saturn S–IV stage was static-fired in first full-duration (seven-minute) test at Douglas Aircraft Co., Sacramento.

October 5: USAF announced special instruments on unidentified military test satellites had confirmed artificial radiation belt densities calculated from earlier INJUN and TELSTAR data. Radiation of new belt, created by U.S. high-altitude nuclear test in July, was composed primarily of electrons. Radiation sharply increases above 400-mi. altitude at the geomagnetic equator and reaches peak intensities of 100 to 1,000 times normal levels at altitude above 1,000 mi. Lt. Col. Albert C. Trakowski (USAF) stated in press conference that Astronaut Walter M. Schirra, Jr., could have been killed if his MA–8 space flight had taken him above 400-mi. altitude. Answering questions about USAF report on the artificial radiation belt, Col. Trakowski said the lifetime of the belt "cannot be forecast at present—it will be long."

- Nevada Extension of the AEC–NASA Space Nuclear Propulsion Office (SNPO) became operational, with Robert P. Helgeson as Chief. Located at Las Vegas, the Nevada Extension assumed financial and contract responsibilities for construction and operation of facilities at Nuclear Rocket Development Station (NRDS), which will conduct static ground testing of nuclear rockets.

- NASA signed $1.55 million contract with Hamilton Standard Div. of United Aircraft Corp. and International Latex Corp. for development of spacesuit for Apollo crewmen. As prime contractor, Hamilton Standard has management responsibility of the overall program and would develop life-support, back-pack system to be worn by crewmen during lunar expeditions; Latex would fabricate the suit with Republic Aviation Corp. furnishing human factors information and environmental testing. The suit would allow crewman greater mobility than previous spacesuits, enabling him to walk, climb, and bend with ease.

- Third test flight of USN Polaris A–3 was partially successful, the missile functioning properly until second stage went out of control near burnout and was destroyed by range safety officer.

- Sperry Gyroscope Co. announced it was patenting a new gyroscope using liquid instead of wheel as the spinning element. Sperry said this was the first such gyroscope to operate successfully and to go into production.

- Soviet Minister of Agriculture K. G. Pysin, visiting U.S.S.R. Embassy in Washington, commented when asked about Soviet droughts that "control of the weather is being researched up there in outer space."

October 6: NASA Astronaut Walter M. Schirra was greeted by cheering crowds in Honolulu, transferring from aircraft carrier U.S.S. *Kearsarge* to USAF jet at Hickam AFB, for trip back to Manned Spacecraft Center, Houston, Texas.

- NASC Chairman Dr. Edward Welsh, in message to American Legion, refuted arguments that the accelerated space program was detrimental to other interests, and said: "Even before we had a space program, we were not spending enough on education, or medicine, or housing, and there is no reason to believe that we would be doing even as much as we are now on those essential projects if it were not for the space program. In fact, our national space effort increases our gross national product and provides more employment and more funds than we would have without it. This is a solid investment which we must make."

October 6: AEC announced low-yield nuclear detonation in the atmosphere near Johnston Island, the 29th announced test in current U.S. series in the Pacific.

- Decision to sell Cape Colony Inn, partially owned by the seven Project Mercury astronauts, was announced by attorney C. Leo DeOrsey of Washington. He said the astronauts' investment in the motel near Cape Canaveral "had become the subject of controversy in some quarters of the press and might become embarrassing to the National Aeronautics and Space Administration. I believe it is in the best interests of all concerned to sell the motel."

- USAF revealed that three "plastinauts," plastic dummies of human astronauts, were undergoing tests at Kirtland AFB preparatory to being sent into space. Plastinauts in space would be used to determine radiation hazards to astronauts.

During early October: J–2 liquid-hydrogen rocket engine successfully completed its first full-duration static firing (more than four minutes) at Rocketdyne Div. of North American Aviation, Inc. J–2 engines will be clustered to power S–II stage of Advanced Saturn vehicle.

October 7: Astronaut Walter M. Schirra, Jr., after arriving in Houston after trip from Pacific landing of six-orbit MA–8 flight, held press conference at Rice University auditorium: "My intention was to use so little [attitude control] fuel that no one could argue that we [had not] had enough fuel aboard SIGMA 7 for eighteen orbits, if we wanted it. I think I proved that point." He said MA–8 was a "textbook flight" and that he was sorry to see it end.

On the issue of space sickness, Astronaut Schirra stated that despite the prolonged weightlessness of this longest U.S. manned space flight, "there was no problem at all. There was no break-off phenomena, there was no uneasiness, there was no queasiness. I felt great.

"The suit temperature problem . . . is a problem which we have solved. I have been much hotter in the tent at Cape Canaveral than I ever, ever thought of being in SIGMA 7."

He reported sighting the "fireflies" and believing they emanated from the spacecraft.

- *New York Times* reported that more than 20 fragments from SPUTNIK IV had been recovered in Wisconsin. Launched May 15, 1960, the Soviet satellite had re-entered earth's atmosphere Sept. 5, 1962, and disintegrated over Wisconsin.

October 8–12: Series of space science lecture-demonstrations held at St. Johns Univ., N.Y., under auspices of NASA-Aerospace Education Council.

October 9: X–15 No. 2, piloted by NASA's John B. McKay, was flown to 129,000-ft. altitude (24.4 mi.) and 3,716-mph speed (mach 5) in test of aircraft's performance without its electronic flight-control system and without its ventral fin. Pilot McKay induced the most severe yawing motion thus far in tests to check the craft's re-entry stability under severe strain. Although pilot McKay later reported "the whole airplane shook for about three minutes," the stability test was considered very successful.

Altitude and speed were greatest ever achieved in X–15 flights without the lower tail-fin.

October 9: Astronaut Walter M. Schirra, Jr., told newsmen he hoped Astronaut Leroy Gordon Cooper would be selected for Project Mercury flight MA–9. Astronaut selection had not yet been made for the 18-orbit 24-hour flight.

- U.N. Space Registry revealed the U.S. had reported total of 66 satellite and spacecraft launchings as of Aug. 15, 1962. Of these, 25 were NASA launchings and 41 were U.S. military launchings. U.S. reports to U.N. Space Registry gave same data for NASA and military launchings: name of object; launch date; and satellite's apogee, perigee, inclination, and period. U.S. reports also identified satellites in one of four categories: development of space-flight technology; space research and exploration; practical applications of space technology; and non-functional objects.

- USAF announced launching of an unidentified satellite using a Thor-Agena launch vehicle.

- White House officials announced the secrecy policy on launchings of military satellites was being re-examined.

- John Rubel, DOD Deputy Director of Research and Engineering, told the Aerospace Luncheon Club in Washington that DOD was supporting two programs in space—one "directed at clear, identifiable military needs and requirements" and the other, less well defined, designed to provide "building blocks" as "insurance against an uncertain future." Replying to charges that military space activities should be expanded, he noted that DOD's military space program exceeded $1.5 billion this year and probably would be greater next year; DOD expenditures on space developments were "remarkably high in relation to viable concepts for military applications in space." He concluded that the military space program was adequate and well balanced, although "we probably err on the side of allowing too generous a margin of safety for the effects of these uncertainties."

- USN Transit navigational satellite system was described by Capt. Alton B. Moody (USNR) of NASA, at National Electronics Conference in Chicago. The system of four Transit satellites in polar orbits, to be operational by end of 1962, was designed primarily to provide naval craft with constant, accurate navigational fix around the world.

October 10: NASA announced MARINER II would miss Venus by 20,900 mi. instead of the 9,000 mi. previously expected, but that the probe would still come "well within the region where the scientific planetary experiments are expected to be very effective." The added distance was caused by midcourse correction maneuver on Sept. 4 increasing MARINER II's velocity by 47 mph instead of the intended 45 mph.

NASA scientists said telemetry data ‌from ‌MARINER II indicated that (1) solar particle radiation appears to be sufficiently strong to distort solar magnetic fields; (2) some atmospheric gas was present in entire path of the probe; (3) density of small particles in deep space was at least 1,000 times less in interplanetary space than in the near vicinity of the earth.

October 10: DOD released "Project Horizon," an Army study of manned lunar flight program aimed at 1965 lunar landing. Written in 1959 as a classified report, the study was declassified in 10 security reviews by three agencies and unclassified version sent to House Science and Astronautics Committee at its request.

- AEC displayed its latest radioisotope electrical power generator, Snap-9A, in exhibit at Whittier (Calif.) Civic Center. Designed to produce 20 watts of electrical power for several years, the nuclear generator was scheduled to power radio transmitters in USN Transit satellite to be launched into polar orbit within the next few months.

October 10–12. NASA held agency-wide representatives conference on automatic data processing in Washington, D.C.

October 11: Opening of the Ecumenical Council in Vatican City was viewed in U.S. with 24-min. TV broadcast relayed via TELSTAR.

- General Dynamics/Convair reported developing a new laser with increased pulse rate. When perfected, new laser would be able to machine and trim any material with greater speed and accuracy than available by any other means.

Ocober 12: USAF Titan II rocket achieved all test objectives in flight from Cape Canaveral. This was fourth success in six test launchings of the rocket, a modified version of which will be used in NASA's Project Gemini beginning in 1964.

- International Association of Machinists reached agreement on new contract with Aerojet-General Corp., after negotiations which had followed IAM walkout October 2–4.

October 13: USAF announced the third in series of Project Firefly upper atmosphere experiments would be conducted during 1962. Approximately 30 research rockets, launched from Santa Rosa Island Aerospace Facility near Eglin AFB, would carry chemical payloads to above 50-mi. altitudes.

- Soviet military newspaper *Krasnaya Zvezda* reported that U.S.S.R. had been first nation to use rockets in combat and cited rockets fired by Soviet airplanes against the Japanese in Mongolia, August 1939.

October 13–14: Open house at NASA Wallops Station, Va.; more than 15,000 visitors toured the launch site facilities.

October 14: NASA announced plans to launch Ranger (5) toward the moon during mid-October. The 775-lb. lunar probe would represent third U.S. attempt to take close-up photographs of the moon, gather information on lunar surface composition, and land instrumented capsule on lunar surface.

- 33-member board of trustees of Pacific Science Center Foundation was announced. The foundation would operate the U.S. Science Pavilion for Century 21 in Seattle as a permanent science center.

October 15: Astronaut Walter M. Schirra, Jr., returned to his hometown, Oradell, N.J., for triumphal welcoming ceremonies in honor of his successful MA–8 six-orbit space flight for NASA Project Mercury. NASA Administrator James E. Webb presented Cdr. Schirra with the NASA Distinguished Service Medal.

- Analysis of scientific measurements by RANGER III lunar probe showed gamma ray intensity in interplanetary space probably is 10 times greater than expected, NASA reported. Measurements were taken by gamma ray spectrometer on RANGER III after it was launched Jan. 26, 1962. Similar instrumentation was planned for Ranger (5), to be launched in mid-October.

October 15: NASA announced five additional lunar spacecraft would be launched in Project Ranger during 1964. The five additions brought total scheduled Rangers to 14.

- Beginning of two-month-long Project Firefly under USAF Cambridge Research Laboratories, the project involving about 30 research rockets carrying various chemical payloads to altitudes of from 50 to 150 mi. Some of the payloads were intended to create higher-than-normal electron density in ionosphere for communications studies. Other payloads were designed to remove electrons from the F-layer to generate a "hole" in the ionosphere. Other released chemicals would interact with ambient elements to produce glow; observation of glow at night would enable tracking and measuring of upper-atmosphere winds. Mixtures exploding at high altitudes would permit observation of shock-wave perturbations of the ionosphere. Vehicles used included approximately 16 Nike-Cajuns, 9 Honest John-Nikes, and 5 Aerobee 150's. Launchings were from Eglin AFB, Fla.

- Dr. Siegfried J. Bauer of NASA Goddard Space Flight Center, one of the scientists who discovered existence of 900-mi.-thick helium layer in the atmosphere, offered explanation of USAF probe's failure to detect helium layer. USAF Cambridge Research Laboratories probe did not detect helium layer because the probe was launched at night, when the layer is so thin that the instruments used could not detect it. According to Dr. Bauer, even in daytime, helium layer will become relatively thin during periods of minimum solar activity and relatively thick during periods of maximum solar activity. Dr. Bauer feels the layer will not be detected by standard ionosphere probing but will necessitate use of special instrumentation.

- Dr. Hugh L. Dryden, NASA Deputy Administrator, described use of instruments in space exploration, in address at 17th annual Instrument-Automation Conference and Exhibit, New York: "There are three main areas. The first is that of observations of spacecraft by instruments on the ground, and the processing of [such] data. . . . The second area is that of measurements made by instruments in the spacecraft on the space environment . . . and on performance of the equipment in the spacecraft. . . . The third area consists of similar measurements of the performance of the rockets and the acceleration, vibration, and noise environment which they impose on the spacecraft. . . ."

- DOD announced steps were being taken to organize an integrated scientific and technical information program within DOD, headed by Dr. Harold Brown, Director of Defense Research and Engineering.

- Soviet press agency Tass announced U.S.S.R. would test "new versions of multistage carrier rockets for space objects" in the Pacific test ranges between October 16 and November 30.

- NASA announced appointment of Robert H. Charles as procurement consultant to the Administrator. Charles previously had served as executive vice president of McDonnell Aircraft Corp.

- Dr. Lawrence E. Lamb, chief of Clinical Sciences Div. of USAF School of Aerospace Medicine, predicted before the Second Interstate Scientific Assembly of D. C. and Virginia Physicians that manned space exploration would yield a wealth of information for practicing physicians during the next ten years.

October 15: RCA reported development of laser that directly converts sunlight into beams of coherent infrared radiation. Such devices would be potentially useful for communications, tracking, and geodetic measurement systems on board future spacecraft and satellites.

• NASA announced William B. Taylor, NASA systems engineer and Assistant Director for Flight Systems, Office of Manned Space Flight, had received citation for meritorious service for his work with Army Corps of Engineers Sept. 1960–April 1962. Taylor was praised for his "outstanding leadership, technical competence, and dedicated service . . ." while serving as Deputy Director of Army Engineers' Geodesy, Intelligence, and Mapping Research and Development Agency (GIMRADA), before he joined NASA in May 1962.

• Governor Nelson Rockefeller, addressing New York State Associated Press Association, proposed establishment of New York state science and technology foundation "to stimulate industrial expansion through research attuned to the space age."

Mid-October: J. Thomas Markley, Apollo project officer at NASA Manned Spacecraft Center, announced details of space facility to be established by NASA at White Sands Missile Range (see Sept. 17). To be used in testing Apollo spacecraft's propulsion and abort systems, WSMR site facilities would include two static test-firing stands, a control center blockhouse, various storage and other utility buildings, and an administrative services area.

October 16: Nike-Apache two-stage sounding rocket carried 65-lb. instrumented payload to 103-mi. altitude, in NASA launching from Wallops Island, Va. Primary objectives of the flight were to measure ionosphere electron density and temperature and to obtain data on the ion density and conductivity of the ionosphere.

• AEC–DOD high-altitude nuclear test failed, the safety officer of Joint Task Force 8 destroying the Thor booster and nuclear payload within minutes after liftoff because of booster's malfunctioning. This was third failure of four attempted high-altitude tests in the current Pacific series.

• Astronaut Walter M. Schirra, Jr., and his family were welcomed by President Kennedy in brief visit at the White House. Also in Washington, Cdr. Schirra appeared before NASA Headquarters employees, later at the Pentagon receiving USN astronaut wings from Secretary of the Navy Fred Korth.

• NASA announced appointment of Dr. Fred L. Niemann as Assistant Director for Technical Programs in NASA North Eastern Office, Cambridge, Mass. Dr. Niemann would conduct technical liaison with contractors, research institutions, and other government agencies in New England.

• NASA announced selection of International Business Machine Corp. to provide ground-based computer system for manned space flight Projects Gemini and Apollo. The computer complex will be part of the Integrated Mission Control Center (IMCC), at NASA Manned Spacecraft Center, Houston.

• George M. Low, NASA Director of Spacecraft and Flight Missions, predicted in Voice of America interview that final decision on manned lunar flight technique would be made soon.   Of the lunar orbit rendezvous (LOR) method, he said: "We in NASA have already reached the firm conclusion that this is the best, cheapest, most reliable and quickest way to reach the moon. . . . The final decision has not been made, but in the meantime, we are proceeding with the lunar orbit rendezvous approach."

October 16: Secretary of the Air Force Eugene M. Zuckert, addressing National Rocket Club, asserted: "Developing the right kind of space program is a continuing job in the Department of Defense. . . . The difficulties encountered do not relate to problems of principle. The problems are questions of specifics.   There are tough decisions of choice to be made.   We have to weigh effects on other activities, make careful estimates as to time, determine rates and select the projects to be pursued. . . . The current rate of investment in space by the Department of Defense reflects our best judgment as to which projects are dictated by specific requirements and those which should be carried on for general development purposes. . . . As projects become more precisely defined and funding and time estimates can be firmed, and as requirements become clearer and more specific, then efforts are focused on those most pressing and most promising. . . ."

October 16–17: U.S.S.R. successfully fired new multistage rockets approximately 7,500 mi. to target area in the Pacific, according to Soviet news agency Tass.

October 16–22: International Rocket Week, with more than three dozen nations participating in intensive cooperative study program.

October 17: U.S.S.R. announced successful launch of COSMOS X into earth orbit (apogee, 208 mi.; perigee, 126 mi.; inclination, 65° from the equator; period, 90.2 min.).   Tass said the satellite's many scientific instruments to measure radiation and other spatial conditions were functioning normally.

- NASA postponed Ranger (5) launch because of weather at Cape Canaveral.

- Among the 22 scientists and engineers honored at Franklin Institute's Medal Day ceremony, Philadelphia, were: Dr. Wernher von Braun, co-recipient of an Elliott Cresson Medal for his contributions to design and development of liquid-propellant rocket engines; Dr. James G. Baker, co-recipient of a Cresson Medal for his contributions to design of cameras for satellite tracking and aerial mapping; Dr. Wilbur H. Goss, recipient of Howard N. Potts Medal for engineering designs and technical supervision leading to development of first successful supersonic ramjet; and Dr. Charles H. Townes and Dr. Arthur L. Schawlow, recipients of a Stuart Ballantine Medal for their concept of an operable solid-stage, optical maser.

- Dr. Wernher von Braun, Director of NASA Marshall Space Flight Center, predicted U.S. will have as many as 24 men operating a moon station within a decade.

- USAF Minuteman missile was destroyed after eight seconds of flight, in silo launch at Cape Canaveral.

- Soviet news agency Tass reported Soviet scientists had developed instrument to photograph any object by its own invisible thermal radiation.

October 18: RANGER V lunar probe launched from Cape Canaveral by Atlas-Agena B, the Agena B stage attaining parking orbit and 25 min. later reigniting to send RANGER V toward the moon. The spacecraft's solar cells did not provide power, making it impossible for reception of flight-path correction signal and rendering its television cameras useless. RANGER V was to have relayed TV pictures of the lunar surface and rough-landed an instrumented capsule containing a seismometer to send back data on moon quakes and meteoritic impact. Jet Propulsion Laboratory scientists tracked the spacecraft for 8 hours, 44 min. before its small reserve battery went dead.

- Transatlantic telephone conversations via TELSTAR opened ceremonies honoring Paul Julius Reuter, founder of international news agency and communications pioneer who in 1850 linked cities of Brussels and Aachen, Germany, using 40 carrier pigeons. TELSTAR, in its 914th orbit, transmitted nearly flawless exchange of telephone messages between Herman Heusch, Lord Mayor of Aachen, and Turner Catledge, managing editor of *New York Times.*

- Japanese Foreign Ministry announced it had filed protest with U.S.S.R. against Soviets' establishing of restricted areas in the Pacific for its rocket tests. Japan said barring ships and aircraft from the areas violated rights of other nations to the open seas.

- President Kennedy presented 1961 Harmon International Aviation Trophies to four pilots: Lt. Col. William R. Payne (USAF); Jacqueline Cochran; and Cdr. Malcolm R. Ross (USNR) and Lt. Cdr. Victor E. Prather (USN) (posthumously).

October 19: Second successful high altitude U.S. nuclear test in Operation Dominic was conducted near Johnston Island in the Pacific, a low-yield explosion at altitude of between 20 and 30 mi. Specially-built rocket with Sergeant motor was used to loft the nuclear warhead.

- NASA would later select additional astronauts for Project Apollo, Dr. Robert Voas of NASA Manned Spacecraft Center told Virginia Psychological Association in Richmond. Dr. Voas said the nine new astronauts recently selected would participate in Project Gemini two-man, orbital rendezvous missions and Project Apollo, culminating in manned landing on the moon.

- DOD announced launch of Anna geodetic satellite would be attempted October 23 at Cape Canaveral.

- NASA ordered a technical review board to conduct reliability analysis of the Ranger program and to study reliability of future Ranger spacecraft. Cause of RANGER V's failure to get power from its panels of solar cells was still not clear.

- Atom Affairs Ministry, West Germany, announced work on third stage of Western Europe's first space rocket would soon begin in West Germany. First stage would be developed from U.K. Blue Streak missile, second stage from French Veronique. Satellite to be launched by the three-stage vehicle would be built by Italy. First model would be completed in 1966.

- NASA Wallops Station announced construction of bleachers located about a half-mi. from the launching area and capable of seating about 200 persons.

October 19: USAF Atlas F missile was successfully fired more than 4,000 mi. from Cape Canaveral in test flight.

October 20: U.S.S.R. launched COSMOS XI into orbit (572-mi. apogee, 152-mi. perigee, 96.1-min. period); one of series of satellites to study radiation belts, propagation of radio waves in the ionosphere, and the sun.

- An Echo I-type balloon launched from White Sands Missile Range ruptured at 21-mi. altitude and fell back to earth 91 min. after launch. The 100-ft.-diameter balloon was to have reached 24-mi. altitude in structural test.

- At executive session of Defense Industry Advisory Council, a subcommittee to investigate implementing changes in Armed Services Procurement Regulation was established.

October 21: RANGER V, its batteries dead, passed within 450 mi. of the moon on its way into orbit of the sun.

- NASA announced U.S. and Swedish experimenters were studying samplings of noctilucent clouds obtained in four Nike-Cajun sounding rocket flights during August. Preliminary analysis indicated that samples taken when noctilucent clouds were observed contained significantly more particles than when no noctilucent clouds were visible. Analysis of origin and structure of the particles may take up to a year. Participants in the project include scientists from University of Stockholm Institute of Meterology, Kiruna (Sweden) Geophysical Observatory, NASA Goddard Space Flight Center, and USAF Cambridge Research Laboratories.

October 22: President Kennedy addressed the Nation concerning the Cuban crisis, pointing to the "unmistakable evidence" of preparation of missile sites, the purpose of which "can be none other than to provide a nuclear strike capability against the Western Hemisphere." President Kennedy demanded removal of Soviet offensive missiles and bombers from Cuba. Two days later, U.S. began naval quarantine of Cuba.

- Report of Soviet dual space flights VOSTOK III and VOSTOK IV, published in *Pravda*, disclosed the minimum distance between the two orbiting spacecraft had been 6.5 km. (about 4 mi.), a revision from 5 km. as earlier estimated. On radiation, *Pravda* said maximum dose for both cosmonauts was 11 millirad. Total dose incurred by Maj. Nikolayev was 43 millirad and by Col. Popovich, 32 millirad; *Pravda* described these dosages as "absolutely safe."

- USAF announced names of 10 additional officers to be trained as space pilots, engineers, and program managers for X–20 (Dyna Soar) manned space flight programs: Capt. Alfred L. Atwell; Capt. Charles A. Bassett; Maj. Tommie D. Benefield; Capt. Michael Collins; Capt. Joe M. Engle; Maj. Neil R. Garland; Capt. Edward G. Givens; Capt. Francis G. Neubeck; Capt. James A. Roman, Capt. Alfred H. Uhalt.

- DOD postponed launch of Anna geodetic satellite to Oct. 24, for undisclosed reason.

- Lunar and Planetary Laboratory of University of Arizona, directed by Dr. Gerard P. Kuiper, reported its analysis of lunar photographs taken by LUNIK III differed with that announced by Soviet scientists. The most extensive feature of moon's far side, photographed in 1959, was named "The Soviet Mountains"; this feature was identified by the Arizona laboratory as an elongated area of bright patches and rays, possibly flat. Another feature was named the "Joliot-Curie Crater" by Soviet scientists, but it was re-identified by the Arizona laboratory as Mare Novum (New Sea), first identified by German astronomer Julius Franz near the turn of the century.

October 22: U.S.S.R. exploded nuclear device at high altitude above central Asia, with "yield of a few hundred kilotons," AEC announced. Unnamed official sources indicated the explosion created a new artificial radiation belt around the earth, probably high within the earth's natural radiation belts.

October 23: Thirteen incorporators named by President Kennedy to set up communications satellite corporation elected Philip L. Graham as chairman. Graham is president of The Washington Post Co. and board chairman of *Newsweek.*

- X–15 No. 3 piloted by Maj. Robert Rushworth (USAF) reached 134,000-ft. altitude and 3,818-mph speed in test to evaluate the craft's re-entry stability without its ventral tail fin.

- International Association of Machinists announced it had requested the Government to seize Lockheed Aircraft Co. plants and to operate them until an "equitable contract" is achieved. IAM said it had "exhausted all avenues of settlement that lay within the confines of free collective bargaining. . . . there remains only the strike weapon, [but] we fully realize the gravity of such a work stoppage in this critical period [of the Cuban crisis]."

October 24: NASA test pilot Joseph A. Walker, key speaker at 40th Anniversary and 1962 Annual Meeting of National Aeronautic Association in Washington, said: "I feel very strongly that the source of our strength in the air in commerce and in National Defense is gradually drying up. Looking at the attrition of small airports, the cost of flight instruction, the attitude of local residents near airports, and the increasing restrictions upon use of airspace and facilities, I cannot help becoming alarmed. . . . The supply of individuals with skills capable of supplying the demand for pilots will inevitably decrease from old age and retirement. . . .

"Two things are apparent. One, that the present outlook is gloomy. Two, that history and foresight give us grounds for expecting that there will be plenty of need for pilots in space as well as in the atmosphere. . . . The need for pilots will not be confined to the cockpit. Engineers with pilot training have a better understanding of the cockpit problems. We need the understanding of the cockpit problem in operations, in maintenance and repair, in management and in planning. Beyond doubt, pilot experience can be an effective tool and augment business administration as well as engineering and accounting. . . .

"The space programs are depending upon qualified, experienced test pilots. It is my belief that they don't happen, they are made, and they are made just about the time of the first solo."

- Problem of communications blackout during spacecraft re-entry ionization has been surmounted, according to Boeing Co.'s X–20 Project Manager A. M. Johnston and engineers at Radio Corp. of

America. RCA-developed equipment using higher radio frequencies would penetrate the blackout sheath 99 per cent of the time during which it engulfed a spacecraft.

October 24: USAF Cambridge Research Laboratories reported its 150-ft. radiotelescope at Sagamore Hill Observatory, Hamilton, Mass., would be ready for operation in early 1963. Telescope would be among three largest radiotelescopes in U.S., the others being one operated by Stanford Univ. at Palo Alto, Calif., and one by NRL at Chesapeake Beach, Md.

- White House memorandum asking U.S. news media to use "caution and discretion" in handling information regarded by DOD as vital to national security during the Cuban crisis. All military commands had been ordered not to release such information, memorandum said, but "such information may come into the possession of news media."

- A single communications satellite in stationary equatorial orbit could connect 92 per cent of world's telephones, Dr. Herbert Trotter, Jr., president of General Telephone & Electronics, Inc., told U.S. Independent Telephone Association in Chicago.

- Launching of DOD geodetic satellite Anna was postponed indefinitely for undisclosed reasons.

- Scientists at Lockheed-Georgia Co. and Georgia Tech announced development of method to prevent space vehicle disintegration upon atmospheric re-entry. New method involved coating the vehicle with liquid refractory (heat resistant) materials: Refractory materials in powder form were mixed with inert gases at temperatures up to 40,000° in spray gun. Heat turned the materials to liquid, which could be sprayed on the foundation metal.

- AEC announced a second 24-hour postponement of high-altitude nuclear test in the Pacific, for technical reason.

October 25: At annual NASA awards ceremony, NASA Administrator James E. Webb presented Group Achievement Awards: to four groups at NASA Manned Spacecraft Center's Assistant Directorate for Engineering and Development, Preflight Operations Division, Mercury Project Office, and Flight Operations Office; and to NASA Goddard Space Flight Center's Directorate for Tracking and Data Systems. NASA Deputy Administrator Dr. Hugh L. Dryden presented Exceptional Scientific Achievement Awards to Robert E. Bourdeau and John C. Lindsay, both of NASA Goddard Space Flight Center. NASA Associate Administrator Robert C. Seamans, Jr., presented Outstanding Leadership Awards to Maxime A. Faget, Assistant Director for Engineering and Development, MSC; George B. Graves, Jr., Assistant Director for Information and Control Systems, MSC; George M. Low, Director of Spacecraft and Flight Missions, Office of Manned Space Flight; and John W. Townsend, Jr., Assistant Director for Space Science at GSFC.

- Maj. Virgil I. Grissom (USAF) was first astronaut to pilot paraglider in test at Edwards, Calif., in development tests for Gemini manned spacecraft landings. Kite-like paraglider was towed aloft by biplane and released to glide downward. Maj. Grissom landed the paraglider upright, although the craft's nose wheel crumpled upon hitting the ground.

October 25: An S–IV stage for the two-stage Saturn began three-week, 3,500-mi. trip via barge from Douglas Aircraft Co. in Santa Monica, Calif., to NASA Marshall Space Flight Center in Huntsville, Ala., for dynamic testing.

• *London Times* reported that Manchester University team conducting photographic survey of the moon, working at Pic-du-Midi Observatory in the French Pyrenees, and similar Japanese team, working at Kwasan Observatory, Kyoto, would be joined by a third observing station in California. Eight-hour time difference between France and Japan enables the two stations to photograph the moon for 16 hours of night; the California station would be a central point in time between the other observatories, making possible the first round-the-clock photographic survey of the moon. Manchester University program is sponsored by USAF, while Kyoto University program operates under NASA research grant.

• $16,280,069 contract for construction of F–1 rocket engine test stands was awarded to Santa Fe Engineers, M. M. Sundt Construction Co., and Stolte, Inc., by Army Corps of Engineers as agent for NASA. Complex of three test stands and control center would be built at Edwards AFB, Calif.

• First live two-way radio broadcast via TELSTAR was conducted by Armed Forces Radio Network. TELSTAR relayed conversations of representatives of Conseil International du Sport Militaire (Military Olympics) between Annapolis, Md., and Germany.

• NASA Administrator James E. Webb announced first nationwide conference of scientists and educators would be held in Chicago November 1–3 to study the relationship between NASA and universities in meeting national space goals.

• United Auto Workers employed by North American Aviation, Inc., voted to defeat union shop proposal. IAM workers at General Dynamics/Convair and UAW workers at Ryan Aeronautical Co. would also vote on the proposal.

• Republic Aviation Corp. selected RCA to design and build data acquisition and communications subsystem for Project Fire, NASA program to study re-entry heating of spacecraft. Two Project Fire spacecraft will be launched late 1963.

• *Washington Post* reported comparative scores of U.S. and Soviet nuclear tests (through Oct. 22, 1962): U.S., 239 tests with about 143-megaton total yield; U.S.S.R., 121 tests with about 305-megaton total yield.

October 26: USAF announced launching of unnamed satellite employing a Thor-Agena launch vehicle from Vandenberg AFB, Calif.

• Titan II rocket launched by USAF from Cape Canaveral scored its fifth successful test flight in seven attempts.

• NASA published proposed revisions to patent regulations, simplifying procedures under which NASA can waive to contractors commercial rights to inventions developed under NASA contracts. Patent waiver regulations would not be effective until after public hearings, scheduled for December. NASA also issued regulations prescribing how NASA-held patents would be made available for industrial or commercial development. NASA announced it would make provisional agreements with industry for nonexclusive commercial development of inventions on which NASA patents were pending. Revised patent policy was designed to promote widespread use of space inventions by industry and the public.

October 26: Third high-altitude nuclear test of Operation Dominic was conducted by U.S. near Johnston Island in the Pacific. The sub-megaton bomb was carried to estimated altitude of between 30 and 40 mi. by Thor missile.

- *New York Times* reported a 300-ft.-diameter radiotelescope at National Radio Astronomy Observatory, Green Bank, W. Va., had recently detected in one day as many radio signals from Jupiter as smaller instruments have recorded in a month. Source of Jupiter's long-wave radio signals was still unresolved.

October 27: EXPLORER XV energetic particles satellite (S–3b) placed in orbit by Thor-Delta vehicle launched from Cape Canaveral. Initial orbital data: apogee, 10,960 mi.; perigee, 193.7 mi.; period, 315.3 min.; inclination, 18.02° to the equator; velocity at apogee, 6,015 mph; velocity at perigee, 21,598 mph. Launched to study artificial radiation belt created by U.S. high-altitude nuclear explosion in July, the 98-lb. satellite was similar to EXPLORER XII and EXPLORER XIV which successfully measured energetic particles of natural radiation belts. Although satellite achieved orbit within predicted estimated range, it was spinning on its axis about 10 times faster than the planned 10-rpm rate. Unnamed spokesman said: "We have a high spin rate. It may well be that we'll receive acceptable data from the main experiments, with some secondary experiments not able to function. . . . We won't know for certain until we have made a thorough study of data over a period of days." EXPLORER XV carried experiments on magnetic field, ion-electron, electron flux, and distribution in pitch angle of electrons; transmitter; optical attitude sensor; and solar cells, including solar-cell damage experiment. Launching marked 13th successful satellite-orbiting by NASA-developed Delta vehicle.

- Dr. Hugh L. Dryden, NASA Deputy Administrator, was among five Government career men selected for 1962 Rockefeller Public Service Awards.

- AEC announced two U.S. nuclear test explosions—one in atmosphere near Johnston Island and one underground in Nevada—and a Soviet test in the atmosphere near Novaya Zemlya.

- Eleven per cent of national budget is spent for research and development—more than twice the combined investment in research of industry and nonprofit institutions, *New York Times* reported.

October 29: Aerobee sounding rocket launched from NASA Wallops Station carried 230-lb. payload to 116-mi. altitude before landing in Atlantic Ocean 59 mi. from launch site. Flight tested performance of three photometers developed for use in Oao (Orbiting Astronomical Observatory) satellite to be launched by NASA in 1964. Other instrumentation obtained data from the comparatively young stars Gamma Cassiopeia and Delta Persei, for use in star-evolution study being conducted by University of Wisconsin. Launch was under management of NASA Goddard Space Flight Center.

- Jet Propulsion Laboratory scientists disclosed they were conducting radar exploration of Venus. Data obtained by the experiment, continuing from October 1 for about three months, would be compared with those obtained by MARINER II spacecraft when it flies in vicinity of Venus.

October 29: NASA officials said five experiments aboard EXPLORER XV were working well but that two others had been adversely affected by satellite's excessive spin rate. The two experiments were designed to determine decay rate of artificially created radiation.

- All industry proposals for 210-ft. antenna at Goldstone Tracking Station were reported rejected by JPL because their prices exceeded sum allocated for the project. JPL was considering relaxing its requirements and requesting bidders to compromise on price.

- W. Randolph Lovelace II, of Lovelace Foundation for Medical Education and Research, told 200 space experts from 14 countries that scientists would have to volunteer for space flight teams. "It takes four years of college plus three years or more of further study to reach the doctor of philosophy level. It is a bit too much to ask our test pilots to undergo this further training." He said that in projects such as two-man Gemini a scientist could make space flights in company of trained flier. Lovelace was addressing an international symposium on environmental problems of man in space, held in Paris.

- USAF awarded $78 million contract to General Dynamics/Astronautics for a standardized Atlas space launch vehicle, to be known as Atlas SLV-III. Contract covered design and development of standard space vehicle version of Atlas ICBM, modifications to launch sites at AMR and PMR, and production and launch of SLV-III boosters.

- Indonesian scientist Dr. Hadinoto said Indonesia hoped to launch her first space rocket in 1964 for scientific purposes, with aid provided by U.S.S.R.

- AEC announced three Soviet nuclear tests in past two days, two of intermediate yield and one of low yield.

October 30: MARINER II Venus probe passed the earth on 65th day of its 110-day flight to vicinity of Venus. At that point in its smaller-than-earth's orbit, the 447-lb. spacecraft was 11.5 million mi. from earth and traveling at 70,500 mph relative to the sun.

- NASA announced realignment of functions within office of Associate Administrator Robert C. Seamans, Jr. D. Brainerd Holmes assumed new duties as a Deputy Associate Administrator while retaining his responsibilities as Director of NASA Office of Manned Space Flight. NASA field installations engaged principally in manned space flight projects (MSFC, MSC, and LOC) would report to Holmes; installations engaged principally in other projects (Ames, LARC, LRC, GSFC, JPL, and Wallops) would report to Thomas F. Dixon, Deputy Associate Administrator for the past year. Previously most field center directors had reported directly to Dr. Seamans on institutional matters beyond program and contractual administration.

- NASA announced signing of detailed contract with Space and Information Division of North American Aviation, Inc., for development and production of Saturn S–II stage. The $319,922,328 contract, largest single contract ever awarded by NASA, covered production of nine live flight stages, one inert flight stage, and several ground-test units for the Advanced Saturn vehicle.

October 30: Oran W. Nicks, NASA Director of Lunar and Planetary Programs, summarized results obtained from MARINER II scientific experiments, at Eighth Tri-Service Conference on Electromagnetic Compatibility in Chicago: "One of the primary scientific objectives in measuring magnetic fields in interplanetary space is to determine the magnitude and direction of the steady field component. . . . The Mariner has indicated a typical steady state transverse component of ten gamma or less, and it has clearly shown how the transverse field becomes large (as high as 25 gamma) during geomagnetic storms. In addition to this, the Mariner 2 has provided information on the steady interplanetary field indicating that the radial component is apparently more stable, even under storm conditions, and that there are quiet times when the radial component from the sun is essentially constant. . . . Measurements from the ion chamber [onboard MARINER II] have indicated an average radiation level as expected, corresponding to about 1.1 milli-roentgen per hour (about 100 times the cosmic ray intensity at the surface of the earth). . . . It has been reported by measurements [of the probe's cosmic dust detector] thus far that the flux of interplanetary dust particles is at least one thousand times less in interplanetary space than in the near vicinity of the earth. The cause of this is unknown. . . . Preliminary data from the [solar plasma detector in the] Mariner indicate abrupt changes in the velocity and intensity of solar winds, and correlated effects on the magnetic fields in space. Measurements show that the particle energies generally range from about 750 electron volts to 2500 electron volts, although some plasma with an energy of 3225 electron volts has been observed. These variations almost certainly were the results of events on the sun, but the exact nature of these events is not yet understood. . . ."

- NASA announced 19 experiments would be included on Pogo satellite (Polar Orbiting Geophysical Observatory) to be orbited by Thor-Agena vehicle launched from Pacific Missile Range. Pogo would be second in Ogo (Orbiting Geophysical Observatory) series and first Ogo in polar orbit; it was scheduled for launch in early 1964.

October 31: Sudden drop in voltage on MARINER II Venus probe was reported by Jet Propulsion Laboratory scientists. JPL turned off the probe's four interplanetary experiments to conserve on-board power for experiments in the vicinity of Venus. Discounting any probability of the solar cells being affected by radiation, experts indicated the problem could have come from faulty voltage regulators.

- ANNA IB geodetic satellite launched into orbit by Thor-Able-Star vehicle from Cape Canaveral (apogee, 727 mi.; perigee, 670 mi.; inclination, 50° to the equator). Mounted on the 350-lb., 36-in. spherical satellite were four high-intensity lights that would flash on and off; by comparing simultaneous observations of the satellite from various points on earth, scientists could measure earth's size, shape, and surface features with accuracy never before possible. The light beacons were to be activated by ground command within three days, after orbital data were precisely analysed. ANNA (Army-Navy-NASA-Air Force) was developed and launched by military services, with Johns Hopkins' Applied Physics Laboratory as prime contractor; NASA cooperated in tracking and in dissemination of orbital data to the international scientific community for worldwide use of the man-made star for geodetic purposes.

October 31: Dr. Thomas L. K. Smull, NASA director of grants and research contracts, told press conference in Chicago that by 1970 one out of every four technically trained persons in U.S. will be engaged in some phase of the space program. Press conference preceded first national conference between NASA and representatives of 300 universities, colleges, and technical schools, Nov. 1-3.

• EXPLORER XIV had transmitted 589 hours of data to NASA Goddard Space Flight Center, which had released about 240 hours of data to the various experimenters.

• Soviet scientist N. Varvarov was reported to have stated U.S.S.R. program to orbit one-man spacecraft had been completed with flights of VOSTOK III and VOSTOK IV.

During October: In letter report to House Committee on Science and Astronautics, NASA Deputy Administrator Dr. Hugh L. Dryden disclosed problems in the Centaur launch vehicle program: Atlas-Centaur capability to carry heavy payloads under severe wind conditions must be studied in greater detail; quality of fabrication must be improved to reach desired level of quality control; guidance system requires further development to achieve necessary accuracy and reliability; tank fabrication requires better welding techniques; Atlas-Centaur inflight separation requires further analysis; nose cone and insulation panels must be further tested to prove their aerodynamic-load resistance. Report said about $100 million would be spent on Centaur in FY 1963 to accelerate development. NASA Lewis Research Center would study use of Centaur in combination with Titan II, Saturn C-1, and Saturn C-1B boosters.

• Program testing astronaut tolerance to space flight conditions was initiated with order of motion-simulator platform from Textron Electronics, Inc., MB Electronics Div. The platform, to be installed at Wright-Patterson AFB, will move in six directions—pitch, roll, yaw, up-and-down, side-to-side, and back-to-forward.

• NASA acquired from the Army a ship basin adjacent to Michoud Operations facility for loading and unloading space vehicles at the Saturn fabrication plant. Plans called for construction of ship dock at the basin and dredging of vessel-turning area.

• Senator Hubert Humphrey, in hearings of the Subcommittee on Reorganization and International Organizations of the Senate Committee on Government Operations, told Melvin S. Day, Director of NASA Office of Scientific and Technical Information: "Mr. Day, I want to commend you for blueprinting one of the finest intra-agency systems in the Federal government today in the information field and, of course, I speak of NASA. While the work is being done by private contractor, it is you who have set the high standards for performance. . . . We feel what has been done within the agency, from our limited examination, is an outstanding job, and we want to compliment you on it."

During October: Geophysical Institute of University of Alaska reported to NASA that specially instrumented sounding rockets would be desirable for measuring extent of solar cosmic rays associated with ionization of the upper atmosphere. Auroral-zone absorption disturbances make ground measurements difficult.

• Three scientists at Boeing Company made seven-day simulated space trip to the moon in a cylinder-shaped capsule 8 ft. in diameter and 6 ft., 4 in. high.

• Atlantic Research Corporation introduced new Archer solid-propellant sounding rocket for use during International Year of the Quiet Sun (IQSY), 1964–65. Archer successfully completed its initial flight to 75–mi. altitude carrying 35-lb. payload, at Pt. Mugu, California. Flight testing would be completed early in 1963.

• USAF announced completion of network of 71 "gravity base measuring stations" throughout the U.S. Exceedingly accurate and sensitive gravity meters at each location would establish a base value of gravity. This knowledge was expected to contribute to missile accuracy, to geophysical exploration for new sources of minerals, and to assist future researchers in determining whether gravity values for a given location change with time.

• USAF prepared proposed FY 1964 budget requesting $23.5 billion, including funds for construction of eight RS–70 reconnaissance bombers and for expanded development of space weapons. Also considered in budget preparation was long-range proposal to use NASA's Gemini two-man spacecraft as a stepping stone toward developing military manned space flight capability.

• Space General Corp. reported development of aspect-insensitive antenna system for satellites and spacecraft. System, which automatically points toward its transmission source on earth, could reduce number of communications satellites in worldwide network "by a factor of four or five," according to Space General.

• International Association of Machinists voted to ratify new three-year contract with General Dynamics Corp., thus leaving Lockheed Aircraft Corp. as the only aerospace company that had not reached contract settlement with union workers.

• George M. Kohler of General Electric Co. suggested to the IAF meeting in Bulgaria that a worldwide, intensive program be established to use asteroids as scientific bases for space exploration. Kohler estimated that between 10 million and 10 billion objects 30 ft. to 300 ft. in diameter pass within 20 million mi. of the earth each year, and that a good number of them probably come much closer—perhaps within several thousand miles. He called for (1) careful search for closer asteroids with special electronic telescopes; (2) landing unmanned spacecraft on asteroids; and (3) steering asteroids into earth orbit, to be hollowed out and used as manned satellite bases.

• Czech scientist Adeneck Dobrichovsky, writing in a Prague technical journal, said the U.S.S.R. would launch a four-man space observatory for two-month orbit in 1965–66.

NOVEMBER 1962

November 1: U.S.S.R. announced it had launched MARS I, a 1,970-lb. space probe on seven-month flight to Mars. Launched with an "improved booster" into parking orbit around the earth, MARS I was hurled into escape trajectory when vehicle engines re-ignited. Tass reported MARS I was flying as planned on its course toward Mars. "All systems are functioning normally and orders sent to the station [i.e., probe] are well received and obeyed." Main tasks of the probe were interplanetary exploration, establishment of interplanetary space radio communications, photographing the Martian surface and relaying the photographs back to earth by radio. "Telemetric, measuring, and scientific instrumentation will be automatically activated in conformity with the flight program by radio commands from the earth," Tass said. The flight had been preceded by an unsuccessful launch attempt the previous week, unidentified DOD sources said. (NASA Administrator James E. Webb stated in September that two Soviet attempts to launch Mars probes had failed in October 1960.)

- NASA reported radiation satellite EXPLORER XV, launched Oct. 27, was spinning at rate of 73 rpm instead of desired 10 rpm because of failure of despin weights to deploy. Preliminary data indicated all experiments were functioning and that all received data were of good quality. Effect of high spin rate on experimental results was under analysis.

- High-altitude nuclear test in U.S. Operation Dominic was made in Pacific test area, the sub-megaton device carried by Thor rocket to estimated altitude of 30 to 40 miles. AEC reported two Soviet nuclear explosions also occurred, one a high-altitude test over central Asia and the other an atmospheric test over Soviet Arctic.

- Development of new type of laser by General Electric and International Business Machine researchers was reported in American Physical Society's *Physical Review Letters* and American Institute of Physics' *Applied Physics Letters*. Called "injection laser," new crystal laser was powered by electrical current; previous solid lasers used strong flash lamps as energy source. At least 10 times more efficient than existing lasers, injection laser was expected to enable engineers to broadcast audio messages on the beams of light and to modulate the brightness of the light according to voice, television, or other pattern. Researchers anticipated that single laser beam could transmit a million telephone messages or a thousand television channels.

- Ten years ago the U.S. exploded a hydrogen device (fusion) at the Eniwetok atoll in the Pacific, the first full-scale thermonuclear explosion in history and the dawn of the so-called "H–bomb era."

- Dr. Lev Davidovich Landau, member of the Soviet Academy of Sciences, was awarded the Nobel Prize in physics for having developed pioneering theories for liquid helium's fluidity at low temperatures.

November 1–3: First NASA-University Conference on the Science and Technology of Space Exploration held in Chicago, with 1,181 educators and scientists representing NASA Headquarters, NASA field installations, and other Government agencies, as well as more than 300 colleges and universities.

November 1: Dr. Hugh L. Dryden, NASA Deputy Administrator, speaking at NASA-University Conference in Chicago, said NASA needed help from universities in four main areas: space sciences, manned space flight, applications of earth satellites to communications and meteorology, and advanced research and technology. "Our educational institutions bear a major responsibility for the success of our national effort to explore space. Our universities and colleges are called upon to produce a body of scientists and engineers of unexcelled competence. Some of these graduates will enter governmental service with NASA and other agencies participating in the space program; some will join private research organizations and industrial corporations.

"But some must remain at the universities where they continue to advance knowledge and produce new talent. This last function should receive high priority. The government laboratory, industry, and research foundation all are users of creative and talented men without reproducing this vital national resource. The university alone is the producer of new engineers and scientists."

• Dr. Homer E. Newell, NASA Director of Space Sciences, addressed NASA-University Conference on the Science and Technology of Space Exploration, Chicago: "During the first four years of the existence of the National Aeronautics and Space Administration, 157 sounding rockets, 55 satellites, and 9 space probes have been launched and have yielded a tremendous amount of information in all of the areas of interest. Many hundreds of research papers have appeared in the open scientific literature and a considerable portion of the scientific community, both national and international, has become involved in the effort to make the most of the opportunities afforded by the space program for the conduct of important research. It is hoped that the university community will continue to find new and challenging opportunities and will continue to increase and strengthen its participation which has already become a major element in the success of the program so far."

• Dr. T. L. K. Smull, Director of NASA Grants and Research Contracts, addressed NASA-University Conference on policies and procedures of NASA-University grants and research contracts: "The basic principle of all NASA policy regarding its relationships with universities is that NASA wishes to work within the structure of the universities in a way that will strengthen the university and at the same time make it possible for NASA to accomplish its mission. . . . NASA hopes to conduct its joint activities in a manner that will preserve and strengthen the universities' educational role. This basic policy is interwoven in the policies and procedures of NASA support of research, facilities, and training.

". . . NASA support of activities in the universities has approximately doubled each year since NASA was organized. During Fiscal Year 1962 some $40 million were committed for these activities. Of this $40 million, $2 million was utilized to initiate the training grant program; $6½ million was used to initiate the facilities grant program; and some $3½ million was utilized for the special purpose research grants. The rest of the funds supported project type research. . . .

"It is expected that the university involvement in NASA activities during the current fiscal year will be more than double that of FY 1962. . . ."

November 2: James E. Webb, NASA Administrator, and the NASA astronauts were honored at Explorers Club's Night of Exploration dinner in New York. Addressing the club on NASA programs and goals, Mr. Webb said: "Curiosity—the driving human thirst for knowledge—is only one reason, of course, for undertaking space exploration on the scale and at the pace which our country has established. . . .

"First, we seek for mankind the benefits inherent in the scientific and technological knowledge and dexterity that will emerge from this dynamic effort to conquer the most hostile environment that man has ever entered—and to use that knowledge and technical skill as an important new resource for human progress.

"Second, we seek to maintain our position as leader of the Free World, through continued superiority in science and technology.

"And, finally, we seek space power as a deterrent to any potential adversary who might attempt to exploit space as an avenue of aggression against us . . . ."

- Transatlantic use of TELSTAR communications satellite was discontinued for about seven weeks. During this period, TELSTAR's orbit brought it too briefly on line of sight with North America and Europe for useful transmission. AT&T said the situation would recur every six months. Domestic transmissions would be suspended only from Nov. 26 to Dec. 2, during which time ground equipment would be "tuned up."

- Unnamed NASA spokesman said MARINER II was sending "about 90 kinds of information" from interplanetary space, despite power drop of Oct. 31. Officials were "pretty optimistic" Venus-bound probe could renew experiments turned off because of drop in voltage of the spacecraft's power system.

- DOD announced ANNA geodetic satellite was functioning properly and its orbit was "near perfect." ANNA's flashing light system would be tested Nov. 4 with single series of five flashes triggered by onboard command. Full-scale operation of all systems would begin within a week with ANNA's flashing light series triggered by ground command from Johns Hopkins Univ. Applied Physics Laboratory, Md.

- Soviet academician M. Sissakian, Presidium member of Soviet Academy of Sciences, was quoted as saying U.S.S.R. hoped to bring MARS I back to earth.

- H. Julian Allen, Assistant Director of NASA Ames Research Center, proposed at NASA-University conference in Chicago that interplanetary spacecraft be cone-shaped and that they enter the atmosphere point-end-first. The originator of concept for returning orbital spacecraft blunt-end-first, Allen stated: "Blunt vehicles are desirable when you are dealing with speeds of the sub-orbital range up to speeds required to escape the effects of the earth's gravity. But when entry speeds exceed this range, as they probably will on return to the earth from trips to other planets, radiation heating drives you away from very blunt bodies, such as the Mercury capsule, to bodies with conical noses. . . . Our concepts of aerodynamic flow, in short, must change as the velocities with which we are dealing increase. . . ."

November 2: Rep. Charles S. Gubser (Calif.) announced NASA and USN had agreed to convert an old dirigible hangar at Moffett Field, Calif., for production of Rift nuclear vehicle stage. Hangar was located about two mi. from prime Rift contractor, Lockheed Missiles and Space Co., and adjacent to NASA Ames Research Center.

• Douglas-built S–IVB stage, originally conceived only as third stage of Saturn C–5 ("Advanced Saturn"), would be modified for use as upper stage of Saturn C–1B, it was reported. "Minor design changes" would be made under initial funding of $2.25 million by NASA.

• Lt. Col. John H. Glenn, Jr. (USMC) was awarded the Alfred A. Cunningham Trophy for being selected as the Marine Aviator of the Year. Mercury Astronaut Glenn orbited the earth three times in first U.S. manned orbital flight MA–6, February 20, 1962.

November 3: MARINER II interplanetary probe reported data showing that interplanetary space was free of cosmic dust and debris found in vicinity of earth, Dr. William H. Pickering, director of Jet Propulsion Laboratory, told press interviewers. "On this basis, one has to think of the earth as moving in its own cloud of dust," Dr. Pickering added.

• AEC announced 31st nuclear explosion by U.S.S.R. in current test series, an atmospheric detonation of intermediate-range yield.

• Four "hot spots" on the moon were reported discovered by Bruce Murray and Robert Wildey of Cal Tech, using new telescope with heat-sensitive, gold-plated mirror to detect infrared radiation. The two space scientists speculated that hot spots could indicate large areas of bare rock exposed on the lunar surface. Spots were discovered during survey of the moon which also revealed the lunar surface gets colder at night than previously believed, −270°F compared to −243°F recorded by earlier heat-measuring devices. Murray said the new evidence could mean there are prominences of heat-retaining rock protruding through thick dust layer on lunar surface.

• NASA announced Col. George M. Knauf (USAF), Deputy Director of Space Medicine in NASA Office of Manned Space Flight, has retired from USAF and accepted appointment to same position as a civilian. Dr. Knauf was responsible for much of development and organization of Project Mercury medical recovery support operation.

November 4: Soviet news agency Tass reported MARS I probe was 606,000 mi. from earth (at 4:00 PM EST) and was increasing that distance by 215,000 mi. a day. Soviet observatories photographed the Mars probe and its carrier rocket on Nov. 3; director of Crimean Astrophysical Observatory said this was the first time moving artificial interplanetary bodies had been photographed from earth.

• Soviet press agency Tass reported the planet Mars has "experienced the effects of meteorites to a much greater degree than the earth."

Emblen Sobotovich, senior scientist at Leningrad Radium Institute, said meteorites were preserved on Mars for long periods of time because Martian atmosphere contained almost no moisture, Tass reported.

November 4: High-altitude nuclear explosion over Johnston Island ended current U.S. atmospheric test series. President Kennedy announced that underground testing in Nevada would continue, and added: "I hope that in the next months we can conclude an effective test ban treaty, so that the world can be free from all testing. . . . We shall devote our best efforts to conclude such a treaty and hope all others will do the same."

- AEC announced U.S.S.R. conducted atmospheric nuclear test with intermediate-range yield.

November 5–7: Symposium on Protection Against Radiation Hazards in Space sponsored by Oak Ridge National Laboratory, American Nuclear Society, and NASA Manned Spacecraft Center, in Gatlinburg, Tenn.

November 5: USAF launched Thor-Agena from Vandenberg AFB, with unidentified satellite to be placed in polar orbit.

- 16th of Project Firefly series of high-altitude rocket launchings was conducted at Eglin AFB, Fla., the two-stage Nike-Cajun rocket releasing chemicals at about 50-mi. altitude. Chemicals caused formation of artificial cloud which ground stations tracked to determine upper atmosphere movements.

- Political Committee of the United Nations passed resolution urging end to all nuclear testing by January 1. No negative votes were cast; 25 nations abstained, including U.S., U.K., France, U.S.S.R. and Soviet-bloc countries. Similar resolution, proposed by U.S. and U.K., emphasized need for international control to enforce a test ban; it was adopted 50–12, with 42 members abstaining.

- NASA Goddard Space Flight Center announced award of $12 million worth of contracts for tracking-network modification in preparation for lengthy manned space flights. Contracts were awarded to: Canogo Electronics Corp., for tracking antenna acquisition aid systems; Radiation, Inc., for digital command encoders; Collins Radio Corp., for RF command systems; and Electro Mechanical Research Corp., for PCM systems.

- Space chimp Enos died at Holloman AFB, of a form of dysentery resistant to antibiotics. Enos had orbited earth two times Nov. 29, 1961, on MA–5 flight.

- Dr. James B. Weddell of North American Aviation Space and Information Systems Div. announced development of new technique to predict solar flares up to 35 days in advance with 70 per cent accuracy. Dr. Weddell announced his new forecasting technique at symposium on radiation hazards, Gatlinburg, Tenn., cosponsored by NASA, Oak Ridge National Laboratory, and American Nuclear Society.

- United Technology Corp. announced development of new technique for throttling rocket engines involving use of helium mixed with liquid fuel. Depending on ratio of gas to liquid, the engine will exert either more or less thrust; system would permit throttling of an engine of 10,000-lb. thrust down to as low as 100 lbs. UTC director of research and advanced technology, William J. Corcoran, pointed out that a practical method of throttling liquid rocket engines had never before been attained.

November 5: Decision on whether NASA or USAF would be assigned development of orbital space stations was reported in process within the National Aeronautics and Space Council.

- Advanced Polaris missile (A–3) exploded 30 seconds after launch from Cape Canaveral.
- No Nobel Peace Prize would be awarded for 1962, Norwegian Parliament committee announced.
- William L. Gill, Chief of Crew Systems Div. Radiation Branch, NASA Manned Spacecraft Center, said walls of Apollo spacecraft would provide most of radiation shielding required for crew. Astronauts would have special shielding devices only for their eyes.

November 6–8: Fourth Annual Liquid Propulsion Symposium held in San Francisco, with more than 500 representatives of Government and industry. Conference was sponsored by NASA, ARPA, and JANAF, with series of restricted sessions not open to public.

November 6: U.S.S.R. reported its interplanetary probe MARS I would pass within 162,000 mi. of Mars on its present course. Press agency Tass added a "precise system of star orientation and special engines" would be used to correct the trajectory so that MARS I would pass within 600 to 6,800 mi. of the planet.

- U.S. transferred to Dominican Republic its rocket tracking station at Sabana, D.R. Formerly used to track missiles launched from Cape Canaveral toward Ascension Island, the facility would be used as a training center.
- First ground-controlled flashes from ANNA geodetic satellite were delayed because delivery of command equipment to Johns Hopkins Applied Physics Laboratory had been delayed.
- Thomas Dixon, Deputy Associate Administrator of NASA, told Liquid Propulsion Symposium in San Francisco that NASA was studying large liquid-propellant rocket engines, beyond the 1.5 million-lb.-thrust F–1 engine, that would be necessary for manned planetary flight. Pointing out the unacceptability of simply scaling up the F–1 engine to produce the required 20 to 30 million-lb. thrust, since it would then measure about 60 ft. high and 45 ft. nozzle diameter, Dixon cited need for "new and imaginative approaches for propulsion in the future . . . . NASA will emphasize advanced research that could culminate in the development of these giant engines for the future." He added that current NASA research projects in advanced propulsion were making progress. "These studies will assess current work on advanced propulsion concepts and point out areas where new concepts need to be investigated."
- NASA and the Japanese Ministry of Posts and Telecommunications signed a Memorandum of Understanding for cooperative testing of NASA-launched communications satellites. Japanese Ministry would make available a ground station for communication via artificial satellites; NASA would arrange for use of experimental communications satellites locally by Japan as well as jointly by U.S.-Japan.
- President Kennedy announced International Association of Machinists (IAM) and Boeing Co. had agreed to extend negotiations for two months so that National Labor Relations Board could poll employees on question of union shop.

November 7: NASA announced selection of Grumman Aircraft Engineering Corp. to build lunar excursion module (LEM) of the three-man Apollo spacecraft. In announcing the selection, NASA Administrator James E. Webb said: "We are affirming our tentative decision of last July . . . using lunar orbit rendezvous as the prime mission mode to accomplish initial manned lunar flight. . . . Studies [of alternate approaches], now completed, along with a great deal of related analyses . . . during the intervening months, make us confident that our present course is the proper one." D. Brainerd Holmes, Deputy Associate Administrator of NASA and Director of Manned Space Flight added that more than 1,000,000 man-hours of some 700 scientists, engineers, and researchers had been devoted to studies of the Apollo mission. ". . . The results of these studies added up to the conclusion that lunar orbit rendezvous is the preferable mode to undertake." Apollo lunar spacecraft will be composed of command module, service module, and lunar excursion module.

- NASA fired two experimental rockets into upper atmosphere within a half-hour of each other, to obtain a comparison of electron density profile with wind profile measured at about the same time. First rocket, two-stage Nike-Cajun, attained 82-mi. altitude with a 55-lb. payload to obtain measurements of electron density and electron temperature in ionosphere at night. Second experiment, using two-stage Nike-Apache, involved use of sodium vapor clouds to measure atmospheric winds and diffusion; rocket reached 34-mi. altitude before ejecting its vapor cloud, which extended to peak altitude of 103 mi. Twin launchings, conducted at NASA Wallops Station, marked first time such experiments were conducted together. Experiments were conducted by NASA Goddard Space Flight Center and Geophysics Corporation of America.

- Advanced USAF missile, Atlas F, was successfully fired more than 4,000 mi. from Cape Canaveral. The missile was 150th Atlas to be launched since the first went aloft June 1957, including 31 launched as space vehicles.

- AEC announced University of California Lawrence Radiation Laboratory and Texas A&M Activation Analysis Laboratory would conduct NASA- and AEC-funded research in methods of determining chemical composition of lunar surface.

First week in November: Armour Research Foundation reported to NASA that surface of the moon may not be covered with layers of dust. Initial Armour studies indicated dust particles become harder and denser in higher-vacuum environment such as that of moon, but studies had not proved that particles eventually become bonded together in rock-like substance as vacuum increased.

November 8: Jet Propulsion Laboratory's Goldstone Tracking Station turned on MARINER II's four interplanetary experiments by command, when the Venus-bound spacecraft was 14.6 million mi. from earth. The scientific instruments—magnetometer, ion chamber and particle-flux detector, cosmic dust detector, and solar spectrometer—were turned off Oct. 31 to conserve power when onboard voltage drop was noted. Experiments were resumed when telemetry from MARINER II indicated onboard power supply was normal again. Signals received at Goldstone showed the resumed experiments were functioning normally and were transmitting excellent data.

November 8: Scientists at Applied Physics Laboratory of Johns Hopkins Univ. sent operating commands to ANNA satellite, and three minutes later ANNA responded by flashing its lights five times to show the message had been received. Satellite's blinking beacons were designed to aid scientists make precise measurements of the earth and to determine its gravitational field.

• NASA Goddard Space Flight Center announced it would conduct experiments using laser in tracking S–66 ionosphere beacon satellite, to be launched in polar orbit early next year. Laser (light amplification by stimulated emission of radiation) would be tested as "potentially vastly more accurate than current tracking methods and requiring no electrical power from the satellite." Laser device, emitting beam $\frac{1}{16}$ of a degree wide, would be mounted on tracking telescope at NASA Wallops Station. Laser signals to satellite would be received on passive reflector and bounced back to earth.

• 30th sounding rocket launched by three-man team at McMurdo Station, Antarctica, the program calling for 60 Arcas rocket launchings at rate of one a week. Purpose of launchings was to obtain information on sudden warming of upper atmosphere that occurs during Antarctic spring.

• AEC released photographs of underground cavity created by first nuclear detonation for peaceful use (Project Gnome). The three-kiloton explosion was made Dec. 10, 1961, near Carlsbad, N.M. Since May 17, scientists have explored the cavern and gathered data on neutron physics, seismic effects, and isotope production. Cavity is 1,200 ft. underground and measures 160–170 ft. in diameter and 60–80 ft. in height.

• French and British ministers agreed to submit to their respective governments plans for joint production of supersonic airliner in mach 2 range (about 1,400 mph). Julian Amery, U.K. Minister of Aviation, and Roger Dusseaulx, French Minister of Public Works and Transport, announced the agreement after conferring in Paris.

• NASA Deputy Administrator Hugh L. Dryden, addressing General Session of U.S. Savings and Loan League in Washington, said: ". . . What we are buying in our national space program is the knowledge, the experience, the skills, the industrial facilities, and the experimental hardware that will make the United States first in every field of space exploration—scientific and practical uses, including military applications. And to accomplish this result we must make advances at the frontiers of knowledge in nearly every field of science and technology which will project new ideas into our whole industrial system. . . .

"The success of the National Space Program hinges on the ability of the American people, through their government, their industry, and their privately endowed institutions, to implement many difficult tasks. Not one or two men will make the landing on the moon, but, figuratively, the entire Nation.

"And our ultimate success also depends, in some measure, on our ability to enlist the cooperation of other nations in this enormous undertaking. An important (but certainly not determining) consideration in our international programs is winning and cementing friendly relations with our allies in the Free World and among the uncommitted nations. This we do by demonstration of our peaceful objectives and our willingness to share what we learn from space exploration with the scientific community the world around. . . .

"The investment in space progress is big and will grow, but the potential returns on the investment are even larger. And because it concerns us all, scientific progress is everyone's responsibility. Every citizen should understand what the space program really is about and what it can do. . . ."

November 8: Speaking before MIT School of Industrial Management, NASA Manned Space Flight Director D. Brainerd Holmes said: "The major national problem today is not a lack of technological or economic resources. It is a lack of trained and courageous minded men. For this reason, the integration of our human, economic, and technological resources is basically a manpower problem. We need a special kind of man.

"This kind of man not only combines general and scientific training and general management with technical experience, but he has the true courage of his convictions. Above all, he must be tireless and devoted and not give in to compromisers and obstructionists. Technical people may not necessarily be the best managers, but in this age of technology if you do not have the technical answers you cannot make good decisions. This is the vital point I want to stress to you today. . . .

"We need group research, group engineering, and group science. But we also badly need individual creativity. In short, we need to utilize effectively all of our available capabilities, regardless of the channels they take. We must integrate group and individual efforts."

- Lt. Col. Joseph A. Connor, Jr. (USAF) of NASA Office of Manned Space Flight received Air Force Legion of Merit for establishing and directing nuclear safety board for TRANSIT IV–A satellite. Lt. Col. Connor's work on the project was done more than a year ago when he was assigned to the AEC.

- Maritime Administration announced award of contract to Vehicle Research Corp. of Pasadena, Calif., for building two models of man-carrying craft that moves over water on cushion of air.

November 9: In emergency landing of X–15 No. 2, piloted by NASA's John McKay, it went out of control when nosewheel broke off, the craft skidding nose down and sideways, then stopping in upside-down position. McKay sustained only superficial injuries; X–15 nose gear, wings, and tail were damaged. Flight was supposed to have reached 120,000-ft. altitude in test of stability during re-entry into earth's atmosphere, but engine failed to develop power when X–15 was released from B–52 mother ship at 45,000 ft. and McKay glided to emergency landing at Mud Lake, Nev. Flight was 31st made in X–15 No. 2, seventh in X–15 by McKay, and first X–15 flight involving any injury to personnel.

- Soviet space probe MARS I was 1,677,704 mi. from earth and "continuing satisfactorily" on its flight to Mars, Prof. Vsevoldod Fedynsky of Moscow Univ. said in Tass interview.

November 9: NASA S–27 Project Officer reported Canadian ALOUETTE topside-sounder satellite was performing as expected.   Launched Sept. 28, ALOUETTE was considered "a very successful experiment since it is producing not only ionospheric data but also information about the earth's magnetic field. . . .   Operation of the satellite continues to be normal. . . ."

- S–I stage of Saturn C–1 space vehicle was static-fired for full duration at full thrust (1.5 million lbs.) for first time, at NASA Marshall Space Flight Center.   Previous full-duration firings had attained 1.3 million-lb. thrust.

- Scientists at Jodrell Bank Experimental Station announced they had measured part of Milky Way's magnetic field, marking first time positive evidence had been obtained of the field's existence. Milky Way's magnetic field measured 25 millionths of a gauss (earth surface's magnetic field measures about half a gauss).   Led by Dr. Rodney D. Davies, research team had spent 20,000 hours over past four years on the experiment, involving radiotelescopic measurement of light from a star cluster 1,000 light years away.

- Floyd L. Thompson, Director of NASA Langley Research Center, announced construction of $12.3-million space radiation effects laboratory would begin in March and would be completed in about two years.   Facility, to be operated by NASA and three -Virginia colleges, would be used to study effects of radiation on space vehicle materials and components; it would simulate radiation hazards astronauts and spacecraft would encounter in space. Facility would be used also in graduate studies program in conjunction with the three colleges—Univ. of Virginia, College of William and Mary, and Virginia Polytechnic Institute.

- Hughes Aircraft Co. reported final assembly of Syncom synchronous communications satellite had been completed.   Spacecraft was undergoing final checkout before delivery to NASA at Cape Canaveral, where it would be launched early 1963.

- Univ. of Illinois radiotelescope at Vermilion River Observatory was dedicated.   The radiotelescope, world's largest, included 400' x 600' x 65' parabolic reflector.   Prof. George W. Swenson, Jr., Univ. of Illinois project engineer for the observatory, said radiotelescope's first job would be to map Milky Way.   The $871,650 telescope was financed mainly with grants from Office of Naval Research (ONR), with supplemental grants from National Science Foundation (NSF) and Univ. of Illinois.

- Use of missiles and earth-orbiting satellites in forest-fire detection was forecast by R. C. Howard of British Columbia Forest Products, Ltd., speaking in Vancouver to annual convention of Canadian Institute of Forestry.   Howard said satellites would report fires to computer center which would evaluate the fire, permitting foresters to dispatch missile with built-in-fire extinguisher.

- *Washington Post* reported that Thrust-Augmented Thor (TAT) booster would be used with Agena D upper stage to orbit first of new Discoverer satellite series.

- British Institute for Strategic Studies estimate reported that the Free World has more than six times as many ICBM's as the Soviet Union, but that the Soviet Union has far more IRBM's than the Free World.

November 10: Baltimore Sun published photograph of ANNA satellite taken by U.S. Coast and Geodetic Survey tracking team, Spesuti Island (Chesapeake Bay), Md. Triggered by Johns Hopkins Applied Physics Lab., flashing lights onboard ANNA were photographed in project to aid geodetic studies; photographs would be made from points around the earth.

- USMC gold medal was presented to Lt. Col. John H. Glenn, Jr. (USMC) at USMC 187th anniversary dinner, Washington. Medal commemorated Astronaut Glenn's Project Mercury orbital space flight Feb. 20, 1962 (MA-6).

November 11: NASA reported MARINER II Venus probe was 15,652,770 mi. from earth and 11,591,751 mi. from Venus.

- USAF announced launching of unidentified satellite with Atlas-Agena from Point Arguello, Calif.

- *Catalogue of Astronomical Photographs* featuring close-up photographs of moon, was released by Lick Observatory. Each photograph showed area 230 mi. by 280 mi. on scale of 30 mi. to the inch (view of lunar surface from 300 mi.). Lick spokesman said precise optical and guidance systems, high magnification, and skilled use of modern astrophotographic techniques were used to obtain the pictures. The 120-in. reflecting telescope at Lick is world's second largest.

- In interview at Soviet Academy of Sciences, Soviet physicist Dr. Sergei N. Vernov stated he believed particle acceleration accounts for outer radiation belt of the earth. Deep-space probes have not detected enough high-energy particles to prove the outer belt is replenished by high-energy particles ejected from the sun. Dr. Vernov believes the particles acquire their energy within the belt—through some as yet unknown accelerating mechanism. He believes there are two such means of acceleration, one sustaining the day-to-day energy level and the other accounting for its increases when a cloud of erupted solar gas reaches vicinity of earth, causing magnetic storms. Dr. Vernov suggested inner belt is sustained by protons formed by neutrons which decay upon bombardment by cosmic rays; some of the high-energy protons are trapped by earth's magnetism to become the inner radiation belt.

- Moscow radio reported Soviet designers were working on airplane that flies by flapping its wings like a bird. The plane was expected to have 10 to 30 times lifting power of similar-size conventional plane.

November 12: NASA Flight Research Center released photographs of "lifting body" experimental wingless craft to test landing techniques in gliding from altitudes as high as 5,000 ft. The 24-ft.-by-13-ft. craft would begin testing early in 1963, first with truck to tow it aloft and later with airplane as towing vehicle. Made mostly of plywood, the 500-lb. craft suggests shape of missile nose cone cut in half lengthwise.

- FAA Administrator N. E. Halaby told International Air Transport Association meeting that British, French, and Soviet competition probably would force the U.S. to develop its own supersonic commercial airliner within 10 years.

November 12: Tave (Thor-Agena Vibration Experiment), flown with Thor-Agena launching ALOUETTE satellite on Sept. 29, measured low-frequency vibrations to Agena stage and spacecraft interfaces during Thor boost phase, it was reported. Tave provided ". . . data verifying the techniques used by Goddard and Lockheed in predicting the launch vibration environment of the Thor-Agena B rocket. Demonstration that these techniques were sound is most important in designing future experiments and structures for the Thor Agena vehicles," Goddard project experimenter James Nagy said. Goddard Space Flight Center had designed, built, tested, and shipped 200-lb. Tave in only 20 days.

November 13: USAF announced it had placed in orbit 1.47-lb. Tetrahedral Research Satellite (TRS), launched piggyback aboard unidentified satellite using Thor-Agena vehicle. TRS was orbited to map radiation in space and radio its findings back to earth. NASA communications and telemetry stations were supporting this USAF project.

• Tass reported interplanetary probe MARS I began broadcasting scientific data on command from earth. The Mars probe, about two million mi. away from earth, was reported functioning normally.

• Rocket-powered instruments will be traveling as far as Jupiter by 1975, Robert J. Parks, director of JPL Planetary Programs, said in interview. Parks said by 1975 the planets Mars and Venus probably would have been studied "quite closely" by instruments.

• Congressman Joseph E. Karth (Minn.), chairman of space sciences subcommittee, said in press interview that planned mission schedule for Saturn launch vehicle "doesn't appear to be the best way to use a vehicle on which we have spent so much money." Mr. Karth said the powerful vehicle might prove useful in "speeding up the interplanetary research program, which may not be ambitious enough."

• Albert J. Evans, NASA Deputy Director of Aeronautical Research, told IATA Public Relations Conference in Washington: "Our understanding of aerodynamics has reached the point where it appears certain that in time an efficient supersonic transport can be developed. . . . There is no question as to whether there will be a supersonic transport; the only question is whose. Right now part of my job is to see that the United States is the first to develop a commercially competitive supersonic transport. . . .

"In the hypersonic region, the X-15 has given tremendous focus to our efforts. The X-15 has such capability that it will be used in furthering certain space research programs. Already we are making radiation measurements—piggy-back experiments along with the hypersonic research program. Our results with the X-15 have been so encouraging that I think we must soon look beyond the transport that goes to Mach 2 or 3, and begin laying the groundwork for the generation to follow. . . ."

• Signals from Venus-bound MARINER II interplanetary probe opened 17th annual meeting of American Rocket Society in Los Angeles.

November 13–14: More than 500 leaders from business, industry, government, and finance participated in New England Regional Conference on Science, Technology, and Space in Boston, sponsored by MIT and NASA. Dr. Howard W. Johnson, Dean of MIT School of Industrial Management, told conference that the "impact of space is largely the impact of space research and development on the economy, one of the important features of the last 20 years. . . ." In 1941, he said, R&D spendings amounted to .7 of 1% of gross national product, or $900 million. Today largely because of space R&D, about 3% of gross national product, or $16.5 billion, goes for R&D.

November 14: NASA announced Maj. L. Gordon Cooper, Jr. (USAF), had been selected for 18-orbit Project Mercury flight MA–9. The day-long space flight would be made no earlier than April 1963. Astronaut Cooper, youngest of original seven astronauts, was back-up pilot and technical adviser for Astronaut Walter M. Schirra in flight MA–8. Back-up pilot for MA–9 would be Cdr. Alan B. Shepard, Jr., who made first U.S. manned space flight (suborbital flight, MR–3) on May 5, 1961.

- In news conference at MIT, Dr. James A. Van Allen said the radiation caused by U.S. atmospheric nuclear test in July would be "undetectable" by July 1963. Dr. Van Allen reported signals from INJUN, TELSTAR, EXPLORERS XIV and XV showed the electronic stream had disappeared within a few days of U.S. explosion and that the electrons at 600-mi. altitude were now undetectable. Electrons at 900-mi. altitude were still creating radioastronomy interference, he acknowledged, but this would be gone by next July.

Speaking at New England Regional Conference on Science, Technology, and Space held at MIT, Dr. Van Allen reviewed suggestions made by panel of scientists which met last July–August under National Academy of Sciences auspices at Iowa City. Panel reviewed NASA program and recommended that NASA (1) send scientists with astronaut training along with test-pilot astronauts on early lunar and planetary flights; (2) establish an academy within a year to train scientist-astronaut volunteers, located at Houston, Tex.; (3) work harder to develop new instruments needed for lunar exploration; (4) ensure that planet Mars is not contaminated with terrestrial organisms until question of existence of Martian life is resolved; (5) make more use of university facilities; and (6) reduce time lag between concept of a space experiment and its execution. Spokesman in Washington said complete report of the panel would be issued by early January.

- West Side Association of Commerce (N.Y.C.) presented National West Side Award to NASA. In accepting the award, NASA Administrator James E. Webb said: "Except in war time, no large national effort has had such a rapid buildup as the space program over the past five years. The end of the buildup is not in sight, but the rate is slowing down and is keyed to the policy . . . that this is a fast paced, driving, prudently managed and efficiently conducted program, but is not a crash program. . . .

"In my view, we are entering a period when our national decisions and the debates which accompany them will not so much relate to whether and when we can achieve pre-eminence in space, but the rate at which we should proceed beyond the time when this pre-eminence is achieved—beyond the time when we have begun manned exploration of the moon. . . .

"It is the contribution of all the forces, in industry, government, and the universities, under the leadership which President Kennedy and Vice President Johnson have given the program, that gives us the basis for the pre-eminence we have in most fields in space and the pre-eminence we will shortly have in all."

November 14: Soviet interplanetary probe MARS I was 2,725,337.592 mi. away from earth at 4 AM EST, Tass reported.

- USAF advanced Atlas missile, Atlas F, was successfully test-fired more than 4,000 mi. down the Pacific Missile Range (PMR).
- Albert R. Hibbs, chief of JPL Arms Control and Disarmament, told American Rocket Society convening in Los Angeles: "Most people on both sides don't feel that space weapons will be too important . . .," but military neglect of space could be dangerous pending ratification of a treaty.
- N. E. Halaby, FAA Administrator, said in speech to Bond Club of New York that recommendations on Federal development of supersonic airliner would be submitted to President Kennedy in about two weeks. He emphasized that U.S. was interested in transport of mach 3 capability (2,000 mph or more).
- Plan for joint British-French development and construction of supersonic airliner was approved by French Cabinet. Under plan, British and French aviation industries would be subsidized by respective governments and would work together on airplane capable of carrying 90 passengers across the Atlantic at mach 2.2 speed (about 1,450 mph)—more than twice as fast as current jet transports.
- Lt. Col. Charles N. Barnes (USAF) of AEC Reactor Development Div. told annual meeting of Association of Military Surgeons of the United States that first attempt to orbit nuclear reactor in joint AEC–USAF Project Snapshot would be made next year. Snap–10A would be launched with Atlas-Agena vehicle, the Agena stage to be orbited with the 950-lb. Snap–10A which would furnish auxiliary electric power for instruments. Intended orbit would be several hundred mi. high. Reactor would be turned off after a year and the accumulated radioactive fission products allowed to decay over period of several hundred years; it would be designed to disintegrate during re-entry into earth's atmosphere.
- Monetary awards for patentable inventions were made to NASA employees under provisions of the Incentive Awards Act of 1954 as follows: George A. Smith of NASA Ames Research Center; Arlen F. Carter, George P. Wood, and Adolf Busemann of NASA Langley Research Center; David G. Evans, Warner L. Stewart, Edward F. Baehr, H. Allen, Jr., C. C. Ciepluch, E. A. Fletcher, Samuel Stein, David M. Straight, and John W. Gregory of NASA Lewis Research Center; William J. D. Escher and Thomas L. Greenwood of NASA Marshall Space Flight Center.
- Soviet AF Maj. Yevgeny M. Andreyev dropped 15-mi. in free fall from balloon before opening his parachute, feat believed to have set new record for free fall without stabilizing devices.

November 15: Venus-bound MARINER II spacecraft set new record for communications, transmitting engineering and scientific data to earth from nearly 18 million mi. in space. Previous transmission record was set by PIONEER V space probe at distance of 17.7 million mi. June 14, 1960. (PIONEER V's signal was tracked to 22.5 million mi., but no scientific data were obtained beyond 17.7 million mi.) Launched by NASA on Aug. 27, MARINER II was providing new information on nature of interplanetary space. It would fly by Venus at distance of 20,900 mi. and measure microwave and infrared emissions from the planet on Dec. 14.

- NASA attempt to launch ionosphere experiment for Commonwealth of Australia failed when Aerobee vehicle malfunctioned 38 sec. after launch from NASA Wallops Station. However, Australian payload electronics, including telemetry instrumentation and lunar aspect sensor, was activated before liftoff and functioned properly during flight. Launch of similar experiment would be attempted in Dec. for Commonwealth of Australia Scientific and Industrial Research Organization.

- NASA Director of Manned Space Flight, D. Brainerd Holmes, addressed National Industrial Conference Board, Chicago, on economic implications of manned space flight program: " . . . In a significant, though incidental way, the space program is already enriching and will continue to enrich our country. It is causing a rapid advancement of industrial technology and the stimulation of our economy. The billions of dollars required for the space effort are not being spent on the moon; they are being spent in our factories and laboratories—for salaries and for new materials and supplies, which in turn represent income for others.

  "Already the space industry is a major industry. It is creating new job opportunities at all levels of skills and abilities. It is improving standards of living.

  "New economically beneficial breakthroughs will be made in many fields. A successful space program will require major and rapid advances in the uses of energy and the development of new materials, fabrics, and lubricants—the very forces that are basic to economic growth.

  "Space research and development is already producing corollary benefits in the form of new products, new methods, and new industrial processes which can be employed in the manufacture of countless articles for human use. The surface has probably barely been scratched. . . .

  "The development of space science and technology will undoubtedly strengthen our whole industrial base and serve as effective insurance against technological obsolescence. . . .''

- NASA Administrator James E. Webb announced three-year grant to Graduate Research Center of the Southwest, Dallas, Tex., for development of advanced scientific experiments in lunar, planetary, and space exploration. Major operating division of GRCSW was Southwest Center for Advanced Studies (SCAS), which would "provide research organization structured specifically to design new space experiments and to provide the scientific guidelines for engineering them."

- AEC announced Dr. Edward Teller would be awarded the Enrico Fermi Award at ceremony on Dec. 3. Citation accompanying the award would praise Dr. Teller "for contributions to chemical and nuclear physics, for his leadership in thermonuclear research, and for efforts to strengthen the national security." Dr. Teller, now a professor at University of California, is most noted for his role in development of thermonuclear bomb.

November 15: USAF announced there would be at least three-month delay in first test-flight of prototype RS–70 (reconnaissance-strike) supersonic aircraft. Delay was attributed to problems in sealing fuel cells.

- Astronomer Dr. E. J. Opik, of Univ. of Maryland, was quoted as saying "the Russians have contributed only five-tenths of one per cent" of the new information on space. "They profess to have sent huge payloads into space 10 to 100 times larger than any the United States is capable of sending. At the same time, however, 99.5 per cent of all information on space has come from the United States' probes." Dr. Opik was cited in 1960 by National Academy of Sciences for "outstanding investigation of meteoric bodies."

- Ambassador Charles Bohlen officiated at the opening of a NASA exhibit at the Palais de la Découverte, Paris, France. The exhibit would stay in Paris until January 15, included models of past, current, and future NASA spacecraft; panels explaining the goals and results of the U.S. space program; a working TELSTAR demonstration; and a Spacemobile lecture and demonstration. The exhibit was enthusiastically received by French scientific and government personages attending the inauguration.

- AFSC announced it was testing electrocardiogram package in free-fall parachute jumps for possible applications in future manned space flight. Worn inside the astronaut's pressure suit, the small, 14-oz. package could monitor heartbeat (and other physical reactions): Conditioning unit in package converts heartbeat to FM radio signal which is sent by transmitter in package to amplifier in cockpit, and then to ground receivers, where it is converted to electrocardiogram. Use of package would provide astronaut freedom of movement by eliminating need for connections from pilot to spacecraft. Developed by Hughes Aircraft Co., the package was being considered for use in X–15 program.

- Astronomer Dr. Carl Sagan told American Rocket Society, meeting in Los Angeles, that there was a statistical likelihood that earth had been visited at least once by intelligent beings from other worlds. In interview, he suggested that successful planetary civilizations endure for perhaps a hundred million years and that, as the societies advance, they reach farther from their respective planets into space. He predicted man would begin interstellar travel in about 100 years—in star ships, powered by hydrogen-fusion ramjet engines and built in space.

- Scientists at Hughes Aircraft Co. displayed Surveyor lunar probe at American Rocket Society, the probe containing redesigned payload weighing about 114 lbs. as compared to originally designed payload weighing 220 lbs. Payload of instruments would be placed on the moon in late 1964. Dr. Leo Stoolman, Surveyor project chief for Hughes, said reasons for redesigning Surveyor instrument package were difficulty in Centaur launch vehicle program and decision to include only instruments that would directly assist U.S. manned lunar landing program.

November 15: Dr. Robert R. Gilruth, Director of NASA Manned Spacecraft Center, told Aircraft Industries Association Quality Control Committee in Houston that designs, procedures, and schedules in manned space flight program had to be flexible in order to absorb continual changes of rapidly advancing technology. Emphasizing importance of equipment malfunctions that occur during systems development or preflight preparations, Dr. Gilruth said: "In manned flight, we cannot afford to regard any of these equipment malfunctions as a 'random' failure. We must regard every malfunction and, in fact, every observed peculiarity in the behavior of a system, as an important warning of potential disaster. Only when the cause is thoroughly understood, and a change to eliminate it has been made, can we proceed with the flight program. . . .

"Rapid corrective response to malfunctions throughout system development and preflight preparations is a critically important requirement of our programs if we are to meet schedules with hardware that is fit to fly. . . ."

- Dept. of Commerce released translation of Soviet report on use of rockets to prevent hailstorms. Report said thousands of acres of grape crops were protected by anti-hail rockets in 1961 and efforts were expanded this year. Rockets were loaded with cloud-seeding chemicals and fired into suspected clouds, the released chemicals crystallizing cloud droplets to (1) prevent existing hailstones' growing larger, and (2) cause the crystallized droplets to "fall as harmless snow."
- Aviatrix Jerrie Cobb told Women's Advertising Club of Washington, D.C.: "We're bypassing the one scientific space feat we could accomplish now—putting the first woman in orbit. . . ."

November 16: Saturn SA-3 reached 104-mi. altitude in ballistic flight from Cape Canaveral, the fully-fueled S–I stage performing as planned. Upper stages were filled with water simulating weight of live stages. At peak altitude, rocket was detonated by radio command and the ballast water was released into the ionosphere, forming massive cloud of ice particles several mi. in diameter. From the cloud experiment (Project Highwater II) scientists hoped to gain data on atmospheric physics. This was third straight test-flight success of NASA's Saturn and first flight with maximum fuel onboard. Vehicle generated 1.3 million-lb. thrust on flight of 4 min., 55 sec.

- NASA Ames Research Center announced construction of four space research facilities, totaling more than $14 million, had been authorized for FY 1963: biosciences laboratory, guidance facility, radiative heat system for Mass Transfer Facility, and helium tunnel. Bioscience laboratory would be used for studies in genetics, radiobiology, immunology, and environmental physiology, and detection of extraterrestrial life. Guidance facility, consisting of three-man capsule with systems and equipment, combined analog and digital computer, and rotating centrifuge to drive capsule, would be used to obtain data on integrated design of guidance, stabilization, control, and crew support systems in advanced manned spacecraft. Radiative heat system would permit Ames scientists to investigate re-entry heating of spacecraft from lunar or planetary missions. Sections of complete heat shields will be tested at the facility. Mach 50 helium tunnel would be used to complement the existing Ames hypersonic free-flight facility, which used gun-launched models to obtain 50 times speed of sound.

November 16: Nike-Cajun sounding rocket launched from Ft. Churchill, Can., under direction of NASA Goddard Space Flight Center. Second stage failed to ignite, so rocket reached altitude of only about 9.5 mi.

- S–IV stage for Saturn space vehicle arrived at NASA Marshall Space Flight Center, Huntsville, Ala., after 23-day journey from Douglas Missiles and Space Systems Div., Santa Monica, Calif. First Saturn stage to be shipped by water from West Coast manufacturing site, the S–IV would be mated with other Saturn stages in dynamic test tower at Marshall, for series of bending and vibration tests. Its external configuration, weight, and other characteristics were same as S–IV flight units.

- Dr. Robert R. Gilruth, Director of NASA Manned Spacecraft Center, told newsmen: "The first unmanned capsule launch [in Project Gemini] has slipped from the third quarter to the fourth quarter of 1963. There is virtually no possibility of a manned flight before 1964 . . . . [Slippage was] simply because of the time it takes to do this very complicated job."

- Dr. Hugh L. Dryden, NASA Deputy Administrator, said in first annual Theodore von Kármán Lecture of the American Rocket Society: "The costs of the presently approved [space] program increase next year to about six billion dollars if current time schedules are maintained. It would not be possible to include the development of a [manned] space station now without still larger resources assigned to the space program. I personally believe that the next large manned space flight project [after Project Apollo manned lunar landing] will be this one rather than extensive exploitation of the moon or manned expeditions to the planets. This might change if the early lunar exploration returned surprises in the form of natural resources of use on earth.

  "I think it will now be appreciated that the present rather arbitrary subdivisions of our program will coalesce, for the manned space station will be useful for both manned and unmanned scientific exploration and could be the site of observation of weather or of communications relay stations . . . ."

- Relationship between very low frequency (VLF) hiss and aurora australis (Southern lights) was confirmed after six months of research by two scientists, it was reported. Ward Helms and John Turtle, working at Byrd Station, Antarctica, believe they have confirmed the theory of Henry Morozumi, Stanford Univ. scientist, when they found peaks of auroral display and VLF hiss were identical in intensity and coincided repeatedly. Helms said he was "not certain that VLF hiss is part of the aurora, per se, but I'm sure that since the aurora and hiss peaks are simultaneous there is reason to think the same particles generating the aurora also generate the hiss."

- Secretary of the Air Force Eugene M. Zuckert announced signing of contract with United Technology Corp. for solid-propulsion rocket motors to be strapped to sides of Titan II rocket in Titan III configuration. He said contract with Martin-Marietta Corp. for airframe assembly and test of the rocket, as well as for structure of transitional upper stage, would be signed soon.

November 16: At American Rocket Society Honors Night Banquet in Los Angeles, Dr. Robert R. Gilruth, Director of NASA Manned Spacecraft Center, was presented the Robert H. Goddard Memorial Award, highest ARS honor, for "general eminence in the field of rocket engineering and space flight." Other 1962 ARS awards: Lt. Col. John H. Glenn, Jr., NASA astronaut, received the ARS Astronautics Award; Vice Adm. William F. Raborn, Deputy Chief of Naval Operations, Special Projects Office, received the James H. Wyld Memorial Award; Samuel K. Hoffman, president of NAA Rocketdyne Div., received ARS Propulsion Award; Dr. Howard S. Seifert, of Stanford Univ. and United Technology Corp., received the G. Edward Pendray Award; Theodore Forrester, director of Electro-Optical Systems, Inc., Ion Physics Dept., received the ARS Research Award; and John R. Winckler, prof. of physics at Univ. of Minnesota, received the first ARS Space Science Award.

- Soviet aircraft designer Artem Mikoyan was quoted as saying U.S.S.R. would have a "rocket plane" within this decade. Aircraft was described as wide-winged, stainless-steel covered, and powered by multistage rocket. After launching, craft would fly ballistic trajectory reaching height of 150–222 km. and later re-enter earth's atmosphere, for intercontinental transport of mail, freight, and, later, passengers. Rocket planes also would be principal transportation between earth and earth satellites.

November 17: Aerobee sounding rocket launched from NASA Wallops Station, the vehicle carrying 211-lb. instrumented payload to 128-mi. altitude in experiment to study behavior of liquid hydrogen exposed to radiant heating and zero gravity. Experiment was conducted for NASA Lewis Research Center, which would use data obtained in development of liquid-hydrogen rocket engines.

- Four USN officers in altitude chamber were injured when electrical spark ignited fire in the chamber, near end of 14-day experiment at USN Air Crew Equipment Laboratory, Philadelphia. Men were participating in NASA experiment to determine effect on human beings of breathing pure oxygen for 14 days at simulated high altitudes.

- USAF Systems Command said re-entry from space flights was being simulated at Flight Control Laboratory, Aeronautical Systems Div., Wright-Patterson AFB, Ohio. The tests were "part of AFSC's long-range Mark IV program to determine the control-display requirements for advanced orbital vehicles."

- Fifth attempt to launch Polaris A–3 missile failed when Polaris second stage veered off course and had to be destroyed.

- DOD released following statement:

  "Lockheed [Aircraft Corp.] stands alone in refusing to follow the course that the other members of the [aerospace] industry considered reasonable. . . . Should a strike take place at Lockheed, Department of Defense programs of critical urgency to the national defense would be affected. It is imperative in the national interest that the Department of Defense make necessary preparations to minimize to the greatest extent possible any loss of production that might follow a work stoppage. Responsible officials of the Department of Defense are therefore considering whether alternate means of production can be utilized so that work on these vital programs can go forward in the event of a stoppage at Lockheed."

November 17: Low yield nuclear device was exploded in atmosphere at U.S.S.R. Semipalatinsk testing grounds, 32nd Soviet test announced by AEC in current series.

November 18: MARINER II Venus probe, launched Aug. 27, was 18,767,-939 mi. from earth and 8,603,582 mi. from Venus, NASA announced.

- Dr. Niels Bohr, pioneer atomic physicist and winner of Nobel Prize and Atoms for Peace Award, died in Copenhagen.

November 19: Dr. Andrie B. Severny, Director of U.S.S.R.'s Crimean Astrophysical Observatory, explained techniques used to photograph MARS I on Nov. 3, when the interplanetary probe was 120,000 mi. from earth. Using 102-in. telescope, largest outside the U.S., astronomers fixed its aim at spot MARS I was calculated to be; automatic controls shifted telescope's aim slightly to sweep along the probe's assumed trajectory. At same time, photographs were made, with each plate exposed twice at 30-sec. intervals. Resulting photographs showed MARS I as well as its booster rocket traveling through space. This was believed to be greatest distance from which man-made objects had been photographed.

- USAF Minuteman ICBM successfully launched from Cape Canaveral silo on 4,000-mi. R&D flight down the Atlantic Missile Range.

- International Association of Machinists announced strike against Lockheed Aircraft Corp. had been set for Nov. 28.

- NASA and Indiana University jointly announced experimental center for industrial applications of aerospace research would be established at Indiana Univ., with initial NASA funding of $150,000. Project would involve the university's business professors, biologists, geologists, physicists, and others, working with research and development people of industry and NASA.

- Bell Helicopter Co. successfully flew unmanned helicopter in demonstrations at Naval Ordance test station, China Lake, Calif. Guided by remote control system, helicopter flew for 72 min. at altitudes up to 2,000 ft. and speeds of about 70 mph; brought back to earth, the craft was refueled, inspected, and relaunched in 15 min. for 48-min. unmanned flight.

- NASA and DOD approved new plan for solid-propellant development. Program considered the solid-propelled boosters as a development concept and called for advanced studies.

November 20: Sparobee sounding rocket launched from NASA Wallops Station with 90-lb. payload to measure electron and neutral particle temperatures at 75 to 225-mi. altitudes. Vehicle reached peak altitude of 214 mi. and payload impacted in Atlantic Ocean about 217 mi. away. Secondary objectives were measurement of ion and neutral particle density and flight testing of newly designed Thermosphere Probe system. Experiment was joint project of University of Michigan and NASA Goddard Space Flight Center.

- NASA named Adm. Walter F. Boone (USN, Ret) as Deputy Associate Administrator for Defense Affairs. Primary responsibility of the newly-created post was to strengthen the flow of technical and management information between NASA and DOD.

November 20: Office of Aerospace Research announced scientists at Air Force Cambridge Research Laboratories (AFCRL) would conduct the "most thorough study of atmospheric ozone ever made," a concentrated one-year series of observations with more than 700 balloons carrying sensitive ozone meters into the upper atmosphere. New 11-station network, from Canal Zone to Greenland, would participate in project. To begin in January 1963, project was expected to yield data on nature of ozone.

- Dr. Eugene B. Konecci, Director of NASA Biotechnology and Human Research, told conference of Manufacturing Chemists' Association that nuclear rocket fuel ". . . may eventually make it an attractive matter to bring lunar ore to the earth for processing. . . ." Nuclear fuel would reduce the cost of each pound of payload.

- USAF Nuclear Effects Research Laboratory dedicated at Kirtland AFB, N.M. Lab was designed to simulate effects of nuclear explosions on ballistic missile systems; study effects of radiation on people and electronic systems; simulate magnetic and radiation effects of high-altitude nuclear explosions; and conceive radiation experiments for space probes.

- Dr. Andrei S. Severny, Soviet astronomer who heads the Crimean Astrophysical Observatory whose 102-in. telescope is the largest outside of California, reported a partial theory on cause of solar flares—solar flare convulsion is brought about by a release of the accumulation of magnetic energy. Based on observations of changes in the strong magnetic fields, Dr. Severny's interpretation of flares is based on technique developed by Dr. Horace D. Babcock and his son at Mount Wilson in California. Skeptical of Severny's thesis, a colleague of the Babcocks, Dr. Robert Howard, was reported in Russia to challenge the Soviet data.

- Lockheed Aircraft Corp. and International Association of Machinists (IAM) union agreed to Federal Mediation Service request that they resume negotiations. The request came after the union had set Nov. 28 as strike deadline.

- Soviet astronomer Vladimir K. Prokofief reported that he had discovered traces of oxygen in the atmosphere of the planet Venus, based on observations from the spectrum of sunlight reflected from the Venutian cloud cover.

November 21: Temperature within MARINER II Venus probe was reported to be rising. Calculations indicated that when the spacecraft reached Venus its battery temperature would be 129°, 9° more than design limit—creating pressures that could rupture the sealed battery.

- NASA confirmed Rift nuclear rocket stage would be manufactured in renovated dirigible hangar at Moffett Field NAS, Calif. To be used to flight-test NASA–AEC Nerva engine, Rift was being developed under technical direction of NASA Marshall Space Flight Center with Lockheed Missiles and Space Co. as prime contractor. Rift would be flight-tested in 1967–68 as top stage of Advanced Saturn (C–5) launch vehicle.

- Difficulties in Centaur launch vehicle development were forcing NASA to program 1964 Mars probe for Atlas-Agena vehicle, *New York Times* reported. Use of Atlas-Agena would mean smaller payload, with about 60 lbs. of instruments instead of 200 lbs.

- USAF launched unidentified Blue Scout probe from PMR.

November 22: Soviet astronomer Dr. Vasily Grigoryevich Fesenkov reported in interview his conclusion that earth is followed by a comet-like tail, the tip of which stretches as far out as the moon's orbit. From lifetime of observations of zodiacal light and gegenschein, Dr. Fesenkov believes solar winds and sunlight pressures form tail from dust cloud enveloping the earth.

November 23: MARINER II reported by JPL to have again lost power from one of its solar panels on November 15 (similar to inoperative period between October 31 and November 7, after which it suddenly returned to life). Temperatures in MARINER II continued to mount within 20° F of what was foreseen (200° F on sunward face of solar panels, 100° F toward the shady side).

• Letter from President Kennedy defending Administration policy of permitting astronauts to sell their personal stories for profit was made public by M. B. Schnapper, editor of Public Affairs Press, in Washington, D.C. Replying to Mr. Schnapper's criticism of the policy, the President said the "policy decision was reached only after most careful review of all the facts and lengthy consultation both within and without the Government. Efforts were directed toward removing specific controversial aspects of the original astronaut agreement, and we feel that this goal has been reasonably achieved."

• In interview, FAA Administrator Najeeb Halaby stated that he would probably recommend the development of a 2,000 mph commercial jet transport, pending results of a full study. Supersonic transport would carry up to 150 passengers between New York and Los Angeles in one hour and 40 minutes. Development costs of perhaps $700 million would, said industry leaders, require Federal subsidy.

• In press interview in Houston, Astronauts Leroy Gordon Cooper (Maj., USAF) and Alan B. Shepard, Jr. (Cdr., USN) said television pictures of Maj. Cooper would be transmitted to ground stations during MA–9 flight by means of miniature transmitter inside the Mercury spacecraft. Pictures could not be viewed live on commercial TV because transmissions would be on different scan speed. Astronaut Cooper also said that a rendezvous maneuver had been considered for MA–9 flight, but decision had been made not to attempt the maneuver. MA–9 is considered to be the last Project Mercury flight, to be followed by Project Gemini.

November 23–December 2: Space Science Fair, held in Cleveland and cosponsored by NASA and Cleveland *Plain Dealer*, included extensive exhibit covering "virtually all topics of space technology" as well as series of lectures for junior and senior high school students by aerospace scientists and engineers Honorary co-chairmen of the fair were NASA Administrator James E. Webb and former NASA Administrator T. Keith Glennan, President of Case Institute of Technology in Cleveland.

November 24: USAF launched unnamed satellite from Vandenberg AFB with Thor-Agena B booster.

• DOD announced selection of contractor for 22 test models of TFX aircraft (tactical fighter, experimental), the Convair Division of General Dynamics, with Grumman Aircraft Engineering Corporation as an associate. Based upon much work at Langley Research

Center, TFX fighter was designed in USAF and USN versions, having variable geometry wings for flight speeds up to 1,650 mph. Contract which would procure up to 1,500 aircraft was said to be bigger than "any fighter aircraft program since World War II in both numbers and dollars."

November 24: DOD issued restrictions, it was reported by *L.A. Times*, on assigning new contracts to strike-threatened Lockheed Aircraft Corp. pending settlement of its labor-management dispute.

• Soviet astronomer Dr. Iosif Samuilovich Shklovsky, head of radio-astronomy department at Sternberg Astronomical Institute, Moscow, said in interview with western press that first place to seek signs of intelligent extraterrestrial life is in Andromeda Nebula. The spiral nebula is closest galaxy to our own Milky Way—about 1.5 million light years away. Reasoning that it was likely an advanced civilization should be able to harness enough energy from its sun to create a powerful radio beacon, and pointing out that no such beams had been detected from other parts of our own galaxy, Dr. Shklovsky said such beams could be detected by a single radiotelescope aimed at the entire Andromeda Nebula.

November 25: MARINER II Venus probe established new communications record, transmitting "excellent quality" data from more than 22.5 million mi. in space. (PIONEER V probe's transmission had ceased with position signal from 22.5 million mi. in 1960; PIONEER V had ceased transmitting scientific data at 17.7 million mi.) NASA said MARINER II signals were relaying good data on the four scientific experiments on board. MARINER II was expected to pass within about 21,000 mi. of Venus on Dec. 14.

• Dr. Glenn T. Seaborg, AEC Chairman, predicted in interview that in 20 years "we'll be well advanced in the use of nuclear power in space both for propulsion of rockets and for auxiliary power within orbiting satellites.

"Around the year 1982 we will have already made, or be seriously planning, a manned journey to one of the near planets. And we'll be using nuclear energy rather routinely in satellites as an auxiliary source of electric power.

"For example, by then, we'll probably be able to have [world] television broadcasts relayed by satellites brought directly into homes, . . . Within 20 years it might be possible to get reception direct from a satellite wherever a television set was located, even in the darkest part of Africa . . . .

"I feel that even such predictions as I have outlined may be too conservative in the sense that there may be spectacular developments that we do not even foresee . . . ."

• U.S. diplomatic source in Geneva disclosed U.S.S.R. had been conducting nuclear tests at the rate of one every two days since Nov. 20, the date previously announced by Soviet Premier Khrushchev as end of current series.

• Pertinent facts of human evolution should be included in preparation for interstellar communication, according to Prof. Robert Ascher of Cornell Univ. dept of anthropology and his wife, Marcia, assistant professor in Ithaca College dept. of mathematics and physics. Models of primitive man's efforts in surmounting barriers can be used as theoretical background for thinking about possible civilizations on other planets.   Problem of watching and listening for signals across space and developing means to reach source of those signals is essentially same as problem of prehistoric man making contact with other races or civilizations. Ascher also declared whenever two civilizations have made contact, conquest by dominant group has followed.   Models of those communications should be constructed and studied, Ascher said, and applied to future interstellar communications.

November 25: Secretary of Defense Robert S. McNamara allotted additional $50 million for development of radar and other components in USAF RS–70 project.   DOD announcement said restudy of the project had been completed and that recommendations had been submitted to President Kennedy last week.

- U.S.S.R. was building an atom accelerator about twice the size of any in existence, it was reported.   The "alternating gradient synchrotron" would be able to accelerate a beam of protons to energies of 50–70 bev (billion electric volts), as compared to 33-bev accelerator at Brookhaven.   Atomic research station, with many supporting facilities, was located near Serpukhov.

- U.S.S.R. was building "largest radiotelescope in existence" near Serpukhov, it was reported.   20-acre, cross-shaped radiotelescope featured two movable antenna arms of 1-km. length with supporting towers of 65-ft. height, the instrument modeled after cross designed by Dr. Bernard Y. Mills of Australia.   Chief charactertic of cross was its ability to collect considerable amount of radio energy emanating from an extremely small area of the heavens. Another instrument at the radioastronomy station was radiotelescope with parabolic antenna 72-ft. in diameter.   Station was under direction of Dr. Viktor V. Vitkevich and operated by Lebede Institute of Physics of Soviet Academy of Sciences.

November 26: NASA Flight Research Center announced investigation of X–15 No. 2 accident on Nov. 9 had revealed probable cause was succession of equipment failures coupled with landing gear collapse.   Inability of rocket engine to attain full power was attributed to faulty governor actuator which prevented engine's receiving sufficient amount of propellant.   Failure in engine's receiving mechanism did not allow landing flaps to come down and resulted in high load on aircraft's landing gear.   Unusually high landing-load caused collapse of main landing gear soon after the X–15 touched down at Mud Lake, Nev.   Pilot John McKay sustained severe bruises, a back injury, and shock; but he had been released from hospital and had returned to work at NASA Flight Research Center.   Accident was under continuing investigation by NASA and USAF engineers, with plans calling for mechanical improvement in flap-lowering mechanism before another X–15 flight.

- Dr. Harold Urey, Nobel prize-winning chemist, predicted U.S. would land a man on the moon in 1969.

- Bell Telephone Laboratories reported TELSTAR communications satellite's refusal to respond to transmitter cut-off commands had been followed by the satellite's turning off its command receiver.   Bell scientists expected to correct TELSTAR communications system by remote control.

November 26: Closed-door sessions of the 13 directors of Communications Satellite Corp. were reported in Drew Pearson's column, discussions on the revolution in communications made by satellites and the international complications involved.

- J. G. Morse of Martin-Marietta Corp. told Atomic Industrial Forum in Washington that the Government should begin immediately to develop nuclear power systems for communications satellites. Morse suggested that only nuclear power in form of radioactive chemical or isotopic generators could satisfy communications satellite needs—systems in power range of 60 to 300 watts, with operational lifetimes of more than 10 years.

- Soviet scientist Nikolai Semyonov, writing in *Pravda*, noted that many U.S. corporations had applied-science laboratories and U.S. universities had laboratories devoted to pure science. ". . . This is profitable for them inasmuch as it is university science which gives them, in the main, those fundamental scientific results which supply the laboratories of corporations, and through them, production." Mr. Semyonov's article was in support of Premier Khrushchev's order that Soviet research institutions be developed on an industry basis.

November 27: Aerobee 150A sounding rocket launched from White Sands Missile Range, N.M., in NASA experiment to gather data on solar radiation intensity in ultraviolet region. Rocket carried spectrophotometer package to 124-mi. altitude. Preliminary experimental results: "High voltage arcing in the instrumentation prohibited collection of scientific data . . . ."

- Dr. Eugene Shoemaker, Assistant Director of Lunar and Planetary Programs, NASA Office of Space Sciences, was appointed chairman of Joint Working Group of NASA Office of Space Sciences and Office of Manned Space Flight. Group would be responsible for (1) recommending detailed program of scientific exploration in future manned flights; (2) defining information desired from unmanned flights to support manned flights; and (3) establishing and maintaining close liaison with NASA field centers, other Government agencies, and universities in development of integrated scientific program for manned flight. Dr. Shoemaker had been employed by U.S. Geological Survey as Chief of Astrogeology Branch in Menlo Park, Calif.

- Experimental new method of producing synthesis of chemicals with low-energy ions was reported by Rocket Power, Inc., a solid-propellant manufacturer. Milton Farber, vice president and director of laboratories, and Dr. Stanley Singer, chief chemist, indicated several new compounds had been produced. Mr. Farber said the ion-synthesis method could produce rocket propellants perhaps 10 per cent more efficient than any now available in U.S. "The key is the speed of the ions. If the beam of ions meets the target chemical at too high a rate, shattering occurs. By controlling the velocity of the ions, they can be made to merge with the chemical, creating a new compound." Research project was sponsored by USAF OAR.

- Representative Chet Holifield, Chairman of Joint Congressional Committee on Atomic Energy, declared in speech that "moon madness" is starving scientific programs other than space exploration. He predicted budgetary troubles for atomic energy programs because of "greater priority" of space programs.

November 27–December 9: Four launchings of Centaur rockets took place from the Argentine Rocket Range, Chamical, as part of worldwide program of simultaneous measurements of winds and turbulence in the ionosphere by means of sodium-cloud experiments. The Argentine launchings were joint cooperative effort of French National Committee on Space Studies and the Argentine National Commission on Space Research.

November 28: Presidential Science Adviser Jerome Wiesner, speaking before American Society of Mechanical Engineers in New York, said special panel to study scientific manpower had concluded that "impending shortages of talented, highly trained scientists and engineers threaten successful fulfillment of vital national commitments." As first step to overcome shortage, Wiesner urged that 8,000 additional college students enroll in autumn 1963 as full-time graduate students in engineering, mathematics, and physics. Panel set goal of 7,500 Ph. D. graduates by 1970 (3,000 Ph. D. degrees were awarded in 1960).

- NASA Manned Spacecraft Center released sketch of Project Apollo suit to be worn by astronauts making four-hour expeditions on lunar surface. Suit contains portable life support system to supply oxygen and pressurization, control temperature and humidity, and protect against solar radiation.

- AEC announced first U.S. "breeder-reactor" had achieved self-sustaining nuclear chain reaction. Producing more fuel than it burns, plutonium-fueled reactor was expected to become valuable and economical power source.

- Dr. Edward Teller, Associate Director of Lawrence Radiation Laboratory, told American Nuclear Society in Washington that peaceful use of atomic explosives had "the potentiality of becoming the first really important and thoroughly economic use of atomic energy." Feasibility of such use was demonstrated last summer with 100-kiloton thermonuclear explosion at 635-ft. depth in Nevada, the explosion removing about 7,500,000 cu. yds. of earth and leaving crater 1,200 ft. in diameter and 320 ft. deep. Dr. Teller said radioactivity can be controlled so that excavated area can be entered a few weeks after the explosion with no more exposure than in an atomic laboratory. He predicted nuclear explosions on the moon could be used to extract water from chemical compounds in lunar rock.

- International Association of Machinists (IAM) workers struck at Lockheed Aircraft Corp. in California, Florida, and Hawaii. President Kennedy, invoking Taft-Hartley Act, created three-man board of inquiry to investigate the dispute; board's recommendations, due Dec. 3, would contribute to President's decision whether to request Federal court order barring strike for 80-day period. Included in Lockheed production items are Agena second-stage rocket, used by USAF and in NASA Ranger and Mariner programs.

- USAF Skybolt missile failed fifth flight-test, dropping into Atlantic Ocean minutes after launch from high-flying airplane.

- Nike-Zeus antimissile was shot outside earth's atmosphere and successfully maneuvered with controlled blasts of rocket exhaust, in test over the PMR.

- Rodney D. Steward was named RL–10 rocket engine program manager at NASA Marshall Space Flight Center. Steward replaced William D. Brown, who resumed full-time duties as deputy chief, Engine Management Office, MSFC.

November 28: Project stabilization agreement—set of standard pay and benefit practices for both union and non-union workers— was adopted at Cape Canaveral, it was reported. Agreement was product of more than a year's negotiation by contractors and unions working at Cape Canaveral, undertaken at suggestion of President's Missile Sites Labor Commission.

November 29: Dr. Edward Welsh, Executive Secretary of the National Aeronautics and Space Council, told Atomic Industrial Forum: "We cannot afford to find ourselves with a space gap due to a lack of advanced planning and of attention to advanced developments. Yet, we will assuredly have such a gap if we do not push enthusiastically the potential of nuclear energy and power at an increasing rate in the field of space. . . . It would indeed be foolish not to develop and use the propulsion system which would carry the greatest weights the greatest distances most economically. The potential of nuclear propulsion in such an objective cannot be overestimated. . . ."

- NASA test pilot John McKay predicted X–15 would soon attain speeds seven times speed of sound at altitudes up to 500,000 ft. during a new series of flights.

- AFSSD broke ground at Edwards AFB for construction of world's largest and most highly instrumented solid-rocket engine test facility. Stand was primarily to test five-segment 120-in. solid rockets for Titan III, but it also had capacity to test segmented solid rockets up to and includng 156-in.

November 30: Franco-American scientific sounding rocket launchings coordinated when two U.S. launchings were made from Wallops Island while France launched one from Algeria and failed to launch one from France. First U.S. rocket (Nike-Cajun) fired at 5:57 AM carried a Langmuir probe to determine electron density and the temperature of the "E" layer of the ionosphere (50–100 miles altitude); the second (Nike-Apache) launched at 6:15 AM, released a sodium vapor cloud to 106-mi. altitude which spread over 100 miles of the eastern seaboard. "Sporadic E" regions of the ionosphere have electron density that reflects much higher radio frequencies, and may result from wind shears. On November 7, GSFC scientists undertook the first sounding rocket firings to examine correlation between "E" region and wind shears.

- Sixth and final powered test of Kiwi B-4A ground-test reactor for Project Rover was successful at Jackass Flats, Nev. Newsmen were permitted to watch this test from a mile away, the first test to which outside observers had been allowed. Kiwi B-4A, looking more like a flyable engine than some of its predecessors, was mounted on a remote-controlled railroad flatcar, consumed tons of liquid hydrogen in the half-power, five-min. test. Part of the joint AEC-NASA program for development of a nuclear-powered rocket, the series of ground reactors tested different reactor designs, fuels, metals, and accessories.

November 30: Dr. Jesse L. Greenstein, prof. of astrophysics at Cal Tech, estimated aggregate of energy far out in universe was equal to amount of energy released by 10,000,000 stars like our sun in their entire lifetimes. "No known physical phenomenon can account for it. It must be the result of some fantastic catastrophe in the universe," Dr. Greenstein said. Electromagnetic energy was detected by radiotelescopes.

- Representative Charles E. Goodell (R–N.Y.), speaking at House Education Subcommittee hearing, criticized NASA's hopes to recruit 1,000 graduate students into space research each year during this decade: "That's 25 per cent of our production of Ph.D.'s in the space-oriented sciences. That's an incredible share of our brainpower to concentrate on one agency." John F. Clark, Associate Director and Chief Scientist, NASA Office of Space Sciences, defended NASA aim as being commensurate with NASA's share of total Government expenditures for research.

- Representative John E. Moss (D–Calif.), Chairman of special House Subcommittee on Government Information, criticized "Government news management which is unique in peacetime." He objected strongly to handling of news about military space launches: "All launches by military agencies, whether they are for a military purpose or to gather information of a scientific nature, are covered by a blanket of secrecy. . . . All information about the billions of dollars spent by the military in space research is channeled through the Pentagon's single public information voice."

  Moss recalled last September when NASA said that six Russian attempts to send space probes to Venus and Mars had failed. Since then, Moss said, "there has been a complete blackout of information about Russian satellite efforts. This is the kind of news management that causes grave concern. . . . If we cover up Russian successes . . . we can certainly cover up our failures. This . . . leads to a dangerous delusion of the American people."

- In press interview, former President D. D. Eisenhower recommended reduction in next year's Federal spending by at least $4 billion, with $2 billion coming out of the $4-billion space budget.

- At meeting of the National Academy of Sciences at Austin, Texas, President Frederick Seitz said that other nations should bear a larger share of space exploration efforts. This would help solve "a big problem for us today—whether the attention being given to aero-space in America is draining too much manpower from the other sciences."

- Secretary of Labor W. Willard Wirtz said in San Francisco that Lockheed Aircraft Corp.'s "extraordinary intransigence" was responsible for deadlocked negotiations with the International Association of Machinists which resulted in two-day strike (IAM strikers returned to work today). Three-man board of inquiry appointed by the President began their investigation of the "union shop" conflict.

During November: In letter to House Committee on Science and Astronautics, NASA Administrator James E. Webb advised Congress of plans to reprogram $10,426,000 in FY 1963 funds to begin construction of facilities at White Sands Missile Range (WSMR).

Of this sum, $1,717,000 was needed for Little Joe launch-complex construction; $9,084,000 was for site development and support facilities.

During November: Goodyear Aircraft Corp. announced development of tunnel to simulate conditions encountered by satellites, missile nose cones, and recovery packages during re-entry into earth's atmosphere.

- Aerojet-General Corp. reported first successful firing of development engine for Apollo spacecraft's service module, full-scale prototype engine with ablative thrust chamber assembly fired for checkout runs up to 50 sec.

- General Dynamics/Astronautics named by NASA as systems integration contractor for Project Fire re-entry test program. Project systems included Atlas D launch vehicle built by GD/A, velocity package built by Chance Vought, and re-entry vehicle built by Republic Aviation.

- NASA selected Hamilton Standard Div. of United Aircraft Corp. as prime contractor to manage and integrate spacesuit program for Project Apollo as well as to design and produce its life-support pack. Spacesuit development and fabrication would be done by International Latex Corp. under subcontract.

- USAF OAR announced it had awarded basic research grants and contracts to 56 universities and 20 research firms in U.S. and Europe in November.

- USAF awarded Radio Corp. of America a contract to develop new communications system which would allow X–20 to maintain communications with ground stations during re-entry. New system would use frequencies in "super high" range and would sustain communications for greater than 97 per cent of re-entry phase.

- General Dynamics/Astronautics announced results of its molecular erosion-rate tests of low-altitude satellites showed erosion due to molecular bombardment was less than one billionth of an inch per year.

- Project Mercury boilerplate capsule successfully landed at Gary Army Air Field, Tex., in soft-landing technique being studied by NASA Manned Spacecraft Center for future manned space vehicles. System utilizes retrorocket for deceleration and steerable paraglider device. Three previous landing tests were made on water.

- Dr. Marvin H. Gold, manager of Propellant Chemicals Div. of Aerojet-General's Solid Rocket Plant, was awarded Navy Meritorious Public Service Citation by ONR for his contributions to high energy rocket propulsion for advanced Polaris.

- NASA Ames Research Center established Space Sciences Div. to be headed by Dr. Charles P. Sonett. New division will conduct research in the areas of geophysics, interplanetary and planetary physics, planetary sciences, astronomy, and astrophysics.

DECEMBER 1962

December 1: Three sounding rockets were launched from NASA Wallops Station in series of GSFC experiments to study structure and composition of upper atmosphere and ionosphere:

Nike-Apache carried 65-lb. payload to 105-mi. altitude, payload designed to measure electron density and temperature and ion density and conductivity of the ionosphere.

Nike-Apache with 70-lb. payload containing pitot-static probe reached 82-mi. altitude, instrumentation designed to measure pressure, temperature, density, and winds in upper atmosphere.

Nike-Cajun with payload to measure winds and temperatures in upper atmosphere, payload consisting of 12 special explosive charges which were detonated at intervals from about 24 to 58 mi.

NASA Nike-Apache was also launched from Eglin Gulf Test Range, Fla., to 128-mi. altitude to measure winds by sodium vapor method.

- Medium-angle camera on TIROS VI meteorological satellite stopped transmitting pictures, during orbit 1,074, but satellite's wide-angle camera was still sending pictures of "excellent quality." Through orbit 1,073, the medium-angle, Tegea-lens camera took 12,337 cloud-cover pictures of which 11,131 or 90.2% were usable for weather analysis.

- President Kennedy was reported by *New York Times* to have requested Bureau of the Budget to look into possibility of obtaining extra funds for U.S. manned lunar flight program. An additional $300,000,000 to $400,000,000 was estimated requirement for fiscal year 1963. President's request came after White House meeting with NASA Administrator James E. Webb and NASA Director of Manned Space Flight D. Brainerd Holmes.

- Unnamed DOD spokesman said DOD was preparing a list of exceptions to its ban against disclosing information on military space shots. New list of exceptions would provide for information about both launch and subsequent performance of USAF, USN, and U.S. Army scientific research vehicles. Specifically included were X–20 (Dyna Soar) manned space vehicle; Project West Ford (orbiting metallic filaments); space projects to gain scientific data; Transit navigational satellite; Project Anna geodetic satellite; and U.S. Army scientific rocket launchings from White Sands Missile Range, N.M.

- Harold Berger, Associate Physicist at Argonne National Laboratory, said recent experiments with neutron beams had demonstrated they not only match penetrating power of x-rays but also reveal structural features which x-rays cannot expose. "It has been known virtually from the time the neutron was discovered 30 years ago that these particles might be used to make pictures through objects opaque to light," Berger said. "But neutron radiography could not be developed until sources of slow neutron beams became available. Such sources were provided by the development of the nuclear reactor and particle accelerator.

The value of neutrons is that their absorption characteristics are quite different from those of x-rays. . . .

"One of the fascinating things that has emerged concerning neutron radiography is that the same neutron beam used to inspect several inches of uranium or lead can be used to inspect specimens such as leaves, insects, and thinner biological specimens."

December 1: "Upper Mantle Project" to explore outermost 600 mi. of earth's solid-rock interior was reported by National Academy of Sciences' Geophysics Research Board. The $31 million international study would take three years.

• Aerospace Industries Association told designers and manufacturers that the "number one target in all technical fields" in coming decade would be improved reliability. "An entirely new methodology must be developed to obtain the necessary reliability of space missions," AIA report said.

December 2: MARINER II Venus probe was 27,293,368 mi. from earth and traveling 81,080 mph through space. Launched Aug. 27, the spacecraft had traveled 158 million mi. on its orbital path toward Venus.

• 20th anniversary of first controlled, self-sustaining nuclear reaction, achieved at Univ. of Chicago in 1942. AEC reported an estimated 518 reactors now in operation or under construction around the world, with 286 in U.S., 39 in U.S.S.R., 39 in U.K., 25 in France, 18 in West Germany, 14 in Italy, 11 in Japan, and 10 in Canada.

• Radiation research results reported by George M. Woodwell of Brookhaven National Laboratory to American Association for the Advancement of Science showed that plants, like animals, vary in sensitivity to radiation. Large plants that reproduce slowly, such as trees, appear more susceptible to radiation than less complex plants, such as algae and fungi. ". . . The hypothesis seems tenable that small organisms with wide ecological amplitudes and high rates of reproduction—in short, weeds and other organisms frequently considered pestiferous because of their persistence under persecution—have survival advantage under conditions of long-term exposure to ionizing radiation over large organisms with longer life cycles."

• Development of two-way portable communicator providing confidential conversations up to 10 mi. apart via infrared beams was announced by Raytheon Co. Pencil-thin light beam is so narrow that it is virtually impossible to intercept or tap, Raytheon said.

December 3: Successful ground tests of Automatic Picture Transmission (APT) subsystem reported by NASA. To be flight-tested on a Tiros satellite next year, APT would eventually enable weather stations to obtain local cloud-cover pictures directly from orbiting weather satellites, thus making possible a world-wide weather satellite service.

• Senator Albert Gore (D–Tenn.) told U.N. General Assembly's Political Committee that U.S. was seeking following goals in space policy:

"To be guided by the general principles already laid down by the United Nations for establishment of a regime of law in outer space, and to negotiate an extension of those principles by international agreement.

"To conclude a treaty banning immediately the testing of any more nuclear weapons in outer space.

"To preclude the placing in orbit of weapons of mass destruction.

"To take all reasonable and practicable steps, including consultation with the world's scientific community, to avoid space experiments with harmful effects.

"To conduct a program which is as open as our security needs will permit and as cooperative as others are willing to make it.

"To press forward with the establishment of an integrated global satellite communication system for commercial needs and a cooperative weather satellite system, both with broad international participation."

December 3: DOD announced it had canceled plans for immediate development of detector-interceptor satellite (unofficially called "Project Saint"), and that the project was being "re-oriented towards achieving longer-term" objectives. Official explanation was that Saint had been "bypassed by technological developments."

- NASA and Air Force Cambridge Research Laboratories launched Nike-Apache sounding rocket from Eglin Gulf Test Range, Fla. Intended to measure winds by sodium vapor method, rocket rose to 117-mi. altitude and emitted a fair vapor trail.

- NASA request for AEC study of nuclear power unit was reported. 40-watt units for NASA interplanetary monitoring probes would be launched at rate of two per year for four years, beginning in 1964.

- President Kennedy presented AEC's Enrico Fermi Award for 1962, carrying a $50,000 prize, to nuclear physicist Dr. Edward Teller, the citation reading: "For contributions to chemical and nuclear physics, for his leadership in the thermonuclear research and for efforts to strengthen national security."

- Temporary restraining order against Lockheed Aircraft Corp. and International Association of Machinists to prevent further IAM strike action was issued by U.S. District Judge Jesse Curtis, effective until Dec. 13. Hearings on Justice Dept. request for temporary injunction, based on report from President Kennedy's three-man board of inquiry, were set for Dec. 10.

- Editorial on tactical fighter (TFX) program, contract for which was awarded to General Dynamics and Grumman Aircraft by DOD, appeared in *Aviation Week and Space Technology:*

"... Underlying the whole TFX concept is one of the solid, basic technical explorations by the researchers of the old National Advisory Committee for Aeronautics (NACA) that did so much to keep this country the international leader in supersonic aircraft development. Without the fundamental research into the variable sweep wing and the detailed development of this principle by the Langley research laboratory group headed by John Stack, the current TFX concepts of both final competitors would have been impossible. ... The full story of the Langley contribution to the TFX program should be hammered home as an example of how these research and development investments eventually pay substantial dividends. ..."

December 4: USAF announced launch of unidentified satellite with Thor-Agena booster from Vandenberg AFB. This was 100th launch with Thor as space booster, 93 of which were considered successful. To date, Thor launched more satellites and probes into space than all other boosters combined, including PIONEER I space probe, EXPLORER VI scientific satellite, PIONEER V space probe, and TIROS I weather satellite as Thor-Able vehicle; and ECHO I balloon-satellite, OSO I, ARIEL, TELSTAR, and four Tiros satellites as Thor-Delta. Thor-Able-Star (Vanguard-derived upper stages) combination launched series of USN navigation satellites, Army communications satellites, and tri-service ANNA geodetic satellite. With Agena upper stage, Thor boosted series of USAF satellites as well as Canadian ALOUETTE satellite for NASA.

- NASA Goddard Space Flight Center launched two Nike-Cajun sounding rockets, one from Wallops Island, Va., and one from Ft. Churchill, Canada, for purpose of comparing data on wind and temperatures in upper atmosphere. Nike-Cajun from Ft. Churchill reached 69-mi. altitude and ejected 12 high-explosive grenades which detonated on schedule; ground instrumentation measured sound energy from exploding grenades and recorded rocket position during flight. Nike-Cajun from Wallops Island performed unsatisfactorily, reaching altitude of only 31.8 mi. instead of 69 mi. predicted; only two grenade explosions were observed.

- NASA announced award of contract to Army Corps of Engineers to define a research program necessary to give the U.S. capability to construct research station on the moon. D. Brainerd Holmes, Director of NASA Office of Manned Space Flight, said NASA does not have a program to establish a "manned lunar laboratory," but that initial studies "are required now which can lead to the existence of a lunar construction capability." Objectives of the six-month, $100,000 study: (1) define R&D effort required to provide U.S. lunar research capability; (2) define needed experimental facilities; and (3) prepare schedules and budgetary estimates for a lunar construction research program. Areas of research would include lunar soil characteristics; lunar soil movement and excavation techniques; construction materials; structural design; power generation; storing and handling of life-supporting atmosphere, water supply, and sanitation; construction tools; and human engineering and training.

- Dr. Victor A. Belaunde of Peru told U.N. General Assembly's Political Committee that it was "imperative that we have a coordinating authority and a finally decisive authority, with decisive jurisdiction" to regulate exploration of space. He said the jurisdiction must be that of the U.N.

- Soviet news agency Tass announced MARS I interplanetary probe was now more than 7 million mi. from earth; all onboard equipment was working normally and the spacecraft was sending back radio reports to earth. MARS I was expected to pass within 120,000 mi. of Mars in June 1963.

- Dr. Robert C. Seamans, Jr., NASA Associate Administrator, told Joint Computer Conference in Philadelphia: ". . . It is no exaggeration to say that we could never have succeeded in orbiting and tracking even the simplest satellite without the availability of computers. . . . And it is certainly true that one of the major pacing elements in our rate of progress toward achieving the national goals in space will be the development of progressively more advanced computers and computer techniques to help cope with the even greater tasks ahead.

". . . [Data processing] systems of the future, as we see them, will involve compatible combinations of computer and ancillary devices, some of which will be spaceborne, some stationed on celestial bodies other than the earth, and some of course will still be here on earth.

"Such computer systems will be required to perform navigational and attitude control tasks; edit and store scientific and engineering data; make calculations of a conventional nature; time sequences of operations; feed displays; and trigger alarms.

"A task for the future is to extend . . . [the computer's pre-launch checkout] function to spacecraft for launching from orbit, for landing on celestial bodies, and for take-off from remote planets for return to earth. . . . What we are looking for are systems which are fully automatic and fast. . . .

"It is apparent that in general a rather complete distribution range of computing burden will exist. In some missions, such as close-to-earth satellite, earthbased computing equipment will perform most of the processing required. Other missions, such as one bound for Pluto, will require the autonomous activity of a highly redundant, self-adaptive, self-checking computer complex aboard the spacecraft.

"A vital need, therefore, is determination of the optimum nature, disposition and use of computing equipment. . . .

"Eventually . . . we may have large numbers of vehicles simultaneously operating in space. . . . In such a situation we will need sophisticated and very comprehensive Flight Control Centers, with compatible computer systems aboard the spacecraft, to handle information on flight status, command control, navigation, and program sequences. . . ."

December 4: U.S. Army Nike-Hercules missile exploded over McGregor Missile Range, Tex., killing French officer and injuring four other persons. Range safety officer reportedly detonated the missile at 500-ft. altitude when it veered off course.

December 5: U.S. and U.S.S.R. announced in U.N. their bilateral agreement to cooperate in space exploration programs of weather observation, magnetic-field study, and satellite communications. In meteorology, agreement called for experimental phase extending through 1964 "during the development of experimental weather satellites" by both countries; second phase would begin approximately 1964–65 with coordinated launchings by both countries of weather satellite system for operational use. In magnetic survey, agreement called for U.S. and U.S.S.R. each to launch a satellite equipped with magnetometers during International Year of the Quiet Sun (IQSY), 1964–65. Both countries agreed to use the intervening years to continue their own magnetic measurement research and to exchange the data obtained. In satellite communications, agreement called for U.S. and U.S.S.R. to cooperate during coming year in communications experiments via U.S. passive communications satellite Echo A-12. Future negotiations would consider cooperative efforts in experimental system of active-repeater satellites. Agreement provided that "the results of these cooperative experiments would be made freely available to all interested states," U.S. Ambassador to the U.N. Adlai Stevenson said. Agreement was product of Geneva negotiations between U.S. team headed by Dr. Hugh L. Dryden, NASA Deputy Administrator, and Soviet team headed by Prof. Anatoli A. Blagonravov, conducted in spring of 1962.

December 5: Nike-Cajun launched from NASA Wallops Station carried electron-density and temperature instrumentation to 80-mi. altitude, as part of current series of upper-atmosphere studies.

- Sodium-vapor experiment launched with Nike-Apache vehicle from NASA Wallops Station was not successful because the payload did not perform properly. Flight was part of current upper-atmosphere studies being conducted by NASA Goddard Space Flight Center.

- Committee for International Year of the Quiet Sun (IQSY), meeting in London, approved worldwide research program to be conducted during 1964–65 period of minimum solar activity. Proposed program included multi-nation participation in such research areas as meteorology, geomagnetism, aurora, airglow, ionosphere, solar activity, cosmic rays, and aeronomy. Project was outgrowth of International Geophysical Year (IGY), 1957–58, considered highly successful both in scientific results and in international cooperation.

- John Dykstra, president of Ford Motor Co., told American Ordnance Association meeting that a great potential defense capability lay untapped in industry—particularly in heavy, mass-producing manufacturing. Asserting that the problem of developing and maintaining broad defense and space capability throughout industry had not been adequately realized or defined, Dykstra suggested reorganizing U.S. productive resources and putting "much more of our heavy manufacturing industry in a state of preparedness to switch readily from peacetime to wartime production. Such preparedness would mean, at a minimum, that heavy industry would be kept current in matters of defense technology in areas of natural interest to it. Such preparedness could also involve a broader distribution of space and defense research and development effort, as well as of production, on a systematic basis. . . .

  "Whatever form an Industrial National Guard might take, I believe the concept is essential. In view of the enormous demands now being made on all our resources of brainpower, we cannot afford to make less than the best possible economic and efficient use of them. . . ."

- Titan I ICBM was successfully fired 4,000 mi. down the PMR in test of recently modified ground support equipment. This was sixth Titan I launch from Vandenberg AFB.

- Rear Adm. Luis de Florez died at age 73; he had been instrumental in 1945 "Project Paperclip" to ensure German scientists and engineers would come to U.S.

- Nine Soviet cosmonauts were lost in space between February 1959 and October 1961, it was reported by Oton Ambroz of North American Newspaper Alliance (NANA). Article, relying heavily on European sources, claimed five cosmonauts were sent on orbital space flights from which they did not return; two cosmonauts, a man and woman, orbited in a single spacecraft May 17, 1961, but did not return; and two cosmonauts, man and woman, were sent on lunar flight Oct. 14, 1961. Conversations of the dual flights were reportedly monitored by Western receivers; signal from a single cosmonaut ("World—S.O.S.—S.O.S.") was received Nov. 28, 1960. Other flights were based on reception of physiological, tracking, and other signals. None of the flights was reported by U.S.S.R. and NASA had no comment on the report.

December 5: Upon returning to Sweden from Moscow, Swedish scientist Bjoern Malmgren reported that two Soviet cosmonauts were being trained in Swedish-built centrifuge for a trip around the moon and back. Malmgren reported Soviet space officials told him the venture was planned for late next year.

- USAF Atlas missile flight-test series was completed with successful Atlas F flight 5,000 mi. down AMR. Witnessing the launch was "father" of the Atlas—Karel J. Bossart, project engineer at GD/A since five-engine Atlas development was initiated in 1946. DOD canceled program in 1947 and subsequently redesigned the ICBM to three-engine configuration, first flight-tested June 11, 1957. Since then 151 Atlases were launched (105 of them from AFMTC) with 108 successes. Future Atlas missiles launched from Cape Canaveral would be involved in space-probe missions or in testing nose cones.

December 6: Nike-Cajun sounding rocket launched from NASA Wallops Station to measure high-altitude winds and temperatures, the payload consisting of 12 special explosive charges which were ejected and detonated at intervals from 24- to 57-mi. altitude.

- Roger Seydoux, French ambassador to U.N., announced that France planned to launch its first satellite by 1965, and that France would continue its space activities "at the rhythm required by the present day world."

- Titan II missile, fired from Cape Canaveral, failed to travel planned 8,000 mi. because of malfunction in second stage.

- Dr. Hugh L. Dryden, NASA Deputy Administrator, was among the five Federal employees presented with the distinguished Rockefeller Public Service Award for 1962 by President Kennedy.

- Dr. Ernst Stuhlinger, Director of Research Project Div. of NASA Marshall Space Flight Center, was awarded 1962 Hermann Oberth Award of the Alabama section, American Rocket Society.

- No funds for Skybolt missile development were included in Secretary of Defense McNamara's proposed FY 1964 budget, it was reported. DOD officially acknowledged it was giving the Skybolt project a "hard look." Each of Skybolt's five flight-tests had been unsuccessful.

- USN advanced Polaris A–3 missile, launched from Cape Canaveral, began twisting after second-stage ignition and had to be destroyed by range safety officer. Large section of the missile, still under power, escaped detonation and suddenly shifted off course, landing in ocean 300–350 mi. northeast of Cape Canaveral and about 150 mi. east of Savannah, Ga. Range safety procedures were under careful review as result of the errant missile's flight.

December 7: Analysis of data from MARINER II interplanetary probe revealed there are 10,000 times as many small dust particles near the earth as there are in interplanetary space, reported W. M. Alexander of NASA Goddard Space Flight Center. This was revision of earlier estimate of 1,000 times as many dust particles near earth as in interplanetary space. Other observations from MARINER II included in the report, which was published in *Science:* solar plasma experiment showed constant presence of solar plasma (or "solar wind"), generally ranging from 850,000 mph to 1,-550,000 mph; magnetometer experiment found "convincing evidence" that magnetic fields of at least a few gamma are nearly always present, except perhaps for occasional, transient lulls.

- MARINER II temperatures continued rising, JPL reported, but all systems aboard the Venus probe were functioning normally. MARINER II temperatures, planned for maximum 140° F, had reached 160° F.
- X–14A VTOL aircraft was being used in research for lunar-landing maneuvers at NASA Ames Research Center, it was reported. X–14A pilot Fred J. Drinkwater, III, said that a "great similarity exists between the flight control system of our [plane] and that proposed for a lunar craft. This makes the X–14A a logical vehicle to investigate problem areas connected with the let-down and landing phases of the lunar mission."
- USAF Minuteman missile was successfully test-fired 3,500 mi. down the AMR, in test "to evaluate the performance of an operational missile."
- British low-yield nuclear explosion made in Nevada, in cooperation with U.S., Britain's Atomic Energy Authority announed. Test was 23rd British nuclear explosion.

December 7–8: President Kennedy visited SAC bases and U.S. nuclear facilities at Los Alamos, N.M., and Jackass Flats, Nev., where he reviewed progress of AEC-NASA Rover nuclear rocket program.

December 8: U.S.S.R., U.S., and 13 other countries proposed that U.N. General Assembly call for urgent efforts to break deadlock over legal problems of space exploration. Compromise resolution was submitted for consideration in General Assembly's Political Committee, which was debating report of the U.N. Committee on the Peaceful Uses of Outer Space.

- Evidence suggesting plant life has existed on earth for 2.7 billion years (a billion years longer than previous estimates) was reported by Caryl P. Haskins, President, Carnegie Institution of Washington, in his annual report. Research by Thomas C. Hoering of Carnegie's geophysical laboratory indicates existence of one-celled algae in world's oldest known sedimentary rocks—Bulawayan limestone of Southern Rhodesia. Hoering isolated chemicals, mostly hydrocarbons, thought to be "the end products of cellular chemicals that have been subjected to reactions for more than two billion years and, as such, have lost their identity." Two of Hoering's colleagues, Philip H. Abelson and Patrick L. Parker, isolated more complex organic compounds known as fatty acids from rocks as old as 500 million years.

December 8: Secretary of Defense Robert J. McNamara announced Operation Dominic nuclear test series was "highly successful." Tests were conducted from April 25 to Nov. 3 by Joint Task Force 8.

- Another Atlas F squadron was declared operational at Walker AFB, Roswell, N.M.

December 9: Gen. Curtis LeMay, USAF Chief of Staff, was quoted as saying: "Eventually I see the development of a manned vehicle that can take off from existing runways, go into orbit, maneuver into a parking orbit, come out of orbit, maneuver while re-entering the earth's atmosphere, and land at any air base in the conventional manner."

And Lt. Gen. Roscoe C. Wilson, former USAF DCS/Development, described the futuristic spacecraft as "a winged, manned vehicle which could go from the earth into orbit and return, solely on its own power. . . . From our investigations, this concept of a future aerospace vehicle appears feasible, and we look forward to it as a follow-on to the Dyna Soar [X-20]."

December 10: NASA launched 186-lb. payload of scientific instruments for Commonwealth of Australia from NASA Wallops Station, but Aerobee 150A vehicle malfunctioned 42 sec. after liftoff and payload did not reach intended altitude. Payload instrumentation, designed to measure VLF radio waves in the ionosphere, functioned successfully throughout the flight.

- Senator Estes Kefauver (D–Tenn.) announced NASA proposed patent regulations would "flout the clear legislative intent underlying the NASA Act of 1958." Urging NASA not to adopt the regulations he contended they would impede the space program, increase monopoly "in an already concentrated industry," and give private industry the fruits of research financed publicly.

Franz Olson, of Aerospace Industries Association, said at the NASA hearing that proposed patent waiver regulations would lead to an increase in inventions through "the proven incentives of the patent system."

Senator Russell B. Long (D–La.) charged the patent proposals would give "patent monopolies" to the industry on developments resulting from publicly financed research. Senator Long said Monopoly Subcommittee of the Senate Small Business Committee would conduct hearings on the proposed regulations in January.

- Incorporators of communications satellite corp., authorized by the Communications Satellite Act of 1962, decided temporary headquarters for the new company would be established in Washington, D.C. Also at the meeting held in N.Y.C., the incorporators discussed name for the corp.; considered names of 50 persons for its management; and discussed budgetary requirements. Next meeting of group was set for Dec. 21, in N.Y.C.

- National Science Foundation announced it would grant $2.5 million for 74 institutes of college teachers next summer. Consisting of about 30 teachers each, the institutes would cover subject matter ranging from history and philosophy of science and mathematics to various aspects of engineering.

December 10: Aviation Week and Space Technology reported NASA was considering substituting piggyback satellite for Agena D target in Project Gemini rendezvous maneuvers. Dr. Joseph F. Shea, NASA Deputy Director of Manned Space Flight Systems, was quoted as saying NASA could save both time and money by such substitution. USAF plans for Blue Gemini project were not yet firm, pending NASA's decision. USAF hoped to buy four of NASA's 12 two-man Gemini capsules for use in its own manned space flight training using Agena D as rendezvous target for two-man Gemini crew. DOD had not yet approved any Blue Gemini funds for FY 1964.

- USAF Minuteman ICBM exploded within seconds after launch attempt from Vandenberg AFB, Calif.
- U.S. Army Pershing missile met all test objectives in flight from Cape Canaveral with pre-programed zig-zag maneuvers designed to test missile guidance system's ability to keep the Pershing on course.
- NASA Flight Research Center announced modifications to X–15 (No. 1) had been completed, the modifications providing window and camera in fuselage for Follow-On Program of high-altitude research. Camera with window would be used to study optical degradation resulting from observations through hypersonic boundary layers and shock waves. This project would begin in early 1963 and would include first five flights of the 35 flights planned in X–15 Follow-On-Program, a two-year program sponsored by NASA and USAF.

December 11: Secretary of Defense Robert S. McNamara met in London with British Defense Minister Peter Thorneycroft to discuss future of Skybolt air-launched missile, among other defense matters. DOD was considering canceling Skybolt development because it would be very costly to develop, might not be successful by proposed operational date of 1965, and was thought no longer needed in view of Minuteman and Polaris successes. U.S.-U.K. agreement of 1960 had provided for joint participation in Skybolt project.

- First USAF Minuteman squadron was declared operational, with 20 of the ICBM's placed in underground silos at Malmstrom AFB, Mont.

December 12: President Kennedy, back in Washington after inspection of defense and nuclear-rocket facilities, said:

"We are going to let these test go on of the [Project Rover] reactor. These tests should be completed by July. If they are successful, then we will put more money into the program, which would involve the Nerva and Rift, both the engine and the regular machine. We will wait until July, however, to see if these tests are successful.

"It should be understood that the nuclear rocket will not come into play until 1970 or '71. It would be useful for further trips to the moon or trips to Mars. But we have a good many areas competing for our available space dollars, and we have to try to channel it into those programs which will bring us a result, first, on our moon landing, and then to consider Mars."

- Unidentified payload launched by USAF with Thor-Agena vehicle from Vandenburg AFB.

December 12: Two tandem 600-ft.-tall research balloons launched from Palestine, Texas, with dummy payload of 6,300 lb., a development test for February 1963 launch of 36-in. telescope with germanium eye to obtain undistorted look at the Martian atmosphere. Scientists from Princeton University, Univ. of California, Vitro Corp., and Schjeldahl Corp. conducted test flight.

- DOD announced Nike-Zeus antimissile missile successfully intercepted Atlas ICBM over the Pacific Ocean. Target Atlas was launched from Vandenberg AFB, Calif., and two Nike-Zeus missiles were fired from Kwajalein Island, the first Nike-Zeus making the successful intercept. This was second successful intercept-test by Army's Nike-Zeus.
- NASA Agena B vehicle program would be transferred from NASA Marshall Space Flight Center to NASA Lewis Research Center, Associate Administrator Robert C. Seamans, Jr., announced. Transfer included Atlas and Thor boosters used with Agena B upper stage. Dr. Seamans said the transfer, which would be completed within three months, would allow "Marshall to wholly concentrate its work on the vital Saturn vehicle development for the manned lunar landing program and for the large unmanned scientific payloads of the future. . . . In addition, it concentrates at Lewis the two Atlas-based vehicles, Agena and Centaur, which are essential to NASA's program of space sciences." Centaur was transferred to Lewis in September.
- "At least 15 per cent of the original synchrotron radiation" created by U.S. high-altitude nuclear blast last July could persist for more than a year, American and Peruvian scientists reported. Measurements made at National Bureau of Standards radio astronomy observatory in Jicamarca, Peru, showed that about one half the radiation had decayed within two months of the explosion, but that the rate of decay was "thought to be decreasing."
- USAF SAC crew launched Atlas ICBM from Vandenberg AFB in routine training exercise.

December 13: RELAY communications satellite launched into orbit by Thor-Delta vehicle from Cape Canaveral (about 800-mi. perigee; 4,600-mi. apogee; 3 hr., 5 min. period). Built by RCA for NASA, RELAY was designed to be first active repeater satellite linking three continents—North America, South America, and Europe. Efforts to turn on RELAY's communications equipment (by NASA test station at Nutley, N.J.) were unsuccessful, the satellite's onboard battery power being too low to operate the transponders. NASA said telemetry data indicated "abnormal drain upon the power supply" was probable cause of low voltage; RELAY communications experiments were postponed indefinitely. Objectives of NASA Project Relay were to test intercontinental microwave communications by low-altitude active repeater satellites, measure energy levels of space radiation in its orbital path, and determine extent of radiation damage to solar cells and electronic components.

- NASA and Canadian Government launched two Black Brant III sounding rockets from NASA Wallops Station, in joint U.S.-Canadian experimental test series to determine vehicle flight performance characteristics and to obtain engineering data on effectiveness of instrumentation. In each flight, 100-lb. payload carried a cosmic ray sensor for measuring altitude, roll-rate magnetometer, and new telemetry transmitter and related antennae. Both vehicles reached 61-mi. altitude and both flights were considered successful.

December 13: Attempts to launch scientific experiment from NASA Wallops Station failed when malfunction occurred in launch vehicle and caused "some damage to the launch tower and building." Experiment was to have determined distribution of certain molecular and atomic species in the upper atmosphere.

- USAF would undertake "development program to advance the technology of large solid propellant rocket motors" under recent NASA-DOD agreement, DOD announced. Program was designed to "keep open the possibility of expedited development" in case large solid-propellant motors should be required for future space boosters. Principal effort would be devoted to providing techniques for and proving feasibility of manufacturing, handling, transporting, and firing motors measuring 260-in. diameter and capable of about six million lbs. thrust.

- Howard Simons, science writer for the *Washington Post*, was named winner of 1962 newspaper science writing award by AAAS. Award was based on Simons' article on structure of matter, entitled "There's Dichotomy Among Neutrinos," appearing in the *Post* July 8, 1962.

December 14: MARINER II passed within 21,600 mi. of planet Venus and made 42-min. instrument-scan of Venutian atmosphere and surface before continuing into perpetual orbit of the sun. In world's first close interplanetary contact, MARINER II measured and returned temperatures and other characteristics of Venutian surface and its atmosphere; telemetry signals would be analyzed and evaluated by JPL, technical manager of interplanetary programs for NASA. In press conference at NASA Headquarters, JPL Director Dr. William H. Pickering said all instruments onboard MARINER II functioned as planned and the mission was virtually a complete success. NASA Administrator James E. Webb hailed the achievement as "an outstanding first in space for this country and for the Free World." NASA Director of Space Sciences Dr. Homer E. Newell pointed out that the probe's internal temperatures ranging from 125° F to 200° F had not prevented the scientific experiments from succeeding; temperatures were expected to continue rising as the spacecraft traveled closer to the sun. JPL would continue tracking the probe as long as possible, even though its scientific mission was considered completed.

Dr. Newell outlined significant interplanetary data already obtained from the four scientific instruments onboard MARINER II. ". . . The plasma probe revealed a steady 'solar wind' at 250–450 miles per second. The magnetometer showed that space contains fields of at least a few gamma and that there are fluctuations by factors of as much as 5 to 10. . . . The cosmic dust detector indicated that the meteoritic particles in space are less numerous than near Earth by a factor of 10,000. . . ." When MARINER II neared Venus, two additional experiments were activated—microwave and infrared radiometers to measure temperatures and identify their sources (surface or atmosphere).

December 14: X–15 No. 3 flown by Major Robert White (USAF) in stability test at speed exceeding mach 5, with lower tail-fin absent and with nose raised to 25° angle above the horizon. Successful flight was the 75th by the rocket research airplane.

- Ground signal station at Nutley, N.J., turned on RELAY satellite's communications transponders and sent signals to the satellite; telemetry indicated RELAY received signals, but RELAY failed to return transmission to Nutley, Andover (Me.), or Pleumeur-Bodou (France) stations. All onboard systems were turned off except command receivers.

- Unidentified USAF payload launched with Thor-Agena vehicle from Vandenberg AFB.

- Soviet news agency Tass distributed announcement by U.S.S.R. Academy of Science giving new details on MARS I interplanetary probe. Announcement said scientists had held 37 radio communications with MARS I during first month of the probe's journey and that more than 600 orders had been transmitted to it. Command, measuring, and computing centers onboard the spacecraft were reported working properly. Announcement said MARS I would pass planet Mars at distance of about 119,000 mi. (193,000 km) and added that its trajectory would have to be corrected sometime during the flight. Data from probe indicated space radiation had increased about 50 to 70% since 1959 Soviet lunar probes made measurements; other data indicated extremely low density of meteor matter at great distances from earth. Announcement said MARS I onboard equipment included television to photograph Martian surface; spectrograph to study ozone absorptions in Martian atmosphere; equipment to measure magnetic fields and radiation in space and around Mars; and radio-telescope to register streams of low-energy protons and electrons.

- Project Stargazer balloon landed after 18½-hr. trip to 82,000-ft. altitude, in southwestern New Mexico, by Capt. Joseph A. Kittinger, Jr. (USAF), and William C. White, astronomer from U.S. Naval Test Station at China Lake. White had clearest view of heavens of any astronomer in history by using telescope mounted on top of gondola. J. Allen Hynek, director of USAF OAR's Project Stargazer, predicted great future for balloon astronomy.

- General Assembly of the United Nations voted unanimously to approve resolution submitted by the U.S., U.S.S.R., and 24 other nations, calling for continued scientific cooperation directed toward using space to improve weather forecasting and intercontinental communications system. The U.N. Assembly also approved continuation of the U.N. Committee on the Peaceful Uses of Outer Space.

- Scientists requested Continental Air Defense Command to shoot down tandem pair of runaway test balloons launched from Palestine, Texas, on December 12th. 600-ft.-long balloons, which had drifted eastward across the Gulf of Mexico were considered a hazard to the airways.

- USAF Minuteman ICBM launched from silo at Cape Canaveral, its re-entry package landing approximately 5,000 mi. down the AMR.

December 15: Joint U.S.-Norwegian-Danish Nike-Cajun sounding
rocket launched from Andoeya, Norway, with instrumented
experiment to probe the ionosphere and the Northern Lights.
Third in a series, the sounding rocket reached an altitude of near
68 mi.

- JPL scientists said MARINER II's earth-seeking sensor had unaccount-
ably gained strength since the probe passed Venus Dec. 14,
enabling Goldstone Tracking Station to receive signals from the
probe for perhaps another month—or distance of nearly 80
million mi. from earth. Internal temperatures close to boiling
point of water were reported from the probe.

- Power supply onboard RELAY communications satellite remained too
low to operate the satellite's instrumentation properly.

- Two-month Project Firefly series of 27 upper atmosphere chemical
releases was completed by USAF Cambridge Research Laboratories,
ending four-point basic research program aimed at learning
more about the ionosphere.

- Article in *Komsomolskaya Pravda* stated that one of MARINER II's
main missions was to learn temperatures of Venutian atmosphere
and surface but that Soviet radioastronomers already had
answered these questions. Experiments by Pulkovo Observatory
determined that Venutian surface was between 300° C and 400°
C, and Venutian atmosphere was within limits of 0° C to 100°
C and therefore, that Venus should be defined as a "red hot
planet."

December 16: EXPLORER XVI (S–55b) launched into orbit by four-stage
Scout vehicle from NASA Wallops Station (initial orbital data:
apogee, 733 mi.; perigee, 466 mi.; period, 104 min.; inclination,
52° to the equator), began measuring micrometeoroids in space.
Satellite was designed to measure micrometeoroid puncture hazards directly by means of samples of spacecraft structural sur-
faces, measure particles having different momentums; and com-
pare the performance of protected and unprotected solar cells.
Cylindrical satellite was 24-in. in diameter and 76-in. in length;
total weight in orbit was 222 lbs., including Scout fourth-stage
motor case which was an integral part of the satellite assembly.

- RELAY satellite's 136-mc beacon was detected by tracking stations
at Santiago, Johannesburg, and Woomera, indicating the beacon
spontaneously turned itself on.

- MARINER II would orbit the sun every 345.9 days, JPL scientists
announced. The spacecraft would make its closest approach to
the sun Dec. 28, at distance of 65,505,935 mi., and would be
farthest from sun June 19, 1963, at 113,813,087 mi. MARINER
II's closest approach to earth would come Sept. 27, 1963, when
it would be 25,765,717 mi. away. All systems aboard the spacecraft were still functioning normally, with temperatures and other
data still being telemetered to earth from more than 37 million
mi. away.

- Radio Moscow quoted Mikhail Yarov-Yarovoy, senior scientist at
U.S.S.R. Astronomical Institute, as saying:
"We learn today that America's space vehicle, the Mariner II,
has passed quite close to Venus. The successful launching and
the reception of radio signals from such a great distance indicate

that American scientists and engineers have achieved more and more in conquering space.

"We congratulate our American colleagues."

December 16: NASA Goddard Space Flight Center scientist Dr. Michael E. Lipschutz discovered diamonds in meteorite that fell in India 90 years ago, the findings published in *Science*. Dr. Lipschutz, an astrochemist, is first lieutenant in U.S. Army on active duty with NASA. His research was made with minute portion of the Dyalpur meteorite, which fell to earth May 8, 1872. Dr. Lipschutz observed two sizes of diamond crystallites in the meteorite—a few large crystals and many small crystals. He concluded that diamonds in Dyalpur meteorite were formed by shock when meteorite's parent body collided with another body in space.

December 17: Although RELAY power supply remained low, Nutley, N.J., ground station was able to obtain about 10 min. of usable telemetry data from the NASA satellite. Engineers were conducting intensive analysis of telemetry data in effort to diagnose cause of RELAY's low battery voltage.

- Walter C. Williams, Associate Director of NASA Manned Spacecraft Center, said "I can't guarantee we're going to be first [to land a man on the moon], but if we don't try, we're going to be last. Either we have to spend more money on space . . . or change our goals." Williams was speaking at dinner commemorating 59th anniversary of Wright brothers' flight, in Beverly Hills, Calif.

- John Stack, former NASA Director of Aeronautical Research and NACA aerodynamicist, was awarded Wright Brothers Memorial Trophy—sterling silver model of Wright Brothers' Kitty Hawk airplane—at Wright Memorial Dinner, Washington. Earlier in the day, ceremonies at Kill Devil Hills near Kitty Hawk, N.C., commemorated 59th anniversary of Wright brothers' historic flight.

- USAF launched Atlas-Agena vehicle with undisclosed payload from Pt. Arguello, the vehicle failing in flight.

- U.S. Army fired Nike-Zeus antimissile missile in successful development test at White Sands Missile Range, N.M.

December 18: USN TRANSIT V-A navigational satellite launched from Point Arguello, Calif., into near-perfect polar orbit (apogee, 395 nautical mi.; perigee, 375 nautical mi.; period, 99.06 min.; inclination, 90.7° to the equator), but satellite's radio receiver failed to function in first five attempts to transmit data to the satellite. TRANSIT V-A was to have been the first operational satellite in system of four satellites, but now it would be useful as test vehicle rather than operational satellite. This was first time in Transit series that the radio command system had failed to work. The satellite was launched into orbit by four-stage Blue Scout rocket by USAF for USN.

- Spontaneous transmission of RELAY I 136-mc beacon signal has been observed at minitrack stations, NASA Relay Project Officer J. Russell Burke reported.

- NASA Administrator James E. Webb, at press conference and luncheon announcing Chicago and Midwest Space Month, said: " . . . I would like to stress one of the most challenging opportunities of the Space Age for American business—and for the consumer. In designing, building, testing, and flying space hardware we are searching for and creating new materials, lubricants, manufacturing processes, and techniques. We are stressing miniaturization, quality control, and foolproof automatic operation over long periods of time. We hope that many of these new products and new techniques can be put to good use on earth as well as in space and that the benefits they yield for business and the American consumer will eventually pay a substantial portion of the cost of the space program. We do not know what new products or uses may develop from space activities of the next few years. But we are hoping that business-men throughout the country will be watching closely for profitable possibilities. And this is one way in which close contacts between business and the scientific community will pay off. The university can act as a transmission belt to make this accumulation of new technology available to industry . . . . "

December 18: USAF launched unidentified Blue Scout space probe from Point Arguello, Calif.

- USAF Atlas ICBM exploded shortly after launch by SAC crew at Vandenberg AFB on training mission. USAF spokesman said no one was injured and no facilities were damaged.

December 19: NASA announced Ranger 6 lunar spacecraft would not be launched but would, instead, be subjected to "an exhaustive test program . . . intended to achieve the high reliability required for Ranger lunar missions." Improvements resulting from test program would be incorporated in Rangers 7–9, launching of which would be delayed "several months." Revised schedule was based on recommendations by Board of Inquiry headed by Cdr. Albert J. Kelley (USN), Director of Electronics and Control in NASA Office of Advanced Research and Technology. After intensive one-month study of Ranger system Board concluded that "certain improvements could be made in Ranger spacecraft design, construction, systems test and checkout which could contribute to increased flight reliability." Kelley board, appointed by NASA Space Sciences Director Dr. Homer E. Newell after RANGER V flight, consisted of officials from NASA Hq., five NASA field installations, and Bellcom.

- U.S. Weather Bureau announced development of infrared spectrometer, to be flight-tested in new balloons during next six months. The 100-lb. "flying thermometer" was planned for use in Nimbus weather satellites.

- NASA Flight Research Center announced it had received A–5A (A3J) aircraft from USN for use in its supersonic transport research. The "Vigilante" would be used primarily in study of problems in terminal area of air-traffic control operations. Working closely with FAA officials, FRC engineers would plan and conduct supersonic flights of Vigilante on assigned Federal airways and into high-air-traffic-density areas; the flights would provide basis for formulation of control plans necessary for safe operation of future commercial supersonic aircraft.

- Titan II ICBM fired from Cape Canaveral in successful 5,000-mi. flight down AMR, DOD announced. R&D flight was "designed to further test the missile's propulsion and guidance system in flight," announcement said.

December 19: Polaris A–2 missile, modified to carry guidance system for advanced A–3 model, was launched on successful 1,500-mi. test flight from Cape Canaveral.

- All three stages of Army's Nike-Zeus antimissile missile fired successfully in test from Point Mugu, Calif., last of Nike-Zeus flight series from that site.
- Final USAF Atlas ICBM squadron declared operational at Plattsburgh AFB, N.Y., bringing total operational Atlas force to 123. Additional composite squadron of Atlases was being maintained at Vandenberg AFB, partly for R&D and partly for training use.

December 19–21: President Kennedy and Prime Minister Harold Macmillan, meeting in Nassau, reached defense agreement designating Polaris missile to replace the Skybolt air-launched missiles as Britain's primary nuclear deterrent weapon. Under terms also to be offered to France, Britain and France would receive Polaris missiles from U.S. and eventually these would form the backbone of a NATO nuclear force.

December 20: X–15 (No. 3) piloted by Joseph A. Walker (NASA) reached 3,886 mph and 157,000-ft. altitude in test of re-entry stability without the craft's ventral fin. Walker said high climbing speed (186 mph faster than planned) was attained because sun's glare temporarily prevented his reading instrument panel; thus, re-entry after engine shut-off was faster than planned. Test was seventh in series without X–15's ventral fin.

- Navy announced that TRANSIT V–A navigation satellite transmitter was now dead, which indicated that its power supply had failed completely. TRANSIT V had previously been unable to receive commands from the ground.
- Management of Project Anna geodetic satellite program was transferred from DOD to NASA. ANNA I–B, launched Oct. 31, would remain a responsibility of DOD, with NASA scientific direction, but further launchings would be planned and executed by NASA.
- DOD announced plans to support NASA Syncom communications satellite operations in early 1963. U.S. Army Satellite Communications Agency (SATCOM), Ft. Monmouth, N.J., would send signals to activate and test the first U.S. synchronous-orbit satellite.
- National Science Foundation announced FY 1963 expenditures for Government-sponsored research and development would total $14.7 billion, 31% higher than last fiscal year. 95% of the estimated total would be spent by four agencies—NASA, DOD, AEC, HEW.
- Eugene W. Wasielewski, Associate Director of NASA Goddard Space Flight Center, told National Rocket Club that ARIEL (U.K.–U.S.) satellite was "just beginning to show signs of trouble." Launched April 26, 1962, the satellite was designed for one year of transmitting life. Wasielewski said some of the experiments aboard the satellite were "not doing as well as some others," that the radio signals from some of the experiments were fading.
- Six cash awards totaling $12,000 were presented by NASA Deputy Administrator Dr. Hugh L. Dryden to 14 employees of NASA Manned Spacecraft Center and NASA Langley Research Center: $4,200 for design of Mercury-type spacecraft to Maxime Faget, Andre J. Meyer, Jr., R. G. Chilton, Jerome B. Hammack, and

C. C. Johnson, all of MSC; W. S. Blanchard, Jr., of LARC; and A. B. Kehlet, formerly of LARC and now in private industry.

$1,500 for development of emergency safety system for manned spacecraft, used in Project Mercury as Mercury escape tower, awarded to Faget and Meyer.

$1,000 for invention of vehicle parachute and equipment jettison system, also incorporated in Project Mercury, awarded to Meyer.

$2,100 for design of contour couch used in manned spacecraft, awarded to Faget, William M. Bland, Jr., and Jack Heberlig, of MSC.

$2,000 for invention of ablation-rate meter awarded to Emedio M. Bracalente and Ferdinand C. Woolson, of LARC. Rate meter had been used to gather data from three small rockets and one four-stage Scout vehicle, all launched from NASA Wallops Station.

$1,200 to George P. Wood, Arlen F. Carter, and Dr. Adolph Busemann of LARC for development of plasma accelerator, considered a milestone in progress toward electrical propulsion of space vehicles and large-scale laboratory simulation of hypersonic and re-entry flight. This award was made under Employees Incentive Awards Act of 1954; other five were made under National Aeronautics and Space Act of 1958.

December 20: USAF Minuteman ICBM launched from Cape Canaveral fell short of its intended range, but "many of the test objectives were achieved," DOD announced. Causes of malfunction would be determined by study of telemetry data.

- Washington *Evening Star* reported responsible officials credited U.S. with 3-to-1 nuclear missile-power advantage over U.S.S.R. The commanding lead, said to be maintainable for decades to come, was chiefly due to new operational status of Minuteman ICBM.

- University of Alabama Research Institute dedicated at Huntsville, Ala.

- Claim of two Soviet scientists of Turkmen Medical Institute that they had discovered micro-organisms of extraterrestrial origin in the Sikhote-Alinst meteorite was disputed by U.S.S.R. Academicians V. Fesenkov, A. Imshenetskiy, and A. Oparin, *Izvestia* reported. Investigation by U.S.S.R. Academy of Sciences' Institute of Microbiology, of which Imshenetskiy was Director, showed that the claim had been based on incorrect premises and poorly conducted research.

December 21: Governments of U.S. and Canada jointly announced cooperative venture to build data acquisition station for Nimbus meteorological satellite at Ingomish, Nova Scotia. Successor to Tiros weather satellite program, Nimbus satellites will record meteorological data, including TV pictures, and relay the data back to earth on command; they will travel in near-polar orbit and thus be able to "see" the entire earth every 24 hours. Ingomish station would be staffed mainly with Canadian personnel and would be completed in 1964.

- Full-size test boilerplate command module of Project Apollo spacecraft was delivered by North American Aviation's Space and Information Div. to Northrop Corp's Ventura Div., which would equip spacecraft with earth-landing system in preparation for drop tests early next year.

December 21: NASA announced selection of 88 colleges and universities to receive graduate training grants for academic year 1963–64. Estimated 800 students in space-related science and engineering would participate in the program, established by NASA to "help achieve the long range objectives of the national space program and meet the Nation's future needs for highly trained scientists and engineers." Project began in 1962 with approximately 100 students receiving graduate training under grants to 10 universities.

- Spokesman said that the comsat corporation created by an act of the Congress would be named the Space Communications Corporation by decision of the board of incorporators. The board also decided to lease Tregaron, estate of Mrs. Herbert May and her former husband, the late Joseph E. Davies, as temporary headquarters in Washington, D.C.
- USAF announced award of $30 million contract to United Technology Corp. for design, development, fabrication, delivery and flight test of large segmented solid-propellant motors.

December 22: COSMOS XII satellite placed in orbit by U.S.S.R. Reported orbital data: 270-mi. apogee, 126.6-mi. perigee, 90.45-min. period, 65° inclination to the equator.

- Revealed that world's first attempt to flight-test an ion rocket engine was made on USAF Scout launching from PMR during the past week (perhaps 12/18/62), according to the *Washington Post*. Malfunction limited value of the test but launching marked beginning of new phase of space flight. Electric propulsion, with low thrust over long periods of time and of light weight, can propel large payloads, once boosted into space, for long distances at greater speeds than heavier chemical rockets with heavy fuels.
- President Kennedy's approval of $5.7 billion budget for NASA in FY 1964 was reported. Of the total nearly $4.5 billion would be for manned flight.
- Soviet Ambassador to India Ivan A. Benediktov announced in Bombay that U.S.S.R. would soon launch a manned spaceship weighing "several dozen tons." No further details were announced.
- Soviet government organ *Izvestia* stated that the Anglo-American agreement concluded at Nassau "turns out that they [U.S. and U.K.] want to drag atomic weapons into NATO in the form of Polaris rockets . . ."
- U.S.S.R. detonated atmospheric nuclear explosion in Novaya Zemlya area, the 36th reported by AEC in current test series. Force of the explosion was in intermediate range.
- Air-launched Skybolt missile was successfully test-flown near Cape Canaveral, only successful flight in six attempts and last test scheduled in Skybolt program. Acting Secretary of Defense Roswell F. Gilpatric stated that the successful test of Skybolt had not changed DOD decision to discontinue the program: "Today's single test did not conclusively demonstrate the capacity of the missile to achieve the target accuracy for which the Skybolt system was designed . . ."
- USAF announced "routine training launch" of Atlas ICBM from Vandenberg AFB by the 576th Strategic Missile Squadron.
- U.S. Army's Nike-Zeus successfully intercepted ICBM target in first test involving decoys, DOD announced. Atlas ICBM with its

decoys was launched from Vandenberg AFB toward Kwajalein Island in the Pacific, from which two Nike-Zeus were launched; one of the antimissile missiles had to be destroyed after launching, but the other successfully completed the mission.

December 23: NASA announced $240,000 grant to Smithsonian Astrophysical Observatory for setting up 16-station "prairie network" with automatic cameras to photograph bright meteors over seven midwestern states and to enable prompt recovery of meteorites. Project would enable scientists to study chemical and organic structures of meteorites before contamination on the surface of the earth.

• Pope John XXIII, addressing diplomats gathered in Consistorial Hall of the Vatican, called upon men of all nations to join in cooperative and peaceful exploration of outer space. "The church applauds man's growing mastery over the forces of nature. Thanks to those men who harbor thoughts of peace, mankind could dedicate itself, in noble rivalry, not only to the great economic and social tasks which confront it but to the continuing exploration of space and to the bold achievements of modern technology . . ."

December 24: Edgar M. Cortright, NASA Deputy Director of Space Sciences, said in press interview that stringent rules on sterilization of spacecraft components constituted one of the greatest problems in Project Ranger. NASA policy to prevent contamination of the moon by terrestrial germs did not require sterilization "to the point of jeopardizing reliability," Cortright said, but rejection rate on many spacecraft components became much higher as result of the policy; some components would not operate properly after being subjected to sterilization processes (including alcohol, ethylene oxide, ultraviolet rays, and heat). Relaxation of strict sterilization rules for Ranger spacecraft was one of the changes being considered in "re-orientation" of Project Ranger, Cortright indicated.

• Japanese Meteorological Observatory reported abnormal atmospheric pressures from a Soviet nuclear blast, apparently one of two recorded also by Uppsala University in Sweden. Japanese agency estimated the strength of the blast at 20 megatons. It said the abnormal pressure lasted from 40 to 100 minutes. Two blasts registered at the seismological institute in Sweden were listed at 19 megatons and 8 megatons. The geodetic institution of the Technological High School at Stockholm, Sweden, registered another gravitational disturbance which may have been a third Russian nuclear test, one which had a force of about 10 megatons. During the past 10 days, the U.S. Atomic Energy Commission had announced that the Russians have conducted at least four nuclear tests.

December 26: At AAAS convention in Philadelphia, Dr. P. J. Coleman, UCLA scientist, reported on the findings of MARINER II in the vicinity of the planet Venus. Coleman reported on the magnetometer experiment which showed no rise in the average value of the magnetic field above the "interplanetary value" during the fly-by of Venus: "The sensitivity, or lower limit, of the field's change that could be observed on the magnetometer was five gamma . . . During the encounter, no changes were observed of this magnitude—five gamma—which might be attributed to Venus." Observations thus far, he said, suggest that planets and satellites such as the moon that rotate slower than the earth have small magnetic fields.

December 26: Dr. Homer E. Newell, NASA Director of Space Sciences, urged in speech before AAAS that scientists be among the next group of astronauts selected for training. "I have a complete and utter conviction that we should take a scientist and make a flyer out of him rather than the other way around."

In interview, Dr. Newell outlined tentative NASA plans for 1963 launchings: 6–8 communications satellites, three weather satellites; 8 scientific satellites; at least one manned orbital flight; initial unmanned flight-test of Gemini capsule; three lunar impacts; Apollo capsule boilerplate tests of emergency flight conditions; three Saturn launch vehicle flight-tests; two electrical rocket-engine flight-tests; three re-entry heating tests.

- FERRET super-sensitive robot-inspector satellite developed by Lockheed Aircraft and RCA for U.S., reported in orbit by *Newsweek.* FERRET was reported capable of tapping microwave telephone messages and pinpointing missile launching sites by their radio guidance signals. *Newsweek* said satellite was aimed particularly at Baikonur, major Soviet rocket-testing base.

- NASA Marshall Space Flight Center's selection of Lockheed Missiles and Space Co. to establish method of assessing hazard potential of operational nuclear vehicle (Saturn C–5/Nerva) was reported. Three-phase study would include: (1) study of influence of impact delay time on hazard magnitude at impact in case of flight failure; (2) study of effect of vehicle trajectory on flight hazards; and (3) study to produce integrated hazards evaluation technique.

- In interview with *Data* magazine, Director of NASA Manned Spacecraft Center, Dr. Robert R. Gilruth, said that Houston Chamber of Commerce had predicted that the "arrival of the Manned Spacecraft Center will influence the economic growth of the area to a degree similar to that associated with the opening of the ship channel here almost 50 years ago. Community leaders are enthusiastic concerning the diversion of Houston's growth from an almost total reliance on the petrochemical industry. Of course, you must also remember that Houston, in terms of population increase, is one of the most dynamic cities of America. The Manned Spacecraft Center represents only a segment of its growth. . . .

"Undoubtedly, there has been some immediate impact on the economy since MSC's arrival in the Houston area. More than 80 aerospace companies have opened local offices here. . . ."

December 27: Establishment of joint AEC-DOD-NASA program for development of space nuclear reactor was announced by the three agencies. Col. Elwood M. Douthett (USAF) would head program office, responsible for research, technology, and component development phases of nuclear-electric power unit Snap–50/Spur. A 300–1,000 kw power unit capable of 10,000-hr. unattended operation, Snap–50/Spur would be unification of two existing projects: AEC's Snap (system for nuclear auxiliary power) and USAF's Spur (space power unit reactor). Unit could be used as power source for "space missions, communications satellites, and other space applications."

December 27: MARINER II calculated to reach its closest point to the sun (65,505,935-mile perihelion) according to JPL scientists.

• Alexander V. Topchiev, vice president of the Soviet Academy of Sciences, died in Moscow. Winner of the 1949 Stalin Prize and twice winner of the Order of Lenin, Topchiev was believed to have played a major role in the development of new rocket fuels and had attended many important international conferences since 1960. He had recently declared that "war is no longer possible" and had advocated East-West collaboration on nuclear space propulsion.

• Goddard Space Flight Center officials reported that they had received "indications the voltage had recovered somewhat" in RELAY satellite, but doubted that its communications mission would be accomplished.

December 27-29: Second Western Conference of the American Geophysical Union at Stanford, Calif. Sidney M. Serebreny of SRI reviewed TIROS I photographs on 14 orbits in May 1960, which showed a blocking pattern in east-central Pacific accompanied by an invasion of tropical air aloft into northerly latitudes and rationale for positioning of the jet stream over the northern Pacific. Life history of tropical cyclones was analyzed from Tiros photographs by James C. Sadler of NSF. And Richard D. Tarble of U.S. Weather Bureau indicated that TIROS I photographs of ice pack and areal snow cover could be used to predict river flow and water supply from snow melt. Other space science papers on meteorology and planetary sciences were presented.

December 28: Preliminary results of particle-flux-detectors onboard MARINER II were reported by Dr. L. A. Frank and Dr. H. R. Anderson at American Geophysical Union meeting, Stanford Univ. The instruments showed absence of particles near Venus, indicating the planet's magnetic field does not extend out as far as trajectory of MARINER II; this observation was confirmed by onboard magnetometer. Cosmic-ray measurements during the probe's interplanetary flight indicated cosmic-ray flux was a constant measurement throughout the flight (approximately 3.0 particles per sq. centimeter per sec.). High-energy solar particles were generally absent except for a single solar-flare event beginning Oct. 23; low-energy solar-particle counter detected this event and at least eight others. Total radiation dose recorded in Oct. 23 event was only about 0.24 roentgens inside ionization chamber's 0.01-in.-thick steel wall, and radiation was very non-penetrating.

• At national conference of American Geophysical Union at Stanford Univ., scientists John D. Anderson and George Null of JPL reported MARINER II fly-by of Venus produced the most accurate estimate yet of the mass of that planet—0.81485 times the mass of the earth, with probable error of 0.015 per cent. Final analysis might slightly alter value and further reduce error.

• Preliminary results of solar-wind experiments by MARINER II were reported by Dr. Conway W. Snyder of JPL at American Geophysical Union meeting.   Experiment measured velocity, density, and temperature of solar plasma using electrostatic analyzer onboard the probe.   Results showed energies of gas atoms in solar wind are very low as compared to cosmic-ray particles, but number of solar-wind particles is about a billion times greater than number of cosmic rays, making total energy content of solar wind much greater than of cosmic rays.   Solar flares eject plasma clouds which may have higher velocity, density, and temperature than undisturbed solar wind—which MARINER II found to be little less than 250 mi. per sec. velocity, 10–20 particles per cu. in. density, and few hundred thousand deg. temperature.   MARINER II also found solar wind appears to be "supersonic"—and supersonic effects such as plasma shockwave ahead of the earth and other planets would be objects of study of future NASA space vehicles.

December 28: Scientists again attempted to turn on RELAY communications equipment, but the satellite's power level was still too low.

- Interaction of winds in lower atmosphere and turbulence in upper atmosphere was reported at AAAS meeting by Univ. of Chicago scientist Colin O. Hines.   Hines said wave motions ("internal gravity waves") from earth appear to generate whirlpool turbulences up to 50 mph and sweeping upward as high as 50 mi.   As strength of upper atmosphere varies, so too does its chemical composition; changes may react back to lower atmosphere. Hines said turbulence seems to provide heat to upper air at rate of one degree a day—and perhaps as high as 10 degrees a day.   He also noted that turbulence ends rather abruptly at altitudes from 60 to 70 mi., the abrupt change altering upper air's chemical composition: whereas presence of turbulence mixes chemicals in atmosphere, absence of turbulence separates chemicals into distinct layers.   Hines concluded further study of turbulence could contribute to better long-distance radio communications which use ionospheric irregularities, partially produced and controlled by the whirlpools.

- Dr. Sheldon J. Korchin of National Institutes of Health and Dr. George Ruff of Univ. of Pennsylvania reported at AAAS conference that American astronauts are typically Protestants, natives of small towns, members of middle or upper-middle class families, and have average IQ of 135.   Summing up results of testing astronauts, two psychiatrists concluded astronauts have normal anxieties but above normal ability to control them.   Average astronaut prefers outdoor sports to indoor; is oriented to action rather than thought; prefers facts to speculation; has abiding belief in his own competence and willingness to strive for perfection.

- Efforts of special Presidential board to settle dispute between Boeing Co. and International Association of Machinists (IAM) were unsuccessful, board chairman Saul Wallen announced, and no further mediation sessions were planned.   Dispute centered around IAM's demand for union shop and Boeing's unwillingness to grant it.

December 29: Planet Venus may rotate clockwise on its axis at a rate of about once in 250 earth days, Richard H. Goldstein of NASA Goldstone Tracking Station announced. Conclusions were based on MARINER II "fly-by" of planet Venus Dec. 14 and of radar experiments conducted by Goldstein at NASA Goldstone Oct. 1–Dec. 17.

- NASA announced 4,000-lb. micrometeorite satellite would be launched with Saturn test vehicles SA–8 and SA–9. Satellite, to unfold in space exposing surface area of 2,000 sq. ft., would be "bonus" experiment with chief missions of flight to be launch vehicle development test and atmospheric re-entry test of Apollo boiler-plate spacecraft.
- Dr. Freeman H. Quimby, Chief of Exobiology Programs in NASA Office of Space Sciences, announced U.S. plans to land life-detecting devices on Mars in 1966. "Perhaps Mars is sterile. But we're working on the postulate there may be some forms of life," Dr. Quimby told the American Association for the Advancement of Science meeting in Philadelphia.
- Addressing American Association for the Advancement of Science, Dr. Loren K. Eiseley, anthropology prof. at Univ. of Pennsylvania, said that space defense efforts are "consuming an enormous amount of our wealth and energy. . . . Let a few generations go by, and so much of society's wealth and employment will be wrapped up in this sort of thing that you will get—as we are beginning to get now—a vested interest in war. . . . [The space/defense effort is] a kind of gigantic tumor. When these emerge, you not alone run the danger of weakening society, but also encounter the fact that these tumors grow and become monstrous; to reduce them to normal size becomes difficult if not impossible. . . ." Dr. Eiseley predicted the massive emphasis on space and armaments might lead to downfall of modern civilization; he compared the preoccupation with space and armaments to Imperial Rome's colonialism, Ancient Egypt's tomb and pyramid building, and India's elaborate caste system which contributed to downfall of those ancient civilizations.
- Soviet news agency Novosti reported first Soviet radioastronomy experiments with planet Venus Nov. 19 and 24. Scientists sent words "peace," "Lenin," and "U.S.S.R." to Venus in radio code and they returned to earth about 4.5 min. later.

December 30: AT&T announced that second Telstar communications satellite would be launched next spring, one which would attempt to avoid or overcome radiation damage which shortened the life of TELSTAR I. TELSTAR I transmitted the first live intercontinental television after launch on July 10, but developed malfunctions in power sources on November 29.

- Speaking at general AAAS meeting in Philadelphia, Dr. Hugh L. Dryden, Deputy NASA Administrator, supported the thesis that science has developed most rapidly in an environment of a national social need, most often for national defense or health. In an atmosphere of highly motivated and widely supported national activities, science and scientists have received enhanced support over broad areas as well as in narrow specialties: Benefits of this have far outweighed unbalances. Such has been our national experience in the technologies of aeronautics, com munications, radar, nuclear energy, and now space. In such periods, the number of free scientists supported to work on problems of their own selection is greater than in the absence of social pressures, although admittedly there is a still greater expansion of team effort.

Dr Dryden said: ". . . It is our aim in NASA to administer the [space] program in such a way as to strengthen science and engineering broadly, to strengthen our universities and our industrial base, in fact to add to our national strength in every possible way. As regards the problem under discussion, NASA has undertaken as a goal the support of about 4,000 graduate students per year in 150 qualified universities to do our part in increasing the supply."

December 30: Dr. Robert M. Petrie, Director of Dominion Astrophysical Observatory in Victoria, B.C., told AAAS in Philadelphia that the Milky Way Galaxy appears to be expanding. Dr. Petrie's 20-year study of 600 "B" stars (hottest and brightest stars in the galaxy) produced evidence that our sun and all other stars are moving away from the galactic center.

- Dr. James A. Van Allen, State University of Iowa physicist, criticized Government report on atmospheric radiation resulting from U.S. high-altitude nuclear explosion in July, which stated radiation levels were much higher than had been predicted and would last longer than had been predicted. Dr. Van Allen charged the Government report was a "hasty and ill-considered" interpretation of the facts; he predicted the bulk of artificially formed radiation would no longer be detectable by summer of 1963. Dr. Van Allen said President's Science Advisory Committee ignored findings from INJUN satellite (which he used to study effects of the explosion) and relied instead on data from TELSTAR satellite, launched after the nuclear test. Scientist James W. Warwick of University of Colorado, basing his conclusions on radio measurements from ground stations in Hawaii and the Philippines, said his studies were in general agreement with those of Van Allen and his coworkers and were "inconsistent" with Government estimates based on TELSTAR data. The two scientists were in Philadelphia for session of American Association for the Advancement of Science.

- Dr. Joseph A. Conner, Jr., of NASA, said in AAAS paper that there will be an arbitrary retirement age for astronauts, not in terms of years but in exposure to penetrating radiation in space.

- Soviet Cosmonaut Pavel Popovich arrived in Havana to participate in fourth anniversary celebration of Fidel Castro's Cuban revolution, Havana Radio announced.

December 31: DOD formally canceled Skybolt missile program, announcing that "the Air Force is taking immediate action to terminate all production in connection with the Skybolt program." Announcement was first public DOD notice since President Kennedy's press conference statement of December 12 on the high cost of Skybolt.

- Chairman of the Joint Congressional Committee on Atomic Energy, Rep. Chet Holifield, stated that he saw a narrowing of the Nation's objectives in space:

"I cannot help but wonder about the sincerity of our entire space effort in view of the trends which I see in our current program. It appears, increasingly, that our goal is gravitating toward a one-shot manned mission to the moon.

"This effort is centered around the development of chemical rockets whose limitations, compared to nuclear propulsion, are severe.

". . . The leader in space cannot afford to be so lacking in vision as to fail to appreciate the need for nuclear power. If we are sincere in our effort to be the first on this new frontier, we must maintain a concerted effort to develop nuclear powered machinery for use in space."

Holifield's views were made known in a statement accompanying publication of testimony taken during September 1962.

December 31: President Kennedy named Theodore von Kármán as the first recipient of the National Medal of Science for leadership in the science and engineering of aeronautics. A native of Hungary, von Kármán became a U.S. citizen in 1936 as director of the Guggenheim Aeronautical Laboratories of Cal Tech. In 1936 he also initiated the rocket research development which led to the creation of the Jet Propulsion Laboratory. He is presently chairman of the Advisory Group for Aeronautical Research and Development for NATO (known as AGARD).

• U.S. Navy announced that the last of its lighter-than-air ships as well as associated spare parts and equipment was being disposed of. Although the Navy "blimp" program had been terminated in October 1961, a few airships had been maintained on a standby mobilization basis.

• White House and State Dept. officials denied report by French newspaper *Paris-Presse* that U.S. and U.S.S.R. negotiators had signed cooperative agreement to send team of Russians and Americans to the moon in 1970 on first manned lunar flight.

During December: Response to *G.E. Forum* survey of editors, writers, and publishers: 60% considered present expenditure for U.S. space program fully warranted; 26% probably fully warranted; 8% probably not; 5%, not warranted; and 1%, no response. Asked what should be level of expenditure (relative to GNP) in immediate future, 25% replied sharply higher; 30%, a little higher; 22%, remain at about present ratio; 6%, a little lower; 6%, sharply lower; 11%, no response. 33% said U.S. would probably exceed U.S.S.R. in space achievements by 1970, while 40% said U.S. would approximately equal U.S.S.R. by 1970. 45% said prime objective of U.S. in space was to exceed U.S.S.R. space achievements—but without duplication and inefficiencies of an all-out crash program. 35% said prime objective was to sustain a prudent and orderly program of scientific progress in space achievements, with little regard for who leads in the various aspects of space technology. 16% chose crash program to exceed the Russians in space. As to justification for U.S. space program, 59% said: "Because of its by-products, both actual and potential, our space program is an excellent dollar investment, *over and above* its value for military security and international prestige." 29% chose this statement: "Our space program requires a high dollar expenditure which must be justified—preponderantly if not wholly—in terms of its value for military security and international prestige."

During December: NASA awarded contract to Ford Motor Co. Aeronutronic Div. for development of camera system to be hard-landed on the moon. To relay photographs of lunar surface back to earth, camera would be used in future Ranger lunar spacecraft and would replace seismometers carried in RANGERS III, IV, and V.

- Hibernation of space crewmen was proposed for long-term space travel, by USAF surgeons Capt. T. K. Cockett and Capt. Cecil C. Beehler of SAM, Brooks AFB, Texas. Under their plan, one astronaut would work while the other would ride in state of suspended animation, with life processes slowed to the minimum. During simulated hibernation the body withstands stresses that would be harmful under normal body conditions, the surgeons said.

- Subcommittee of House Committee on Science and Astronautics issued report on solid-propellant rocket motors for Nova-class vehicles: "If the cancellation of this development program [solids for Nova] is based on the cost factor alone, then the committee feels that NASA should review the events of the past. For it was this approach in the development of our ICBM that gave the U.S.S.R. their superbooster and the resultant lead in space exploration which the U.S. is still struggling to overcome. The United States cannot afford a second setback of this magnitude. . . .

  "The committee is struck by the large amount of manpower and time that both the DOD and NASA have devoted to coordination in trying to reach an agreement and the extended delay in starting even the most basic development program." Committee recommended giving NASA the funding responsibility for continuation of solid-propelled motor effort, since there exists no direct military requirement for the large boosters.

- Modulation from ARIEL I satellite appeared on the carrier approximately 10 days out of the 31. Data received indicated the x-ray experiment had failed.

- Dr. C. Stark Draper, director of MIT Instrumentation Laboratory, named 1962 recipient of Louis W. Hill Space Transportation Award by IAS. Dr. Draper was cited "for his leadership in the development of inertial guidance and automatic control equipment and techniques vital to the success of space flights."

- Vice Adm. William F. Raborn (USN) wrote in *U.S. Naval Institute Proceedings* that ability to control weather might bring about greater changes in warfare than did explosion of first nuclear bomb, adding: "The capability to change the direction of destructive storms and guide them toward enemy concentration may exist in the future arsenal of the Navy tactical commander. . . . We already have taken our first steps toward developing an environmental warfare capability. We are using satellite weather from TIROS II for current tactical operations and more accurate long-range weather predictions. . . . Some experiments in fog dissipation have shown promise, and some exploratory research has been conducted on ways to change the heading of major storms. . . ."

Admiral Raborn said USN was planning 10-year study of the atmosphere ("ATMOS") which would be coordinated with separate research on oceans.

During December: Soviet Prof. V. Nikiforov, in article entitled "Chemistry in the Cosmos and on the Moon" which appeared in Soviet journal *Aviation and Cosmonautics,* suggested use of a chemical foam envelope on the moon to protect cosmonauts from great temperature changes, solar radiation, and materials on the lunar surface.

• Sir Bernard Lovell of Jodrell Bank in an interview with *London Daily Telegraph:* ". . . Although the Russians may have [space] superiority purely in the sense of rocketry, the Americans have tremendous superiority over the Russians in their ability to instrument their space vehicles and in the extraction of space information.

"If you made an assessment of the scientific information we have obtained about the earth's environment from space vehicles, you could find that this was very heavily biased in favour of the Americans. They have mounted more sophisticated experiments in their satellites and space probes and they have made them work much better."

During 1962: Dr. Louis Smullni of MIT's Lincoln Laboratories succeeded in detecting a laser reflection from the moon, although at too low a signal level and too long a pulse length for quantitative measurements. This was considered a first step in use of laser reflection for scientific research.

• White Sands Missile Range, N.M., was busiest missile test center, recording 2,615 "hot" tests.

• Man's conquest of space accelerated on a broad front. The U.S. achieved manned orbital flight three times, provided man with his first close-up of another planet with MARINER II's fly-by of Venus, brought global communications a step closer with the orbiting of the first active repeater communications satellite TELSTAR I, and saw the X–15 exceed its design speed and altitude (4,104 mph and 58.7 mi.). U.S.S.R. achieved first dual manned space flight in an effort that racked up an impressive total of 112 orbits, and launched a space probe toward Mars.

During the year the U.S. successfully launched a total of 61 satellites, deep-space probes, and probes, the U.S.S.R. 17, according-according to the U.N. Public Registry. Of the U.S. total, 20 were launched by NASA, 41 by DOD.

APPENDIX A

SATELLITES, SPACE PROBES, AND MANNED SPACE FLIGHTS

A CHRONICLE FOR 1962

The following tabulation was drafted from open sources by Alfred Rosenthal, Historian of the NASA Goddard Space Flight Center, and finalized from available open materials by Dr. Frank W. Anderson, Jr., Assistant NASA Historian. Sources included the United Nations Public Registry, the *Satellite Situation Reports* issued by the Space Operations Control Center at Goddard Space Flight Center, public information releases of the Department of Defense, NASA, and other agencies, and the President's Report to the Congress on *U.S. Aeronautics and Space Activities—1962*. Russian data are from the U.N. Public Registry, translations of Tass News Agency statements in the Soviet press, and international news services' reports. The history of space exploration requires the recording of as much of the full story as possible for the events swiftly pass by.

Orbital data are either initial or those reported in the first biweekly *Satellite Situation Report* after launch. Date of launch is that of local launch site time. After all available sources had been checked, there still remained six gaps in the International Code Name Greek-alphabet listing of orbital bodies: 1962 Alpha-Phi, 1962 Beta-Zeta, 1962 Beta-Theta, 1962 Beta-Iota, 1962 Beta-Nu, and 1962 Beta-Xi. Three of these can be accounted for as known but undesignated Soviet launches; the others remain unknown.

Even though the tempo of space flights continued to increase, valid interpretation of the scientific results of many satellites and probes is as yet limited by the normal "lag-time" required to process and evaluate telemetry and other data.

285

Launch Date	Name	International Designation	Vehicle	Payload Data	Apogee (st. mi.)	Perigee (st. mi.)	Period (minutes)	Inclination	Remarks
Jan. 26	RANGER III (United States).	1962 Alpha	Atlas-Agena B.	Total weight: 727 lbs. Objective: Roughland survivable capsule on moon; take TV pictures of moon. Payload: 17'-wide and 10'-tall (cruise position, with solar panels extended) structure. Hexagonal base containing conical mid-course motor, retrorocket, spherical lunar capsule, and omnidirectional antenna. Instrumentation included vidicon camera, gamma-ray spectrometer, radar reflectivity experiment, and seismometer; 3 transmitters; 8,680 solar cells; 1 silver-zinc battery, 6 silver-cadmium batteries.	1,163 A U.	0.9839 A U.	406.4 days.	0.39°	Lunar impact not achieved; excessive acceleration by 1st stage booster caused payload to pass 22,862 mi. in front of moon 1/28/62; no attempt made to land lunar capsule. TV signals were too weak to provide usable pictures because high-gain antenna did not home on earth. Still in solar orbit.
Feb. 8	TIROS IV (United States).	1962 Beta	Thor-Delta.	Total weight: 285 lbs. Objective: Develop hardware and techniques of weather satellite system; obtain cloud photos and heat data for use in meteorology. Payload: 42" x 19" cylinder, containing 2 TV camera systems, infrared sensors, heat-budget sensors; attitude controls; 3 transmitters and 2 tracking beacons; 9,260 solar cells; 63 cadmium batteries.	525	471	100.4	48.3°	All systems provided good data. TV photos used in weather analyses in support of FRIEND-SHIP 7 manned orbital flight. Transmitted 30,000 photos as of 6/14/62, after which photos no longer usable for weather forecasting. Still in orbit.
Feb. 20	FRIENDSHIP 7 Mercury-Atlas VI (MA-6) (United States).	1962 Gamma	Atlas D.	Total weight: 2,900 lbs. Objective: Orbit and recover manned spacecraft; evaluate manned spacecraft performance; investigate man's capabilities in space environment. Payload: Bell-shaped, 9½' x 6' (diameter at base) capsule, containing, in addition to astronaut, 2 cameras; life-support, aeromedical monitoring, and attitude control systems; HF and UHF transmitters, radar recovery beacons; plus retrofire and recovery apparatus.	159.5	97.6	88.2	32.5°	First U.S. manned orbital flight. After 3 orbits, capsule containing Astronaut John H. Glenn, Jr., re-entered, parachuted into Atlantic east of Grand Turk Island, Bahamas, was picked up and taken aboard U.S.S. Noa. Valuable data gained on capsule and systems, weightlessness, consumption of solid and liquid foods, disorientation exercises, etc. Glenn piloted spacecraft in 2nd and 3rd orbits, reported "Glenn effect," greenish glowing spots passing slowly outside capsule.

Date	Name	Designation	Launch vehicle	Objective / Payload					Remarks
Feb. 21	USAF Spacecraft (United States).	1962 Delta	Thor-Agena B.	Not available.	232.64	104.30	89.7	81.97°	Re-entered 3/4/62.
Feb. 27	DISCOVERER XXXVIII (United States).	1962 Epsilon	Thor-Agena B.	Total weight: 2,100 lbs. Other payload data not available.	191.39	129.26	89.7	82.2°	Capsule re-entered and was recovered in mid-air over the Pacific, 3/3/62; 8th mid-air recovery in the Discoverer series. Remainder of satellite re-entered 3/21/62.
Mar. 7	OSO I (United States).	1962 Zeta	Thor-Delta.	Total weight: 440 lbs. Objective: Measure solar electromagnetic radiation in the ultraviolet, X-ray, and gamma-ray portions of the spectrum; investigate dust particles in space; test spacecraft design. Payload: Nine-sided box topped by semicircular sail mounting 1,860 solar cells. Instruments to conduct 13 experiments in solar radiation; others to investigate dust particles in space and thermal radiation characteristics of spacecraft surface materials; 2 transmitters; nickel cadmium batteries; complex attitude control system to keep instruments focused on sun. Not available.	369.8	343.5	96.15	33.8°	Control system reacquired the sun over 1,000 times (77 days) before missing, 5/22/62; had sent data on some 75 solar flares and subflares. Intermittent operation until 5/24/62, then became active again. Still in orbit.
Mar. 7	USAF Spacecraft (United States).	1962 Eta	Atlas-Agena B.	Not available.	413.76	141.93	93.9	90.93°	Re-entered 11/3/62.
Mar. 16	COSMOS I (U.S.S.R.).	1962 Theta	Not available.	Total weight: Not available. Objective: Study of earth's radiation belt, cosmic rays and their acceleration, shortwave emissions from sun and other bodies; earth's cloud systems; meteoric impacts. Payload: Not available.	540	131.4	95.4	48.9°	Experiments reported to have transmitted as planned. Re-entered 5/25/62.
Mar. 29	P-21A Probe (United States).		Scout.	Total weight: 94 lbs. Objective: Experiments on continuous-wave propagation to study electron density and associated parameters of ionosphere; swept-frequency measurements of electron density; positive ion experiment to determine ion concentration at night. Payload: 8-sided cone, containing continuous-wave propagation experiment, swept-frequency probe, ion traps, transmitters.					Probe went to 3,910 mi. in 97-min. Good data on night time ionosphere for comparison with P-21 daytime data. Described newly discovered helium layer under nighttime conditions for first time. Ionospheric data needed to improve earth-space communications.
Apr. 6	COSMOS II (U.S.S.R.).	1962 Iota	Not available.	Total weight: Not available. Objective: Investigate ionosphere, cosmic rays, meteoric dust, earth's magnetic field, cloud formation and distribution, and solar radiation. Payload: Not available.		122.7	100.3	48.98°	Experiments reported to have transmitted as planned. Still in orbit.

Launch Date	Name	International Designation	Vehicle	Payload Data	Apogee (st. mi.)	Perigee (st. mi.)	Period (minutes)	Inclination	Remarks
Apr. 9	USAF Spacecraft (United States).	1962 Kappa	Atlas-Agena B.	Not available.	2,099.6	1,748.5	153.0	86.68°	Still in orbit.
Apr. 17	USAF Spacecraft (United States).	1962 Lambda	Thor-Agena B.	Not available.	332.68	97.99	91.5	73.53°	Re-entered 5/28/62.
Apr. 23	RANGER IV (United States).	1962 Mu	Atlas-Agena B.	Total weight: 730 lbs. Objective: Roughland survivable capsule on moon; take TV pictures of moon. Payload: 17'-wide and 10'-tall (cruise position, with solar panels extended) structure. Hexagonal base containing conical mid-course motor, retrorocket, spherical lunar capsule, and omni-directional antenna. Instrumentation included vidicon camera, gamma-ray spectrometer, radar reflectivity experiment, and seismometer; 3 transmitters; 8,680 solar cells; 1 silver-zinc battery, 6 silver-cadmium batteries.				6.18° to ecliptic.	Timer failure caused loss of control and data 2 hrs. after launch. RANGER IV traveled 231,486 mi. in 64 hrs., impacted on moon 4/26/62 at a point calculated as 229.3° E longitude and 15.5° S latitude. First U.S. spacecraft to impact on the moon.
Apr. 24	COSMOS III (U.S.S.R.).	1962 Nu	Not available.	Total weight: Not available. Objective: Investigate ionosphere, cosmic rays, meteoric dust, earth's magnetic field, cloud formations and distribution, and solar radiation. Payload: Not available.	414.5	130.8	92.5	49°	Experiments reported to have transmitted as planned. Re-entered 10/17/62.
Apr. 26	COSMOS IV (U.S.S.R.).	1962 Xi	Not available.	Total weight: Not available. Objective: Investigate earth's radiation belts for effects on manned space flight; meteoric particles; weather phenomena; and factors affecting communications. Payload: Not available.	206	180	90.6	65°	Flight stated to be a continuation of program of scientific investigations announced 3/16/62 with launching of COSMOS I; experiments reported to have transmitted as planned. Satellite recovered "at a predetermined point" in U.S.S.R. on command, 4/29/62, after 1,250,000-mi. flight.
Apr. 26	ARIEL I (United States-United Kingdom).	1962 Omicron	Thor-Delta.	Total weight: 132 lbs. Objective: Investigate ionosphere and its relationships with the sun. Payload: 23" x 10¼36" cylinder, containing electron density sensor, electron temperature gauge, solar aspect sensor, cosmic-ray detector, ion mass sphere, Lyman-Alpha gauges, tape recorder, x-ray sensors, transmitters.	754.2	242.1	100.9	53.87°	All experiments except Lyman-Alpha transmitted normally. Discovered new ion belt at 450-500 mi. Transmission intermittent after radiation from U.S. high-altitude nuclear test damaged solar cells. First international satellite. Still in orbit.

Date	Name (Country)	1962 Designation	Launch Vehicle	Objective / Weight					Remarks
Apr. 28	USAF Spacecraft (United States).	1962 Pi	Atlas-Agena B.	Not available.	Not avail-able.	Not avail-able.	Not avail-able.	Not avail-able.	Re-entered 4/28/62.
Apr. 28	USAF Spacecraft (United States).	1962 Rho	Thor-Agena B.	Not available.	226.18	109.98	90.0	73.07°	Re-entered 5/28/62.
May 15	USAF Spacecraft (United States).	1962 Sigma	Thor-Agena B.	Not available.	401.4	180.19	94.0	82.46°	Still in orbit.
May 24	AURORA 7 Mercury-Atlas VII (MA-7) (United States).	1962 Tau	Atlas D.	Total weight: 2,975 lbs. Objective: Orbit and recover manned spacecraft; evaluate manned spacecraft performance; investigate man's capabilities in space environment. Payload: Bell-shaped 9½' x 6' (diameter at base) capsule, containing, in addition to astronaut, cameras; life-support, aeromedical monitoring, and attitude control systems; HF and UHF transmitters, radar recovery beacons; retrofire and recovery apparatus; tethered balloon for study of drag and visibility in space.	166.8	100	88.3	32.5°	Re-entered after 3 orbits; spacecraft and Astronaut M. Scott Carpenter picked up 135 mi. east of Puerto Rico (200 mi. beyond intended area) by U.S.S. Pierce. Carpenter performed many experiments, including star observations, studying "Glenn effect," and behavior of liquid under weightless conditions; proved feasibility of drifting flight with its great saving in fuel. Second U.S. manned orbital flight.
May 28	COSMOS V (U.S.S.R.).	1962 Upsilon	Not avail-able.	Total weight: Not available. Objective: Orbital scientific satellite to continue "program announced March 16 of this year," (launch of COSMOS I).	908.3	124.6	101.5	49.0°	Experiments reported to have operated as planned. Still in orbit.
May 30	USAF Spacecraft (United States).	1962 Phi	Thor-Agena B.	Not available.	198	122	89.7	74.1°	Re-entered 6/11/62.
May 31	USAF Probe (United States).		Blue Scout.	Not available.					No information available.
Jun. 1	USAF Spacecraft (United States).	1962 Chi 1	Thor-Agena B.	Not available.	241.09	131.1	90.5	74.26°	Re-entered 6/28/62. Carried OSCAR II as piggyback satellite.
	and OSCAR II (United States).	1902 Chi 2			240	129	90.5	74.3°	10-lb. American Radio Relay League satellite, transmitting the word "Hi", for use by amateur radio operators. Re-entered 6/21/62.
Jun. 17	USAF Spacecraft (United States).	1962 Psi	Atlas-Agena B.	Not available.	Not avail-able.	Not avail-able.	Not avail-able.	Not avail-able.	Re-entered 6/18/62.
Jun. 18	USAF Spacecraft (United States).	1962 Omega	Thor-Agena D.	Not available.	243.5	234.2	92.3	82°	First orbital flight of Agena D. Still in orbit.

Launch Date	Name	International Designation	Vehicle	Payload Data	Apogee (st. mi.)	Perigee (st. mi.)	Period (minutes)	Inclination	Remarks
Jun. 10	TIROS V (United States).	1962 Alpha-Alpha	Thor-Delta.	Total weight: 286 lbs. Objective: Develop hardware and techniques of weather satellite system; obtain cloud photos and heat data for use in meteorology. Payload: 42" x 22" cylinder, containing 2 TV camera systems, infrared sensors; attitude controls; 4 transmitters and 2 tracking beacons; 9,260 solar cells; 63 nickel-cadmium batteries.	604	367	100.5	58.1°	Infrared system did not pass preflight checks but satellite was launched with this inoperative because of need to cover August-September hurricane season. Inclination of orbit increased to 58° to widen global weather coverage (tip of Greenland to fringes of Antarctica). Cameras transmitted excellent photos, although medium-angle (Tegea-lens) camera ceased functioning 7/2/62, after transmitting 4,700 photos. Still in orbit, still transmitting.
Jun. 23	USAF Spacecraft (United States).	1962 Alpha-Beta	Thor-Agena B.	Not available.	143.54	130.49	88.9	75.09°	Re-entered 7/7/62.
Jun. 27	USAF Spacecraft (United States).	1962 Alpha-Gamma	Thor-Agena D.	Not available.	428.14	130.49	93.6	76.04°	Re-entered 9/14/62.
Jun. 30	COSMOS VI (U.S.S.R.).	1962 Alpha-Delta	Not available.	Total weight: Not available. Objective: Continuation of Cosmos scientific satellite series. Payload: Not available.	186.8	162.4	90.1	48.9°	Experiments reported to have functioned as planned. Re-entered 8/8/62.
Jul. 10	TELSTAR I (United States).	1962 Alpha-Epsilon	Thor-Delta.	Total weight: 170 lbs. Objective: Orbit active repeater communications satellite to test broad-band microwave communication in space; study radiation effects; test satellite tracking techniques. Payload: 34½" sphere, containing communications receivers and transmitters; tracking beacons; experiments to study earth's radiation belts and their effect on semiconductors; mirrors for optical tracking; 3,600 solar cells and 19 nickel-cadmium batteries.	3,503	593	157.8	44.79°	First private-corporation satellite, built by AT&T, launched by NASA. Successfully transmitted international TV (between U.S. and France and U.K.) and telephone conversations. Command circuit failed 11/23/62 because of radiation effects, was by-passed from the ground; satellite resumed transmission 1/4/63. Still in orbit, still transmitting.
Jul. 18	USAF Spacecraft (United States).	1962 Alpha-Zeta	Atlas-Agena B.	Not available.	Not available.	Not available.	Not available.	Not available.	Re-entered 7/25/62.
Jul. 20	USAF Spacecraft (United States).	1962 Alpha-Eta	Thor-Agena B.	Not available.	218.11	122.41	90.0	70.29°	Re-entered 8/14/62.
Jul. 24	USAF Probe (United States).		Blue Scout, Jr.	Not available.					No information available.

Date	Name (Country)	Designation	Vehicle	Payload / Objective	Apogee	Perigee	Period	Inclination	Remarks
Jul. 27	USAF Spacecraft (United States).	1962 Alpha-Theta	Thor-Agena B.	Not available.	251.04	129.25	90.7	71.06°	Re-entered 8/24/62.
Jul. 28	COSMOS VII (U.S.S.R.).	1962 Alpha-Iota	Not available.	Total weight: Not available. Objective: Orbit scientific satellite to investigate charged particles in ionosphere, corpuscular streams and low-energy particles, earth's magnetic field and radiation belts, cloud formations, and cosmic rays. Payload: Not available.	218.2	125.2	90.1	65°	Experiments reported to have functioned as planned. Re-entered 8/1/62.
Aug. 1	USAF Spacecraft (United States).	1962 Alpha-Kappa	Thor-Agena D.	Not available.	227.43	120.55	90.2	82.25°	Re-entered 8/26/62.
Aug. 5	USAF Spacecraft (United States).	1962 Alpha-Lambda	Atlas-Agena B.	Not available.	Not available.	Not available.	Not available.	Not available.	Re-entered 8/6/62.
Aug. 11	VOSTOK III (U.S.S.R.).	1962 Alpha-Mu	Not available.	Total weight: Approx. 10,000 lbs. Objective: Orbit manned satellite and make precision recovery; study effects of space flight on man. Payload: Unofficially reported to include, in addition to cosmonaut, TV transmitter, aeromedical monitoring equipment, 2 recorders, transmitters and tracking beacons.	156, later 137.2	105.6, later 107.4	88.5, later 88.13	65°, later 64°50'	Spacecraft carrying Cosmonaut Maj. Andrian Nikolayev transmitted live TV to Soviet ground stations, showing cosmonaut movements during weightlessness. Cosmonaut communicated several times with VOSTOK IV, orbited a day later. On re-entry, Nikolayev left spacecraft via ejection capsule, parachuted to earth near Karagand, Kazakhstan, U.S.S.R., after 64 orbits (95 hrs. 25 min.)
Aug. 12	VOSTOK IV (U.S.S.R.).	1962 Alpha-Nu	Not available.	Total weight: Approx. 10,000 lbs. Objective: Orbit manned satellite in close conjunction with another manned satellite already in orbit. Investigate effects of space flight on man. Make precision recovery. Payload: Unofficially reported to include, in addition to cosmonaut, TV transmitter, aeromedical monitoring equipment, 2 recorders, transmitters and tracking beacons.	158	111	88.5	65°	First tandem flight of manned spacecraft, reported to have been as close as 4.3 mi. to each other at one point in the flight. Cosmonaut Lt. Col. Pavel Popovich in VOSTOK IV was in visual and radio contact with Cosmonaut Nikolayev in VOSTOK III, "floated" in weightless condition for 3½ hrs. On re-entry, Popovich left spacecraft via ejection capsule, parachuted to earth near Karagand, Kazakhstan, U.S.S.R., after 48 orbits (70 hrs. 29 min.). Tandem flight (1) demonstrated high order of Russian launch and vehicle technology; (2) multiplied several times the amount of space aeromedical data.

Launch Date	Name	International Designation	Vehicle	Payload Data	Apogee (st. mi.)	Perigee (st. mi.)	Period (minutes)	Inclination	Remarks
Aug. 18	COSMOS VIII (U.S.S.R.).	1962 Alpha-Xi	Not available.	Total weight: Not available. Objective: Continuation of Cosmos scientific satellite series. Payload: Not available.	374.4	158.7	92.9	49°	All systems reported to have "operated as planned." Still in orbit.
Aug. 23	USAF Spacecraft (United States).	1962 Alpha-Omicron	Blue Scout.	Not available.	526.3	388.35	99.6	98.62°	Still in orbit.
Aug. 25	Venus Probe (U.S.S.R.).	1962 Alpha-Pi	Not available.	Not available.	Not available.	Not available.	Not available.	Not available.	Reported by U.S. (9/5/62) to have been an attempted Venus probe that failed to leave parking orbit. Re-entered 8/28/62.
Aug. 27	MARINER II (United States).	1962 Alpha-Rho	Atlas-Agena B.	Total weight: 447 lbs. Objective: Conduct fly-by of planet Venus with scientific probe; in flight, gather data on surface temperature, atmosphere, magnetic field, and charged particle intensity in vicinity of Venus; gather data on interplanetary magnetic field, charged particle intensity and distribution, density and direction of cosmic dust, and intensity of low-energy protons from the sun. Payload: 16'6" x 11'11" (in cruise position, with solar panels and antenna extended) tubular frame with hexagonal base. Experiments include infrared and microwave radiometers, magnetometer, ion chamber and particle flux detector, cosmic dust detector, and solar plasma spectrometer; equipment includes receiver-transmitter, omnidirectional, high-gain, and command antennas, attitude controls, midcourse motor, 9,800 solar cells, and rechargeable silver-zinc battery.					First successful interplanetary probe. Four experiments transmitted data throughout flight to Venus (except from 10/31–11/8, when instruments were turned off during on-board power drop). All 6 experiments transmitted during Venus fly-by on 12/14/62; closest distance to Venus, 21,594 mi. On passing Venus, MARINER II went into heliocentric orbit; set new space communication record by transmitting usable data to a point 54.3 million mi. from earth. Preliminary evaluation of data indicates: Venus has little or no magnetic field, which suggests very slow rotation rate. Surface of Venus may be considerably hotter than the 600°F previously assumed. Interplanetary data measured solar wind at about 250 mi. per sec. during quiet sun periods; cosmic-ray flux at about 3 particles per sq. cm. per sec.
Aug. 28	USAF Spacecraft (United States)	1962 Alpha-Sigma	Thor-Agena D.	Not available.	250.42	114.33	90.4	65.16°	Re-entered 9/10/62.

Date	Name (Country)	Designation	Launch vehicle	Objective / Payload					Remarks
Sep. 1	Venus Probe (U.S.S.R.).	1962 Alpha-Tau	Not available.	Not available.	Not available.	Not available.	Not available.	Not available.	Reported by U.S. (9/5/62) to have been an attempted Venus probe that failed to leave parking orbit. Information on orbital status not available. Still in orbit.
Sep. 1	USAF Spacecraft (United States).	1962 Alpha-Upsilon	Thor-Agena B.	Not available.	415.69	181.44	94.3	82.82°	Not available.
Sep. 17	USAF Spacecraft (United States).	1962 Alpha-Chi	Thor-Agena B.	Not available.	421	129	91.4	81.94°	Re-entered 11/19/62.
Sep. 18	TIROS VI (United States).	1962 Alpha-Psi	Thor-Delta.	Total weight: 281 lbs. Objective: Develop hardware and techniques of weather satellite system; obtain cloud photos for use in meteorology. Payload: 42″ x 22″ cylinder, containing 2 TV camera systems; attitude controls; 4 transmitters and 2 tracking beacons; 9,120 solar cells; 63 nickel-cadmium batteries.	442	425	98.7	58.2°	Launch moved up from Nov. to ensure maximum coverage of the storm season and of MA-8 launching. Infrared sensor was left out of TIROS VI. Photo quality as good as any in Tiros series. Medium-angle camera (Tegea lens) stopped functioning 12/1/62, after transmitting 12,337 cloud-cover photos. Still in orbit, still transmitting photos from wide-angle camera.
Sep. 27	COSMOS IX (U.S.S.R.).	1962 Alpha-Omega	Not available.	Total weight: Not available. Objective: Orbit scientific satellite for study of ionosphere, radiation belts, and effect of meteoroids on spacecraft. Payload: Not available.	221	188	90	65°	Equipment reported to have functioned as planned. Information on orbital status not available.
Sep. 28	ALOUETTE I (United States) (Canada).	1962 Beta-Alpha	Thor-Agena B.	Total weight: 320 lbs. Objective: Measure electron density distribution in the ionosphere; variations in electron density distribution with time of day and with latitude under varying magnetic and auroral conditions; galactic noise; energetic particle flux. Payload: 2′10″ x 3′6″ egg-shaped satellite, containing swept-frequency pulsed sounder, electron density ionization and Whistler experiments, 2 crossed antennas extending 150′ and 75′; solar panels. Not available.	638	620	105.4	80.46°	First NASA satellite launch from PMR; first satellite designed and built by country other than U.S. or U.S.S.R. Experiments transmitted as programed. Still in orbit, still transmitting.
Sep. 29	USAF Spacecraft (United States).	1962 Beta-Beta	Thor-Agena.	Not available.	228	124	89.1	65.4°	Re-entered 10/14/62.

Launch Date	Name	International Designation	Vehicle	Payload Data	Apogee (st. mi.)	Perigee (st. mi.)	Period (minutes)	Inclination	Remarks
Oct. 2	EXPLORER XIV (United States).	1962 Beta-Gamma	Thor-Delta.	Total weight: 89 lbs. Objective: Describe the trapped corpuscular radiation, solar particles, cosmic rays, and solar wind; correlate particle phenomena with magnetic field observations taken in a highly elliptical orbit over an extended time period. Payload: 27" x 19" cone with magnetometer boom and 4 solar paddles; instruments include cosmic-ray detector, ion-electron detector, proton analyzer, trapped particle radiation experiment; transmitter; optical attitude sensor; solar cells, including solar-cell damage experiment.	61,190	174.2	36.4 hrs.	32.9°	Experiments transmitted good data. Highly elliptical orbit would enable satellite to sample virtually all levels of the magnetosphere as the magnetic field shape and strength changed and strength throughout a year's passage of the earth around the sun. Data to be compared with those from EXPLORER XII. Still in orbit; ceased transmitting 1/10/63; resumed transmitting 1/29/63.
Oct. 3	SIGMA 7 Mercury-Atlas VIII (MA-8) (United States).	1962 Beta-Delta	Atlas D.	Total weight: 3,030 lbs. Objective: Orbit and recover manned spacecraft; evaluate manned spacecraft performance; investigate man's capabilities in space environment. Payload: Bell-shaped 9½'x6'(diameter at base) capsule, containing, in addition to astronaut, cameras; life-support, aeromedical monitoring, and attitude control systems; HF and UHF transmitters, radar recovery beacons; retrofire and recovery apparatus; experiment to compare various types of heat shielding.	176	100	89	32.5°	First U.S. manned orbital flight of more than 3 orbits; first manned spaceflight recovery in the Pacific. Astronaut Walter M. Schirra, Jr., flew 6 orbits, most of it in "drifting" mode, re-entered and landed within 13,000 ft. of predicted impact point 295 mi. NE of Midway Island, was picked up by U.S.S. *Kearsarge*.
Oct. 9	USAF Spacecraft (United States).	1962 Beta-Epsilon	Thor-Agena.	Not available.	238.5	126.75	90.5	81.96°	Re-entered 11/16/62.
Oct. 17	COSMOS X (U.S.S.R.).	Not available	Not available.	Total weight: Not available. Objective: Continuation of Cosmos scientific satellite series.	208	126	90.2	65°	Equipment reported to have functioned normally. Information on orbital status not available.
Oct. 18	RANGER V (United States).	1962 Beta-Eta	Atlas-Agena B.	Payload: Not available. Total weight: 755 lbs. Objective: Roughland survivable capsule on moon; take TV pictures of moon. Payload: 17'-wide and 10'-tall (cruise position, with solar panels extended) structure. Hexagonal base containing conical midcourse motor, retrorocket, spherical lunar capsule, and omnidirectional antenna. In-	Not available.	Not available.	Not available.	Not available.	Power malfunction prevented solar cells from feeding power to instrumentation making payload inoperative from 2 hrs. after launch. No midcourse correction possible; no TV photos; no ejection of lunar capsule. Passed within 450 mi. of moon 10/20/62, went into solar orbit.

Date	Name	Designation	Launch vehicle	Description					Remarks
Oct. 20	COSMOS XI (U.S.S.R.).	Not available.	Not available.	strumentation includes vidicon camera, gamma-ray spectrometer, radar reflectivity experiment, and seismometer; 3 transmitters; 8,680 solar cells; 1 silver-zinc battery; 6 silver-cadmium batteries. Total weight: Not available. Objective: Continuation of Cosmos scientific satellite series.	572	152	96.1	45°	Instruments reported to have functioned as planned. Information on orbital status not available.
Oct. 26	USAF Spacecraft (United States).	1962 Beta-Kappa	Thor-Agena.	Payload: Not available. Not available.					Still in orbit.
Oct. 27	EXPLORER XV (United States).	1962 Beta-Lambda	Thor-Delta.	Total weight: 100 lbs. Objective: Inject scientific satellite into elliptical orbit to study artificial radiation belt created by U.S. high-altitude nuclear explosion 7/9/62. Payload: 27" x 19" cone with magnetometer boom and 4 solar paddles; experiments were on magnetic field, ion-electron, electron flux, and distribution in pitch angle of electrons; transmitter; optical attitude sensor; solar cells, including solar-cell damage experiment.	3,457.92	109.96	147.6	71.4°	Went into orbit with spin rate 10 times the planned one, because of failure of despin weights to deploy; spin rate renders data from the 2 electron directional detectors almost unusable. Still in orbit, still transmitting.
					10,960	193.7	5.2 hrs	18.02°	
Oct. 31	ANNA IB (United States).	1962 Beta-Mu	Thor-Able-Star.	Total weight: 350 lbs. Objective: Orbit geodetic satellite to calibrate instruments at U.S. tracking stations around the world and obtain geodetic data. Payload: 36" sphere, containing 4 flashing-light beacons, Secor radio ranging system, Doppler radio system, transmitters, solar cells.	632	583	107.7	50.5°	Still in orbit. All 3 tracking systems on board satellite still transmitting.
Nov. 1	MARS I (U.S.S.R.).	Not available.	Not available.	Total weight: 1,980 lbs. Objective: Conduct fly-by of planet Mars; gather interplanetary data on trip to Mars and Martian data on fly-by; take TV pictures of Mars and relay them to earth. Payload: 10' x 12' (in cruise position, with solar panels and radiators extended) structure in 2 compartments; orbital compartment contains mid-course motor, solar cells, radiators, and antennas; planetary compartment contains TV camera, spectrograph, spectrograph, magnetometers, gas discharge and scintillating gauges, radiotelescope, low-energy-proton-electron-ion traps, and micrometeorite counters. (Source: Tass).					Trip to Mars to take about 7 mos. Soviet calculations indicated that on original trajectory, satellite would pass 119,924 mi. from Mars. No mid-course correction reported as of 12/31/62. Still transmitting.

Launch Date	Name	International Designation	Vehicle	Payload Data	Apogee (st. mi.)	Perigee (st. mi.)	Period (minutes)	Inclination	Remarks
Nov. 5	USAF Spacecraft (United States).	1962 Beta-Omicron	Thor-Agena.	Not available.	249.79	129.86	90.7	75°	Re-entered 12/3/62.
Nov. 11	USAF Spacecraft (United States).	1962 Beta-Pi	Atlas-Agena.	Not available.	Not available.	Not available.	Not available.	Not available.	Re-entered 11/12/62.
Nov. 21	USAF Probe (United States).		Blue Scout.	Not available.					Information not available.
Nov. 24	USAF Spacecraft (United States).	1962 Beta-Rho	Thor-Agena B.	Not available.	196.98	128.63	89.8	65.15°	Re-entered 12/13/62.
Dec. [4	USAF Spacecraft (United States).	1962 Beta-Sigma	Thor-Agena.	Not available.	174.61	119.31	89.16	65°	Marked 100th Thor launch as space booster, of which 93 were successful. Re-entered 12/8/62.
Dec. 12	USAF Spacecraft (United States). and	1962 Beta-Tau 1	Thor-Agena.	Not available.	1,731.22	144.79	116.2	70.36°	Still in orbit.
	INJUN 3	1962 Beta-Tau 2		Not available.	1,726.87	154.11	116.3	70.34°	Still in orbit.
	and	1962 Beta-Tau 3		Not available.	1,699.45	136.81	115.6	70.29°	Still in orbit.
	and	1962 Beta-Tau 4		Not available.	1,728.65	144.78	116.2	70.35°	Still in orbit.
Dec. 13	RELAY I (United States).	1962 Beta-Upsilon	Thor-Delta.	Total weight: 172 lbs. Objective: Orbit active repeater communications satellite; test intercontinental microwave communications by low-altitude active repeater satellites; measure energy levels of space radiation and amount of damage it does to solar cells and electrical components. Payload: 33″ x 29″ octagonal prism, tapered at one end, containing 2 transponders, duplicated telemetry and command system, attitude control system, tracking beacon; solar-cell damage, diode, and radiation monitoring experiments; 8,215 solar cells; 3 nickel-cadmium batteries; 5 external antennas.	4,611.18	828.91	184.9	47.48°	Power failure in satellite made it impossible to receive communications from RELAY I in early days of the flight; all satellite systems except command receiver were turned off in hope that solar cells would build up battery charge; on 1/3/63, first intercontinental test patterns were transmitted from satellite. Still in orbit; still transmitting.
Dec. 14	USAF Spacecraft (United States).	1962 Beta-Phi	Thor-Agena.	Not available.	237.37	124.9	90.4	70.97°	Still in orbit.

Date	Name	Designation	Launch vehicle	Payload					Remarks
Dec. 16	EXPLORER XVI (United States).	1962 Beta-Chi	Scout.	Total weight: 222 lbs. Objective: Orbit satellite to make direct measurements of micrometeoroid puncture hazard to structural skin samples of particles having different momentums; comparing performance of protected and unprotected solar cells exposed to space radiation. Payload: 24″ x 76″ cylinder, containing 160 pressurized cells for meteoroid puncture detection, 60 foil gauge detectors, 46 wire grid detectors, 2 cadmium-sulfide-cell detectors, and impact-detecting transducers; radio tracking beacon; 2 telemetry systems; solar cells; batteries.	735.11	463.56	104.3	52.04°	Third NASA attempt to orbit micrometeoroid satellite, 1st to attain durable orbit (EXPLORER XIII re-entered after 3 days; first attempt on 6/30/61 did not achieve orbit). EXPLORER XVI still in orbit, still transmitting.
Dec. 18	USAF Probe (United States).	Not available.	Blue Scout.	Not available.					No information available.
Dec. 18	TRANSIT V-A (United States).	1962 Beta-Psi	Blue Scout.	Total weight: 140 lbs. Objective: Orbit first of 4 operational navigation satellites. Payload: 43″ x 31″ drum, containing oscillators, continuous transmitters, attitude control system, memory system, clock, Snap auxiliary power system for transmitters.	488.42	403.29	99.2	89.98°	Intended to be world's first operational navigation satellite, but failure of command receiver meant no signals could be sent to satellite's memory system and that its attitude controls could not be activated; this reduced its utility from operational to research. Still in orbit; ceased transmitting 12/21/62.
Dec. 22	COSMOS XII (U.S.S.R.).	Not available.	Not available.	Total weight: Not available. Objective: Continuation of Cosmos scientific satellite series. Payload: Scientific equipment for exploring space; transmitter; telemetry; radio system for precise measurement of orbital elements.	270	126.6	90.45	65°	Equipment reported to be functioning as planned. Still in orbit.
During 1962.	TRS I (United States).	Not available.		Total weight: 1.5 lbs. Objective: Orbit satellite to map artificial radiation belt. Payload: 6½″ pyramid, containing radiation sensors and transmitter.	Not available.	Not available.	Not available.	Not available.	Launched piggyback aboard a larger satellite launched by USAF Thor-Agena. Information on orbital status not available.

APPENDIX B

CHRONOLOGY OF MAJOR NASA LAUNCHINGS

OCTOBER 1, 1958, THROUGH DECEMBER 31, 1962

This chronology of major NASA launchings is intended to provide an accurate and ready historical reference, one compiling and verifying information previously scattered over several sources. It spans NASA's first four years. It includes launchings of all vehicles larger than sounding rockets launched either by NASA or under "NASA direction" (e.g., ABMA served as launching agent for the early Jupiter C and Juno II shots, and AFBMD for the Thor-Able shots).

An attempt has been made to classify the performance of both the launch vehicle and the payload and of total results in terms of primary mission. Three categories have been used for vehicle performance and mission results—successful (S), partially successful (P), and unsuccessful (U). A fourth category, unknown (Unk), has been provided for payloads where launch vehicle malfunctions did not give the payload a chance to exercise its main experiments. These divisions are necessarily arbitrary, since many of the results cannot be neatly categorized. Also they ignore the fact that a great deal was learned from shots which may have been classified as unsuccessful.

A few unique items require separate treatment. Their dates have been kept in sequence, but their history has been relegated to footnotes.

Dates of launchings are referenced to local time at the launch site.

The original table was prepared by Robert Rosholt of the NASA Historical Office, who made a comparative analysis of the several general chronologies prepared by the NASA Historical Office, the first six semiannual reports of NASA, and the Space Activities Summary prepared by the NASA Office of Public Information. Additional information and guidance was received from Headquarters Offices, from Launch Operations Center, and from the very useful comments received on the October 1962 version. For further information on each item, see Appendix A of *Aeronautics and Astronautics (1915–1960)*, *Aeronautical and Astronautical Events of 1961*, and this volume.

Prepared January 1963 (AFEH).

Date	Name	NASA Code	General Mission	Launch Vehicle (Site)	Vehicle S	Vehicle P	Vehicle U	Payload S	Payload P	Payload U	Payload Unk	MR S	MR P	MR U	Remarks
1958 Oct 11	PIONEER I	—	Scientific lunar probe.	Thor-Able (AMR).		x		x					x		Uneven separation of 2nd and 3rd stages; reached 70,700 miles. Verified Van Allen Belt.
Oct 22	BEACON	—	Scientific earth satellite.	Jupiter C (AMR).			x				x			x	Premature upper-stage separation.
Nov 8	PIONEER II	—	Scientific lunar probe.	Thor-Able (AMR).			x				x			x	3rd-stage failure; reached 963 miles; its brief data indicated equatorial region had higher flux and energy levels than previously thought.
Dec 6	PIONEER III	—	Scientific lunar probe.	Juno II (AMR).		x		x					x		Premature cutoff on 1st stage; reached 63,580 miles. Radiation belt discoveries.
1959 Feb 17	VANGUARD II	—	Scientific earth satellite.	Vanguard (AMR).		x			x				x		Excess satellite wobble. Cloud cover data not used.
Mar 3	PIONEER IV	—	Scientific lunar probe.	Juno II (AMR).		x		x					x		2nd and 3rd stage propulsion and pitch malfunction. Communicated to 407,000 miles.
Apr 13	Vanguard	—	Scientific earth satellite.	Vanguard (AMR).			x				x			x	2nd-stage failure.
May 28[1] / Jun 22	Vanguard	—	Scientific earth satellite.	Vanguard (AMR).			x				x			x	2nd-stage failure.
Jul 16	Explorer	S-1	Scientific earth satellite.	Juno II (AMR).			x				x			x	Destroyed after 5½ seconds.
Aug 7	EXPLORER VI	S-2	Scientific earth satellite.	Thor-Able (AMR).	x						x	x			Mapped Van Allen belt. Photographed cloud cover.
Aug 14	Beacon	—	Scientific earth satellite.	Juno II (AMR).			x				x			x	Premature fuel depletion in 1st stage; upper-stage malfunction.
Aug 21[2] / Sep 9	Big Joe	—	Suborbital Mercury capsule test.	Atlas-Big Joe (AMR).	x			x				x			Capsule recovered after re-entry test.
Sep 16[3] / Sep 18	VANGUARD III	—	Scientific earth satellite.	Vanguard (AMR).	x			x				x			Magnetic fields, radiation belt, and micrometeorite findings.
Sep 24[4] / Oct 4	Little Joe 1	LJ-6	Suborbital Mercury capsule test.	Little Joe (WS)	x							x			Qualify booster for use with Mercury test program.
Oct 13	EXPLORER VII	S-1a	Scientific earth satellite.	Juno II (AMR).	x			x				x			Radiation and magnetic storm findings.

Date	Name	Desig.	Purpose	Vehicle	Remarks
Oct 28	Shotput I	----	Suborbital communications test.	Augmented Sergeant (WS)	Canister ejection successful, 100-foot sphere inflation unsuccessful. Successful test of Delta stage.
Nov 4	Little Joe 2	LJ-1A	Suborbital Mercury capsule.	Little Joe (WS)	Capsule escape test. Escape rocket had a delayed thrust buildup.
Nov 26	Pioneer	----	Scientific lunar probe.	Atlas-Able (AMR)	Shroud failure after 45 seconds.
Dec 4	Little Joe 3	LJ-2	Suborbital Mercury capsule test.	Little Joe (WS)	Escape system and biomedical test; monkey (Sam) used.
1960 Jan 16	Shotput II	----	Suborbital communications test.	Augmented Sergeant (WS)	Canister ejection successful, sphere inflation unsuccessful.
Jan 21	Little Joe 4	LJ-1B	Suborbital Mercury capsule test.	Little Joe (WS)	Escape system and biomedical test; monkey (Miss Sam) used.
Feb 27	Shotput III	----	Suborbital communications test.	Augmented Sergeant (WS)	Canister ejection successful, sphere inflation unsuccessful.
Mar 11	PIONEER V	----	Scientific deep space probe.	Thor-Able (AMR)	Communicated data from 17,700,000 ml., position signal from 22,500,000 ml. Failure in upper stages.
Mar 23	Explorer	S-46	Scientific earth satellite.	Juno II (AMR)	Failure in upper stages.
Apr 1	Shotput IV	----	Suborbital communications test.	Augmented Sergeant (WS)	Twelve-sentence voice message relayed successfully.
Apr 1	TIROS I	A-1	Meteorological earth satellite.	Thor-Able (AMR)	First true meteorological satellite; photographed cloud cover.
Apr 18	Scout	----	Launch vehicle development test.	Scout X (WS)	Structural failure prevented 3rd-stage ignition (2nd and 4th stages were dummies).
May 9 / May 13	Echo	A-10	Communications earth satellite.	Thor-Delta (AMR)	Failure in upper stages.
May 31	Shotput V	----	Suborbital communications test.	Augmented Sergeant (WS)	Inflation successful despite excess spin.
Jul 1	Scout	----	Launch vehicle development test.	Scout (WS)	Ground tracking failure led to erroneous destruction by Range Safety Officer.
Jul 29	Mercury	MA-1	Suborbital Mercury capsule test.	Mercury-Atlas (AMR)	Atlas exploded; capsule re-entry qualification test.
Aug 12	ECHO I	A-11	Communications earth satellite.	Thor-Delta (AMR)	First passive communications satellite; 100-foot sphere used for passive communications and air density experiments.
Sep 25	Pioneer	P-30	Scientific lunar orbiter.	Atlas-Able (AMR)	2nd-stage failure.
Oct 4	Scout	----	Launch vehicle development test.	Scout (WS)	Air Force Special Weapons Center payload included.
Nov 3	EXPLORER VIII	S-30	Scientific earth satellite.	Juno II (AMR)	Ion, electron, and micrometeoroid measurements.
Nov 8	Little Joe 5	LJ-5	Suborbital Mercury capsule test.	Little Joe (WS)	Mercury escape system qualification; premature escape-rocket firing.
Nov 21 [a]					

See footnotes at end of table, p.305.

Date	Name	NASA Code	General Mission	Launch Vehicle (Site)	Performance — Vehicle			Performance — Payload				Mission Results			Remarks
					S	P	U	S	P	U	Unk	S	P	U	
1960															
Nov 23	TIROS II	A-2	Meteorological earth satellite.	Thor-Delta (AMR).	x			x				x			Combined infrared measurements with photography. Wide-angle photographs substandard.
Dec 4	Explorer	S-56	Scientific earth satellite.	Scout (WS).			x				x			x	2nd-stage failure; combined vehicle test and Beacon inflatable sphere.
Dec 15	Pioneer	P-31	Scientific lunar orbiter.	Atlas-Able (AMR).			x				x			x	Exploded after 70 seconds.
Dec 19	Mercury	MR-1A	Suborbital Mercury capsule test.	Mercury-Redstone (AMR).	x			x				x			235-mile flight.
1961															
Jan 31	Mercury	MR-2	Suborbital Mercury capsule test.	Mercury-Redstone (AMR).		x		x				x			Booster oversped; chimpanzee (Ham) in 16-minute flight.
Feb 16	EXPLORER IX	S-56a	Scientific earth satellite.	Scout (WS).	x				x			x			Repeat of 12/4/60 shot. Satellite tracking transmitter did not function, but optical tracking provided atmospheric density data.
Feb 21	Mercury	MA-2	Suborbital Mercury capsule test.	Mercury-Atlas (AMR).	x			x				x			1,425-mile flight.
Feb 24	Explorer	S-45	Scientific earth satellite.	Juno II (AMR).			x				x			x	2nd-stage malfunction prevented 3rd and 4th-stage firing.
Mar 18	Little Joe 5A	LJ-5A	Suborbital Mercury capsule test.	Little Joe (WS).	x				x				x		Mercury escape system qualification; premature escape-rocket firing.
Mar 24	Mercury	MR-BD	Vehicle test for Mercury flight.	Mercury-Redstone (AMR).	x			x				x			Booster development test necessitated by MR-2 flight.
Mar 25	EXPLORER X	P-14	Scientific satellite-probe.	Thor-Delta (AMR).	x			x				x			Magnetometer probe. Highly eccentric orbit (145,000-mile apogee).
Apr 25	Mercury	MA-3	Orbital Mercury capsule test.	Mercury-Atlas (AMR).			x				x			x	Failure in 1st stage. Abort successful.
Apr 27	EXPLORER XI	S-15	Scientific earth satellite.	Juno II (AMR).	x			x				x			Gamma-ray experiment.
Apr 28	Little Joe 5B	LJ-5B	Suborbital Mercury capsule test.	Little Joe (WS).		x		x				x			One booster engine fired late. Repeat of Mercury escape system test.
May 5	FREEDOM 7	MR-3	Suborbital manned Mercury flight.	Mercury-Redstone (AMR).	x			x				x			First U.S. suborbital manned spaceflight; Shepard flight.

Date	Name	Desig.	Purpose	Vehicle	Remarks
May 24	Explorer	S-45a	Scientific earth satellite.	Juno II (AMR).	2nd-stage failure.
Jun 30	Explorer	S-55	Scientific earth satellite.	Scout (WS).	3rd-stage failure. Vehicle test and micrometeorite experiment.
Jul 12	TIROS III	A-3	Meteorological earth satellite.	Thor-Delta (AMR).	One camera system failed by end of July.
Jul 21	LIBERTY BELL 7	MR-4	Suborbital manned Mercury flight.	Mercury-Redstone (AMR).	Grissom flight.
Aug 15	EXPLORER XII	S-3	Scientific earth satellite.	Thor-Delta (AMR).	Various experiments including energetic particles profile.
Aug 23	RANGER I	P-32	Scientific lunar probe.	Atlas-Agena B (AMR).	First space test of Ranger instrumentation only partial success, since Agena failed to restart. Remained in parking orbit. Orientation, communications, and electronics worked well for more than 100 orbits.
Aug 25	EXPLORER XIII	S-55a	Scientific earth satellite.	Scout (WS).	Premature re-entry due to tip-off. Vehicle test and micrometeoroid experiment.
Sep 13	Mercury	MA-4	Orbital Mercury capsule test.	Mercury-Atlas (AMR).	One orbit. Capsule recovered. Tracking network checked.
Oct 19	P-21 PROBE	P-21	Scientific geoprobe.	Scout (WS).	Reached 4,261 miles. Electron density measurement, vehicle test.
Oct 27	Saturn	SA-1	Launch vehicle development test.	Saturn C-1 (AMR).	1st stage only.
Nov 1	Mercury	MS-1	Orbital Mercury network test.	Mercury-Scout (AMR).	Destroyed after 30 seconds; Air Force-launched.
Nov 18	RANGER II	P-33	Scientific lunar probe.	Atlas-Agena B (AMR).	Space test of Ranger instrumentation unsuccessful since Agena failed to restart. Remained in parking orbit.
Nov 29	Mercury	MA-5	Orbital Mercury capsule test.	Mercury-Atlas (AMR).	Two orbits. Enos, the chimpanzee, was recovered.
1962					
Jan 16	Echo (test)	AVT-1	Suborbital communications test.	Thor (AMR).	Canister ejection and opening successful, but 135-foot sphere ruptured.
Jan 26	RANGER III	P-34	Scientific lunar lander.	Atlas-Agena B (AMR).	First Ranger attempt at moon; Atlas over-accelerated; missed moon by 22,862 miles; gamma-ray sensor worked.
Feb 8	TIROS IV	A-9	Meteorological earth satellite.	Thor-Delta (AMR).	Performed as planned.
Feb 20	FRIENDSHIP 7	MA-6	Orbital manned Mercury flight.	Mercury-Atlas (AMR).	First U.S. manned orbital flight; Glenn flight, 3 orbits.
Mar 1	Re-entry	----	28,000 ft/sec re-entry test.	Scout (WS).	Re-entry speed lower than planned.
Mar 7	OSO I	S-16	Scientific earth satellite.	Thor-Delta (AMR).	Transmitted data on 8 solar flares.
Mar 29	P-21A PROBE	P-21a	Scientific geoprobe.	Scout (WS).	Reached 3,910 miles.

See footnotes at end of table, p. 305

Date	Name	NASA Code	General Mission	Launch Vehicle (Site)	Vehicle S	Vehicle P	Vehicle U	Payload S	Payload P	Payload U	Payload Unk	Mission Results S	Mission Results P	Mission Results U	Remarks
1962															
Apr 23	RANGER IV	P-35	Scientific lunar lander.	Atlas-Agena B (AMR).	x					x			x		First U.S. spacecraft to land on moon; crashed onto moon, timer failure causing loss of control over spacecraft; no midcourse correction, TV, or lunar-capsule separation.
Apr 25	Saturn	SA-2	Launch vehicle development test.	Saturn C-1 (AMR).	x							x			1st stage only. Project Highwater utilized dummy upper stages.
Apr 26	ARIEL I	S-51	Scientific earth satellite.	Thor-Delta (AMR).	x			x				x			First international satellite; explored ionosphere (UK #1).
May 8	Centaur	F-1	Launch vehicle development test.	Atlas-Centaur (AMR).		x				x				x	Centaur exploded before separation.
May 24	AURORA 7	MA-7	Orbital manned Mercury flight.	Mercury-Atlas (AMR).	x			x				x			Carpenter flight, 3 orbits.
Jun 19	TIROS V	A-50	Meteorological earth satellite.	Thor-Delta (AMR).	x			x				x			Orbit more elliptical than planned; infrared system failed prior to launch.
Jul 10	TELSTAR I	A-40	Communications earth satellite.	Thor-Delta (AMR).	x			x				x			First active communications satellite; owned by AT&T, launched by NASA.
Jul 18	Echo (test)	AVT-2	Suborbital communications test.	Thor (AMR).	x			x				x			Inflation successful; radar indicated sphere surface not as smooth as planned.
Jul 22	MARINER I	P-37	Scientific Venus probe.	Atlas-Agena B (AMR).		x					x			x	Atlas deviated from course, was destroyed by Range Safety Officer.
Aug 27	MARINER II	P-38	Scientific Venus probe.	Atlas-Agena B (AMR).	x			x				x			First spacecraft to scan another planet; passed 21,100 mi. from Venus on December 14; extended space communications record to 54.7 million mi.
Aug 31	Re-entry		28,000 ft/sec re-entry test.	Scout (WS).	x			x					x		Tardy 3rd-stage ignition; desired speed not achieved.
Sep 18	TIROS VI	A-51	Meteorological earth satellite.	Thor-Delta (AMR).	x			x				x			Performed as planned.
Sep 28	ALOUETTE I	S-27	Scientific earth satellite.	Thor-Agena B (PMR).	x			x				x			Ionosphere sounder satellite built by Canada. First NASA satellite launch from PMR.
Oct 2	EXPLORER XIV	S-3a	Scientific earth satellite.	Thor-Delta (AMR).	x			x				x			Energetic particles satellite. Highly eccentric orbit.
Oct 3	SIGMA 7	MA-8	Orbital manned Mercury flight.	Mercury-Atlas (AMR).	x			x				x			Schirra flight, 6 orbits; first astronaut recovery in Pacific.
Oct 18	RANGER V	P-36	Scientific lunar lander.	Atlas-Agena B (AMR).	x					x				x	Payload, including mid-course guidance, did not function because spacecraft failed to get power from solar cells. Passed within 450 miles of the moon.

Date	Name	No.	Purpose	Launch vehicle										Remarks
Oct 27	EXPLORER XV	S-3b	Scientific earth satellite.	Thor-Delta (AMR).	x					x			x	Difficult to analyze data from 2 of 7 experiments because of high spin rate. To study artificial radiation belt.
Nov 16	Saturn	SA-3	Launch vehicle development test.	Saturn C-1 (AMR).	x			x		x			x	1st stage only. Project Highwater utilized dummy upper stages.
Dec 13	RELAY I	A-15	Communications earth satellite.	Thor-Delta (AMR).	x					x		x	x	First launch with uprated Delta. Power supply voltage originally too low for communications experiments; voltage built up and early in Jan. 63 transatlantic TV transmissions began.
Dec 16	EXPLORER XVI	S-55b	Scientific earth satellite.	Scout (WS).	x			x		x			x	Micrometeoroid satellite.

[1] May 28, 1959. Included in the nose cone of an Army-launched (from AMR) Jupiter IRBM were medical experiments sponsored by NASA. Two monkeys, Able and Baker, were successfully recovered after a 1,700-mile flight.

[2] Aug 21, 1959. While a Little Joe was being readied for firing (at WS) a malfunction caused the Mercury escape rocket to fire. The vehicle was undamaged but the capsule was lost in the ocean.

[3] Sep 16, 1959. An Army Jupiter IRBM, containing a NASA biological experiment, was destroyed by the Range Safety Officer shortly after launch (from AMR).

[4] Sep 24, 1959. An Atlas-Able vehicle, scheduled to launch a Pioneer lunar-orbit payload, exploded on the launching pad (at AMR) while being static-tested.

[5] May 9, 1960. The first production model of the Mercury capsule was tested in a "pad abort" test (at WS). The escape rocket was used rather than a launch vehicle.

[6] Nov 21, 1960. Upon being fired (from AMR), a Mercury-Redstone (MR-1) rose one inch, stopped firing, and settled back on the launching pad; was fired again on December 19. The Mercury capsule rocket also fired. The capsule was recovered and also reused on December 19.

INDEX